ROBERT BAKEWELL

Britain's Foremost Livestock Breeder

In memory of much missed former
members of the New Dishley Society

ROBERT BAKEWELL

Britain's Foremost Livestock Breeder

The New Dishley Society

Editor: Tim Bacon

BREWIN BOOKS

BREWIN BOOKS
19 Enfield Ind. Estate,
Redditch,
Worcestershire,
B97 6BY
www.brewinbooks.com

First published by Brewin Books 2025

A CIP catalogue record for this book is available from the British Library.

ISBN: 978-1-85858-782-0

Printed and bound in Great Britain
by Short Run Press Ltd.

Contents

APPENDICES

List of Sponsors

Tim Bacon; James R. Beechey; Sue Brown; Ray, Clive, Tim and Digby Davies, Westwood Hereford Cattle and Shropshire Sheep; Mrs J.G. Harris; Mrs Sue Hatton, Leicester Longwools; Dr Hilary Matthews; Mr John Matthews; The New Dishley Society; Roger Pegg; Rita Poulson; Janet Spavold; Trevor Spavold; John Stanley, F.R.Ag.S.; Patricia Stanley, F.R.Ag.S.; John D.M. Thomson, Longtown Auction Mart.

Acknowledgements

We would like to thank many people for their help and encouragement in the making of this book, but especially:

Acknowledgements for the new book
The New Dishley Society book research and writing team:
Tim Bacon – Editor
Jim Beechey
Sue Brown
Karen Ellis
Brenton Heazlewood
Hilary Matthews
John Matthews
Rita Poulson
Elaine Shirley
Janet Spavold
John W. Stanley F.R.Ag.S.
Patricia A. Stanley F.R.Ag.S.

Additional input
Chris Coleman for the history of the Speeton Flock
Sandra Coote, Secretary of the Roscommon Sheep Breeders Association
Mr Keith Elliot, Measham Museum
Lawrence Weaver
Julian Wiseman, Emeritus Professor of Animal Production, for illustrations
Beamish, The Living Museum of the North
Northampton Museum and Art Gallery for the Portrait of Valentine Barford
Ordnance Survey
Private collectors
Squire Gerard de Lisle
University of Leicester, Special Collection online

Acknowledgements for the original book
J.H.S. Thompson FRICS, chief land agent to the Fitzwilliam Estates
M.J. Rogers FRICS, former chief land agent to the National Trust
The National Trust, Calke Abbey, Derbyshire (Mr K. Usher, administrator)
I.G. & J.E. Mattley, Phenotype Books, Penrith, Cumbria

The Royal Agricultural Society of England and their librarian at the National Agricultural Centre, Stoneleigh, Philip Sheppey
John Heathcote Ball
Jim Beechey
Rita Callwood
Geoff and Judith Gilby
Dick and Claire Clarke
Miss A. Harpur Crewe
Iona Antiques, London
Graftons of Market Harborough
The publishers Farming Press: Roger Smith, Julanne Arnold, and their staff
Kay Edge for her editorial help
David Edgar for his many valuable suggestions
And finally, the three men in my life, John, Ben and Joe, for their tolerance and understanding

List of Illustrations

Black And White

Colour Gallery

All images which do not credit ownership below them are in private ownership and permission to use them has been sought and obtained.

Robert Bakewell at Dishley Grange, by Boultbee.

Introduction

Set out below is the introduction by David Edgar for Pat Stanley's original book. We have kept this as it describes much about the revised book which has been built around Pat's text and acts as our core writing. What the society has endeavoured to show is the extent of Robert Bakewell's influence on livestock breeding and farming more widely. We have used the opportunity to explain the situation surrounding Bakewell's bankruptcy which, to the best of our knowledge, has not previously been explored. We have also set out in detail the Bakewell genealogy to help prevent individuals mistakenly claiming a relationship to this man, who was a confirmed bachelor. We must sadly add that we are not sure David Edgar's opening sentence is quite so true today as it was in 1995!

Original Introduction by David Edgar

Even in these increasingly urban times, almost everyone has heard of Robert Bakewell (1725-95). He was the great improver of farm livestock and he still finds a place in most school history books, with Jethro Tull (seed drills), Turnip Townshend (Norfolk four-course rotation) and Coke of Norfolk, later Lord Leicester of Holkham. Along with collaborators and rivals in their day, these great men of British farming showed the way to expand agricultural production which, with gradually increasing food imports over most of the period, enabled the British population to increase from 7 million in 1760 when food imports were virtually nil to 30 million in 1881 when food imports from the New World became a major factor.

Jethro Tull (1674-1741) showed the way to grow heavier yields of traditional crops; Turnip Townshend (1674-1738) increased the total production of land over a rotation even if at least two of his four crops – wheat, roots (turnips), barley and seeds (hay) – had to be processed through animals to be of value to man; Bakewell provided the improved livestock to process the crops for human use; and Coke of Norfolk (1752-1842) brought all these elements together in his new commercial farming. More or less.

Robert Bakewell is best remembered for developing the New Leicester Sheep, the breed claimed to have been infused into virtually every one of the 40 or more native British sheep breeds, and many abroad. Leicesters are also a more direct parent of breeds such as the Hexham or Bluefaced Leicester and the Border Leicester, which are themselves of major significance in the cross-bred lambs they produce out of yet other breeds – the Swaledale, the

Cheviot, the Blackface and many more. In one way or another, the Leicester is also closely related to other large longwoolled breeds – the Wensleydale, the Lincoln Longwool and the Cotswold, for example.

Longhorn cattle, another of the major breeds on which Bakewell worked, have had a more mixed reception down the years. The Longhorn was eclipsed soon after Bakewell's death, and this was, to an extent, Bakewell's own fault. For it seems that he so stimulated thinking about cattle breeding that farmers improved the fairly new Shorthorn so greatly and rapidly that by early in the 20th century the Longhorn was considered to be of no more than historical interest. Yet the Longhorn survives and is increasing, in response to new or revised ideas on cattle breeding and beef production as we move towards the 21st century.

There is always an element of stubbornness in any farmer who maintains a breed which is not the most popular of its day. Yet there are a number of characteristics which make the Longhorn attractive even in the 1990s, and after three decades of beef breeds being imported from Europe and elsewhere. One is its docility, noted by virtually all its breeders who are, in any case, not likely to keep a wild breed with horns such as the Longhorn has. Another is its record of easy calving, an important feature for any farmer who will almost certainly have a large number of other chores and has no wish to miss his sleep.

A third characteristic of merit is that the Longhorn appears to thrive even on relatively poor-quality grassland such as that in which rock or other factors make it difficult to plough and improve the quality of the sward. Fourth, and by no means least, the Longhorn does seem to have a large percentage of lean meat in its carcase and therefore commends itself to modern Jack Spratt consumers.

Mrs Pat Stanley is, with her husband John, a breeder of Longhorn cattle on a farm in the Charnwood Forest area of Leicestershire. This forest area was a major cradle of the breed as it came into recorded history in the early 1700s. It is also an area on the edge of which the great Robert Bakewell farmed, at Dishley on the outskirts of Loughborough, and was also home to some of his collaborating and rival farmers. Moreover, Pat Stanley grew up on a neighbouring farm on the Calke Estate, in south Derbyshire, which has made its own considerable mark on the history of the Longhorn.

Her method has been to absorb, over many years, all the traditions, written and otherwise, of this great breed of the English Midlands and she here reproduces many of the major writings about the breed over the years, interspersed with her own comments. In doing so she has created what is a substantial history of the Longhorn breed and a new interpretation of the work of the master breeder Robert Bakewell.

Her story is of major interest not only in its discussion of breeding and feeding cattle and other livestock but also in the comments which flow from years past on the issues of the day as seen by the writers she quotes, or from their subjects whom they quote. Attitudes to farm workers, to rural transport and to religion as it affected people at the grass roots all have something to offer to the social historian. Some of the language of earlier writings, and the art of 'likeness' at various times, provide yet more interest in her work.

Pat Stanley focuses at one point on the extra meat that Bakewell's ideas produced for a population that consequently had the nutrition from which to grow. The master historian of *English Farming, Past and Present*, Lord Ernle, made a similar point about the 18th century; 'a time rapidly approaching when beef and mutton were to be more important than the power of draught and the fineness of wool'. Bakewell was the agricultural opportunist who saw the impending change and knew how it should be met. By providing meat for the millions, he contributed as much to the wealth of the country as Arkwright or Watt. There is some foundation for the statement that many monuments have been reared in Westminster Abbey to the memory of men who less deserved the honour than Robert Bakewell.

This book should go some way to reminding us of the important contribution made to the development of the British nation by Bakewell, and by many of his contemporaries who are also discussed.

C. DAVID EDGAR

PART ONE

Robert Bakewell of Dishley Grange

Robert Bakewell (from an engraving by F. Englehart, after the painting by Boultbee).

CHAPTER ONE

Robert Bakewell, Farmer and Innovator

Pat Stanley

> Leicestershire lays claim to a full share of celebrity for being the cradle and nursery of some of the great modern improvements in Agriculture. These improvements, more especially those connected with the breeding of animals, were in great degree projected and executed by Robert Bakewell, whose talents burst the fetters of rustic ignorance and long established usage; who exalted the ordinary drudgery of ordinary cultivation into the dignity of a science of the highest national benefit and importance, and discovered and demonstrated the principals by which form, size, strength and beauty of our most useful farm animals may be brought to the highest perfection.
>
> He seems to have solved that difficult and valuable problem by what means the maximum of wholesome and nutritious food can be obtained for the use of man in the shortest space of time and on the minimum vegetable substance.
>
> *(Local directory, 1860)*

On 23rd May 1725, a son was born to Robert and Rebecca Bakewell of Dishley Grange near Loughborough, Leicestershire, whose life and deeds were to have far-reaching and everlasting effects on agriculture worldwide. The child, the third Robert Bakewell of Dishley, was to become one of the most important agriculturalists of all time. He was truly unique. The like of such men as he number but few in our great history, for his ideas were original and very much in advance of his time.

His early manhood was spent, with every encouragement from his forward-thinking father, travelling both in this country and abroad, absorbing knowledge like a sapling oak taking up nourishment from the earth to produce a king among trees. His thirst for knowledge remained unquenched throughout his life. He was a man of great qualities, amongst which were to be found in abundant quantity, enthusiasm, perseverance, observation, judgement and, above all, great kindness to both men and beasts.

Bakewell was to provide the answer to the increased food demand which was an integral part of the Industrial Revolution, often held to have started in the year he took over Dishley Grange from his father in 1760. His objective was to improve every class of farm livestock, providing the nation with animals exclusively for meat rather than what had been available up to that time, namely aged draught animals retiring from the yoke.

Although his greatest success was to be with sheep, untold numbers of which were touched by the Dishley blood, no history would be complete without the story of the man who was justly named by J. Neville Fitt in 1876 as the 'Great High Priest' of Longhorn cattle.[1]

Bakewell was the descendant of an old and highly-respected farming family; its earliest records are traceable back to the late 1550s. There are also several other Bakewell families in the English Midlands, who may or may not have early connections to this one; there is no documentary evidence early enough to prove, for example, that the Bakewell family centred around Castle Donington is related to the family of Robert Bakewell of Dishley; this has led to confusion and mistakes over ancestry, made worse by the fact that the different families tended to use the same first names that were widely popular at the time. In addition, some previous writers have quoted from the Book of Bakewell, a beautifully produced, illustrated and bound volume in a private collection purporting to trace the history of Robert Bakewell's family over 600 years from the reign of Henry II. It is a supreme example of Victorian antiquarianism, lavishing love and care on a family project. But the resources for research available now were certainly not available then, and the standards of historical writing have changed; documented proof is expected now and the looser conclusions of the antiquarian are no longer acceptable. The family histories for the branches of the Bakewell family are detailed in Chapter 2 and the family trees can be found at Appendices 1.1-1.7.

In 1709 Bakewell's grandfather took Dishley on a 99-year lease from Sir Ambrose Phillips. Bakewell's grandfather (1643-1716) was the first of three Robert Bakewells of Dishley. The second (1685-1773) had, according to the obituary notice in the *Gentleman's Magazine*, 'the reputation of being one of the most ingenious and able farmers of the neighbourhood',[2] and it is from him that Robert (4), our Robert Bakewell, gained much of his pioneering spirit and enterprise. In 1809 the agricultural writer William Pitt remarked that 'Bakewell was encouraged by his father beyond all parallel and without limitation'.[3] It can therefore be concluded that Bakewell served a sound apprenticeship under the guiding hand of his very able father until, at the age of 35, he took over total responsibility for the running of the farm and from then on became increasingly famous beyond the boundaries of Leicestershire for his revolutionary methods of agriculture and superior specimens of farm livestock.

Longhorn bull and cow from an old undated engraving.

His holding, when visited by Arthur Young in 1771, was of 440 acres, 110 of which were arable and the rest in grass. The arable usually included 15 acres of wheat, 25 acres of spring corn and 30 acres of turnips. On this acreage he ran 60 horses, 400 large sheep and 150 'beasts' (cattle) of all sizes and various breeds.[4]

Mr R.E. Prothero (later Lord Ernle, 1851-1937) described Bakewell as resembling the typical yeoman who figured on Staffordshire pottery, 'a tall, broad-shouldered man, of brown-red complexion, clad in loose brown coat and scarlet waistcoat, leather breeches and top boots'.[5] This complements very well the image we have of Bakewell sitting astride his horse at Dishley as portrayed by the 18th-century artist John Boultbee (see front cover).

Writing in 1805, John Lawrence said that 'Bakewell had a kind face which showed a combination of intelligence and shrewdness. His manners had a rustic yet polite and pleasing frankness. He spoke with economy, and always to the point, and had a store of anecdotes and stories'. As to any political leanings, Bakewell probably supported the Whigs but was not politically active. According to Lawrence 'He lived and died one of the warmest supporters and staunchest defenders of liberty'.[6]

Dishley Church, the first written mention of which was in 1220, by the Bishop of Lincoln, was, and is, within the Dishley Grange grounds. It had never been very flourishing, and services were held there infrequently. Bakewell therefore followed his parents to the Unitarian Chapel on Warners Lane in nearby Loughborough, which was built in 1743. He became a trustee, like his father, in 1774, and in 1761 was also made a trustee of Mountsorrel Chapel, near Loughborough. Even throughout his later financial embarrassment, he continued with monetary assistance to these chapels. Religion was a serious part of his life and he would never conduct business on a Sunday, regardless of how important a visitor might be or how far they might have travelled. Later, when he formed the Dishley Society, it was written into the rules that no member could show rams on a Sunday.

Anyone visiting Dishley was shown the most generous hospitality, the door being always open to friends and strangers alike when they shared in Bakewell's passion for the progression of agriculture. His household was run for him by his capable sister, Hannah, who never married and predeceased him in 1793 at the age of 60 years. Hannah's domain, the great Mecca to which all agricultural students and enthusiasts flocked, including 'Russian princes, French and German Royal dukes, British peers and sightseers of every degree',[7] was a building of ancient construction. Illustrations of it appear in the background of some of the paintings of Bakewell's prize animals by John Boultbee.

One such view, when enlarged from a late 18th-century engraving, appears to show two parallel blocks of buildings on either side of a yard. The further or northern block looks to be of medieval origin, showing narrow, possibly lancet windows. The engraver has made no attempt to hatch the walls, as in the brick end of the timbered southern block, which could lead to the assumption that it was stone built. The massive end chimney indicates an original use as a brewhouse, laundry or something similar, when occupied as a principal grange to Garendon Abbey.

Following the dissolution of the monasteries in 1539, the grange buildings, like many others, were used for farming purposes, the lack of a farmhouse being overcome by the adaptation of the bailiff's quarters, or domestic buildings, which were easily converted. The curious curved structure at the eastern end, as yet unexplained, may be represented by an existing single-storey building with a bowed end, standing, as near as can be judged, in the same position, with a long stone wall which may be the rear wall of the northern block.

Robert Bakewell's house, the home of several generations of the Bakewell family, must have been close to these buildings, a fact borne out by their inclusion in the landscapes of his animal portraits. The substantial-looking building had obviously been adapted over the 200 years since its monastic ownership. Additions had been made at various stages of its life for the convenience of the occupants, giving it a look lacking regularity and compactness, but lending it a naïve style, which gave it pastoral simplicity and charm.

However, very little structural evidence from Bakewell's time has survived. His dovecote, capable of housing up to 640 pairs of birds, still stands guard over Bakewell's final resting place, along with other members of his family, inside the tiny, ruined Dishley Church. Of the farm buildings, only a couple of examples which would have been standing during his life are still in existence. Most of the others were probably demolished during a period of agricultural prosperity in the 1850s and their materials reused in a programme of modernisation. Further buildings were lost in the 1960s.

J. Neville Fitt, writing in 1876, portrayed a most poignant scene: 'Not long ago I chanced to be in the neighbourhood of Dishley, and I turned aside to see the grave of the man who had done so much for the peaceful industry of his country. I found it in the little, unused and dilapidated church close to the Grange, where his princely hospitality was dispensed, now only a dovecote and a building place for the fowls of the air'.[8]

Bakewell was surrounded by faithful servants who had been with him for many years, some of them from his father's time. One of his principles was that he would not employ anyone for a period of less than four years, at the end of which time it was rarely found that anyone would desire to leave. The minimum four-year term of service imposed by Bakewell upon members of staff may have been a contributory factor in the prevention of the secrets of his breeding policy becoming common knowledge. A contented and happy worker naturally becomes attached and loyal to his employer and an honest one will never betray a trust that has been shown to him.

Bakewell's animals were renowned for their extreme docility, probably in response to the kind treatment they received at Dishley. In 1771 Arthur Young described with amusement the way in which bulls would just stand still in the field to be examined by visitors: 'A lad with a stick three feet long, and as big as his finger, will conduct a bull, away from his companions, and guide him from one end of the farm to the other. All this gentleness is merely the effect of management, and the mischief often done by bulls is undoubtedly owing to practices very contrary, or else to a total neglect'.[9]

Dishley Grange as it looked in 1790.

In 1790, William Marshall gave another account of the Longhorn bulls which, although of frightening size and appearance, were so gentle that they could be, and were, handled by children – one even submitting to being ridden. 'At an age when most of his brethren are either foaming or bellowing with rage and madness, old C, a son of the old parent Comely, had all the gentleness of a lamb, both in his look and action. He would lick the hand of his feeder, and, if petted or scratched, would bow himself down almost to his knees'.[10]

John Lawrence, writing in 1805, related that he'd heard it said that 'One of Bakewell's work bulls was in the constant habit of dragging a cart by himself, without a driver, to a brewhouse two miles distant and returning home with a load of grains'.[11] This is the only reference so far found inferring that Bakewell used bulls for draught. It may have been just one incident of an exceptionally quiet animal having been used for this purpose so many times that he was able to complete the trip unaided. There are many instances of horses behaving in this way. No doubt we have all heard the many old stories of horses returning their owners to their homes after too much ale has been consumed!

This was all in complete contrast to the behaviour usually shown to animals at this period. Bakewell was always outraged when he witnessed the commonplace acts of cruelty perpetrated against all forms of farm livestock at the time. Drovers travelling the country, taking animals to markets and fairs, were considered to be amongst the foremost culprits. This behaviour was for the most part ignored by a population whose own lives were hard and living conditions intolerable, without wasting their thoughts and time on the suffering of 'dumb' animals.

Engravings by Garrard of Mr Honeyborn's New Leicester bull and cow of pure Dishley breeding.

Marshall further remarked that Bakewell was the only person of his acquaintance to leave calves on their mothers, a practice which had the effect of quick growth. 'The growth of calves reared in this way is strikingly rapid'. Here we see Bakewell leading the field by giving his cattle a good start in life, not only with the milk of the dam but also by providing calves with extra feed which they obtained from mobile creep feeders which he devised.

What a marked difference there must have been between Bakewell's and the neighbouring stock. On the one hand, pedigree animals on well-managed pasture with supplementary feeding, and on the other, local mongrel cattle on unimproved pastures, generally foraging for themselves. Not only was Bakewell paying particular attention to his animals' hereditary make-up, but also to their growth and well-being in the initial stages of life. The rest of this book proceeds to explore these and other aspects in greater detail.

Endnotes

1 Fitt, Neville J., 'Longhorn Cattle: Their History and Peculiarities', *Journal of the Royal Agricultural Society of England*, 2nd Series, Vol XII (London: John Murray, 1876), p463.

2 'Obituary of remarkable Persons; with biographical Anecdotes', *The Gentleman's Magazine* (November 1795), p969.

3 Pitt, W., *A General View of the Agriculture of the County of Leicester published by Order of the Board of Agriculture* (London: Richard Phillips, 1809), p217.

4 Young, A., The Farmer's Tour through the east of England, Vol 1 (1771), p123.

5 Prothero, R.E., English Farming Past and Present (Cambridge University Press, 2013 edn.), p184.

6 Lawrence, J., (ed.) 'Memoirs of Eminent Men who died between 1756-1798', *Necrology*, quoted in Housman, W., 'Robert Bakewell', *Journal of the Royal Society of England*, 3rd Series, Vol V (London: John Murray, 1894), p3-4.

7 Pawson, H.C., Robert Bakewell Pioneer Livestock Breeder (1957), p41. Pawson did not reference his quote.

8 Fitt, Neville J., 'Longhorn Cattle: Their History and Peculiarities', *Journal of the Royal Agricultural Society of England*, 2nd Series, Vol XII (London: John Murray, 1876), p463.

9 Young, A., The Farmer's Tour Through The East of England, Vol I (1771), pp113-4.

10 Copeland, S.W., 'Section XXII The Long Horned Cattle', *Agriculture Ancient and Modern*: a historical account of its principles and practice, exemplified in their rise, progress, and development, Vol 1 (London: Virtue & Company, 1866-7), p375-376.

11 Lawrence, J., *A General Treatise on Cattle, the Ox, the Sheep and the Swine* (London: Sherwood, Neely & Jones, 2nd edn. 1809), p326.

CHAPTER TWO

The Bakewell Family

Sue Brown

Our Robert Bakewell came from a family well versed in agriculture and the livestock business of the day. This chapter traces the direct family of Robert Bakewell of Dishley, from his earliest identifiable ancestor Thomas who died in the 1570s, Thomas's son John and John's son Robert, the first of four Robert Bakewells in direct succession, the fourth being Robert the Agriculturalist 1725-95. They have been numbered to help identify which Robert is being discussed at any one time. Also, the descendants of the four sons of Robert (1) have been researched in detail, partly to show their connections and partly to avoid any confusion with other Bakewell families in that part of north-west Leicestershire. Unless they are of direct relevance to our story, their details can be found in the Bakewell Family Trees in the Appendices. By following the family's evolution and the social and economic conditions of the time, we gain a better appreciation of Robert the Agriculturalist's development. It shows a family that had a business acumen, that understood the market for their livestock and was prepared to innovate, all characteristics that we shall discover in our Robert (4). We will also see how his relatives would help him at various times in different ways; there were a lot of relatives.

A word needs to be said here regarding the earlier writer about Robert the Agriculturalist – H. Cecil Pawson. Pawson described Bakewell as having an 'illustrious ancestry', but he was misled in this, as the lineage supplied by 'an historian' and used by Pawson in good faith, was not accurate, its research being far below his own high standards.[1]

The research that Pawson was unfortunately given for the early Bakewell family appears to follow 'the well-known method of arranging people of one surname in the order in which their names are found in deeds and records and connecting them with lines of descent'.[2] This latter comment was written in 1904 but holds true for many of the late-Victorian and early 20th-century fanciful compilations of family forebears. Not only are the early generations a flight of fancy but even the Normanton and Castle Donington Bakewells are confused, with the author claiming that John Bakewell, born at Normanton in 1639, married twice and was the ancestor of the Castle Donington Bakewells.

Robert Bakewell the Agriculturalist's family was rooted in the deep past of Leicestershire farming, probably from an ancestor who came from the town of Bakewell to the area in the 14th or 15th century and took his name from there. There were several Bakewell families in the north-west area of Leicestershire in the 16th century; all may be related but too far back for the records to show; only DNA could prove a connection. The problem with that would be finding a living male descendant of the Normanton branch, should there be any.

The probability is that the Castle Donington Bakewells descend from the marriage of a Thomas Bakewell of Snarestone and Joane Heyes who married in 1598 at Swepstone; their probable son John Bakewell, a baker of Snarestone, married in 1633, had a messuage at Snarestone. He then lived at Castle Donington and left his messuage at Snarestone to his eldest son John (whose baptism is not in the register) in his will written in 1656.[3] That John was a baker, like his father; he died in 1695, and there is an inventory for him.[4] This has previously been completely ignored and the two families conflated, thus misleading nearly all the Bakewell researchers into the Castle Donington family. This has been a welcome opportunity to put the record straight; Robert Bakewell's family deserve no less. The various Bakewell Family Trees can be seen in Appendices 1.1-1.7.

Leicestershire and Normanton le Heath

> Warwickshire and Northamptonshire are not so full of antiquities, large towns, and gentlemens seats, but this county of Leicester is as empty. The whole county seems to be taken up in country business ... particularly in breeding and feeding cattle; the largest sheep and horses in England are found here, and hence it comes to pass too, that they are in consequence a vast magazine of wool for the rest of the nation; even most of the gentlemen are grasiers, and in some places the grasiers are so rich, that they grow gentlemen: 'tis not an uncommon thing for grasiers here to rent farms from 500l. to two thousand pounds a year rent.[5]
>
> (Daniel Defoe)

The small village of Normanton le Heath in north-west Leicestershire is where this particular Bakewell family had its origins so far back as can be traced with any accuracy. As the name suggests the original landscape was heathland, part of the larger area of heathland that stretched from Donington on the Heath, Ibstock, Heather, Packington on the Heath, right across to Donisthorpe and Oakthorpe in south Derbyshire.

In the late 1500s, when the earliest information on this family is to be found, life in English villages was changing. From the time of the Black Death in 1348-9 onwards the close ties of the English peasantry with their manorial lord had loosened. The Peasants' Revolt of 1381 challenged the status quo, and on the manor tightly regulated services in return for land to farm were being replaced by money rent and written into the court rolls, their tenure known as copyhold. The vast majority of these farming peasants were husbandmen, in other words, tenant farmers renting their cottages and land for a set sum. By Tudor times some had become wealthy enough to buy some land, perhaps only a few acres at a time, but they then moved up the social scale and could be classed as yeomen, or even further into the gentry if they were very successful. Farming in the Midlands was mostly on the open field system; villages had two, three or four huge fields divided into furlongs or selions (blocks of land) with ridge and furrow with a reverse S bend, the remains of which are still visible on the landscape where deep, post-WW2 ploughing has not obliterated it.

The standard holding was by this time known as a yardland or virgate, the two terms being interchangeable. A yardland was reckoned to be sufficient to support a family; a holding of a half yardland probably meant that the farmer also worked at a supporting trade. The size of a yardland varied from manor to manor according to the quality and workability of the land, but was often between 20 and 30 acres. This was sometimes split so that only half or even a quarter yardland was held. With the paucity of information on Normanton, we do not know what size the yardland was there. But we do have information from Ashby de la Zouch where the yardland was 24 acres and the adjacent village of Packington where it was 28 acres, mostly split into single roods, there being four roods to the acre. This meant that a man holding a yardland of 28 acres could have his land contained in 112 separate ridges or 'lands' spread all over that parish. This was a highly inefficient way of farming, with a team of six to eight oxen needed to plough the land, depending on how heavy the soil was. Ploughing was a communal effort, men combining to produce the six to eight oxen needed to plough; it was only the better-off peasants who had more than one ox.

The manor of Normanton was acquired by Thomas Hord of Bridgnorth in Shropshire in 1522; at his death, his daughter Frances inherited; she married Thomas Fermor or Farmer of Somerton in Oxfordshire as his first wife. It is very unlikely that the local Farmer families living in Normanton and the surrounding parishes were connected to the lord of the manor who belonged to a well-known Oxfordshire family, despite the similarity of the surname. This is because the Fermor or Farmer family were catholic and had to submit a record of their income and land holdings. This has led to the survival of a detailed listing of the tenants of Normanton in 1717 and the extent of their holdings. Of the 16 Normanton tenants, one held three and a half yardlands and one held a quarter of a yardland; most of the rest held one or two yardlands with part of another. All were on long-term leases of 99 years for three lives, with names of tenant and heirs noted as well as the entry fine, the heriot payable on death and the rent. There were also eight cottages 'on the waste' with just the tenants' names mentioned.[6]

Much of the land in this part of north-west Leicestershire was rough heathland; an important part of the farming economy in both Leicestershire and Derbyshire was raising livestock, particularly horses. Several farmers from Normanton, particularly the Farmer and Carver families, were buying and selling horses at the important St Peter's Day Horse Fair at Market Bosworth, just over six miles away, during the years 1603-32.[7] This is probably why Lawrence Farmer and his heirs had to provide a coach mare to the value of £40 written into their lease as the heriot payable on their holding when the leaseholder died. Regrettably, none of the Bakewell family appeared in the short period of time for which the fair records survive, but they were certainly raising livestock at this date.

Information about Normanton records for the 1500s is sparse, there being no manorial records. The parish registers only survive from 1697 onwards, which is late. The bishops' transcripts, which were contemporary copies of the registers sent regularly to the bishop, have a patchy survival from 1569, so while there is some information on baptisms, marriages and burials before 1697, much is also missing. The main sources for the family at this period are probate documents and other material. Before 1752, the new year was reckoned to start on 25th March, consequently the dates 1st January to 24th March are usually for the following year; for instance 7th March 1650/1 (Old Style) becomes 7th March 1651 (New Style). All years have been shown here as New Style.

The period from around the 1620s onwards was a time of change in farming practice in this area of Leicestershire. The old ridge and furrow open fields were being enclosed by agreement of the landowners,[8] and the landscape was being divided into smaller fields, each man allotted roughly the same as he had held under the open field system. This was a vast improvement for livestock farmers as it meant that they could control which animals theirs mated with and could do what they liked with their land; they could sow different crops from their neighbours. The downside was that the ploughing was no longer a communal effort, nor was the sowing, harvesting and harrowing; however, the time saved in not having to take the plough and animals to the distant and well spread-out strips of land made a huge difference.

Thomas Bakewell of Normanton (great x 3 grandfather) and his family

The earliest known member of this family is Thomas 'Baykwell' who died at Normanton in 1575. He was almost certainly a husbandman, who held his land either as copyhold or leased from the lord of the manor. In 1564

Normanton was stated to have 12 families so it was a small village even for that time.[9] Thomas was not listed in the 1571 Lay Subsidy,[10] a tax levied on the wealthier portion of the population; only three people from Normanton were so listed. Nor was there any mention of the name in the earlier 1549 Lay Subsidy,[11] which included people of lesser wealth and which has a much larger proportion of names. This suggests that Thomas Bakewell had moved to the village after 1549 when he was younger. If Thomas was the first generation of the family at Normanton then it is not known where he came from, although the adjacent village of Packington is a possibility. There was a Bakewell family there from the late 1400s; William Bakewell of Packington was one of the better-off villagers there who paid four shillings on four pounds' worth of goods in the 1524 subsidy.[12] Unfortunately, there are no wills for that family at that date, only inventories, although they were amongst the leading farmers in that parish at the time.

Thomas Bakewell of Normanton had sufficient estate to make his will on 26th February 1575.[13] He mentioned no wife, implying that by that time he was a widower. His family consisted of two married sons, Robert, who had un-named children, was left 10 sheep; his two youngest were to receive a ewe each, implying that there were other children too. His other son John was also married; his children were named and left bequests which are mentioned below.

The inventory of his goods was taken on 16th (probably a scribal error for 26th) February 1575,[14] and came to a total of £16 7s 5d.[15] Although this seems a very small sum to us, he had goods worth making an inventory for and had a reasonably comfortable, if humble, lifestyle for the period. He had some of the basics: the pot, caldrons, pewter, an old mattress but no bed, although he has the bed coverings, sheets – mostly 'harden', so quite coarse and hardwearing; pillows and pillowbears (pillow slips), towels, a couple of coffers to store his goods in, a table with a form and three 'quesshynges' (cushions) to make life a little more comfortable.

What is missing from this inventory, quite apart from a bed, is any form of irons for the fire and cooking. This strongly suggests that he was lodged with someone else, which would most likely be one of his sons, probably John, whose children are named and given specific goods. Thomas was reasonably elderly for that time, perhaps born in the period 1510-30, which would make him aged between 55 and 75. The value of Thomas's household goods came to £6 7s 5d but the main value of his estate was in his livestock, in this case sheep – one score (20) male sheep, probably castrated, were worth £3 13s, he also had 40 ewes and 10 others valued at £6 6s 8d, a total of approximately £10, around 61% of the value of his other goods. At that time he would have kept sheep for the wool they produced, although there is nothing in his inventory to suggest any spinning or other processing. That quantity of sheep, 70 in total, would have necessitated renting some enclosed pasture for them to graze, as this number was far more than would be allowed on the land he may have held by copyhold from the manorial court. 'Beast stints' as they were called, were strictly allotted to prevent overgrazing on the common land in the period before enclosure.

John's children were not left sheep, apart from William who was probably the youngest (and last named). Robert was the first named and was to receive a Coffer; Lawrence,[16] his brother, was left the table and form; Alice got the main share of the household goods, with the five pieces of pewter, two bordeclothes (tablecloths), 3 pairs of sheets, a pillow bear, bolster and mattress, a pair of blankets and the chafing dish (a ventilated pottery warming dish using charcoal); these were goods she could bring to her future marriage.

There were also bequests of a sheep to John's two servants; Anne Jackson was to have a ewe sheep and Christopher a hogge (adult) sheep. Again, this suggests that Thomas was living with John; he knew his son's servants and liked them well enough to leave them the start of their own flock.

John was his executor and residuary legatee so he would have inherited the remainder of his father's farming interest and the remainder of the sheep, around 55 animals.

John Bakewell of Normanton (great x 2 grandfather) and his family

The little we know of John comes from his own will made 26th November 1599, by which time he was a sick man.[17] His last son was also John, who was under 24. The other children mentioned in Thomas's will were all still living and married with children apart from Robert. Lawrence had a wife named Anne whose children from a former marriage were left 12d each. Alice had married Richard Capenhurst of Netherseal; William was farming at Chilcote. Lawrence and William were given a yearling heifer to invest for their children, whilst Alice was left a two-year-old one. As they were married they had already had their share as was customary and they were established with families of their own. John, the youngest, was to have £30 in money when he came to the age of 24; he was still unmarried. He was also to have household stuff or husbandry gear to the value of 20 nobles; this would be about £6 13s 4d (£1,630 in today's money).[18] John was of Packington in the 1639 will of his nephew William Bakewell of Chilcote,[19] and had earlier witnessed the will of William Starkey also of Packington in 1632.[20]

The poor neighbours were not forgotten, seven of them were to have 4d apiece.

John's wife Jone and son Robert were joint executors and residuary legatees. This suggests that Robert was probably the eldest, or at any rate the most capable. There was also provision for young John to choose more goods, such as either the best cart or wain if his brother Robert died before his mother, but Robert's half of the goods were to remain to any children he might have had by then.

These goods given in John's will are really the only things we can draw upon for John's estate, as unfortunately no inventory has survived for him. He seems to have been farming on a larger scale than his father with 'best' wain or cart being specified, implying that there was more than one. The cattle, like the sheep given by his father, would give the recipients a chance to start their own herds.

His widow Jone Bakewell was buried a few years later on 5th August 1606.

Robert Bakewell (1) of Normanton, (great grandfather) and his family

When his mother died in 1606, Robert must have been around 30 to 35; he has already been mentioned as living in 1575 and we know from later Chancery records that he had a brother Lawrence, which fits with the above Robert living in 1574/5 and his father and grandfather's wills. He was evidently a responsible and respectable member of the small community of Normanton and held the office of churchwarden in 1607 and 1608 and later in 1637-8. He was also literate and signed the contemporary copy of the Bishops' transcript which is all that persists of the parish register entries for Normanton until the first surviving register starts in 1695.[21] His signature has also enabled a degree of certainty about some of his activities that have come to light.

This also tells us that he was the same Robert who made his will in 1669 and described himself as 'well stricken with age'; too true as he must have

1606 signature from Bishops' transcript.

Signature from his 1669 will.

been in his late 90s and getting on for 100 by then, so he must have had a good constitution. He was also much wealthier than his contemporaries as his inventory value was given at over £700, actually £666 when corrected, but still at least double that of the next wealthiest person in Normanton at that date, or indeed many of the surrounding villages.[22] So how did he become that wealthy?

His longevity is important, as there was more than one Robert Bakewell in Normanton during the first half of the 17th century and the signature of the Robert (1) who could be described as Robert 'senior', born in the 1570s, helps to differentiate between the two (or perhaps three Roberts). A 1634 reference to another Robert has been noted as aged about 40 and as 'junior', the son of

Robert Bakewell 'senior' in an ecclesiastical court case about John Hubbocke the parson and his drinking habits.[23] This Robert 'junior' would seem to be the one baptised in 1593 with no parent's name mentioned.[24] He is probably the Robert who made his mark – a sideways B, rather than sign his name, but he was not the son of Robert (1) 'senior', who would by then be aged around 60. The observation came from a witness in an ecclesiastical case, albeit a parson but not of Normanton, who may have assumed that the elder and younger Roberts were father and son. The two Roberts were more likely to be uncle and nephew as there seem to be only two Bakewell families in Normanton at that time, those of Robert (1) and Lawrence his brother.

Robert (1) married Margaret Webster at Netherseal on 25th May 1617.[25] Margaret was the daughter of William Webster; she was left a sheep in the will of her uncle, John Webster of Overseal, in 1612.[26] Margaret also seems to have been called Mary during much of her marriage. By the time Robert married, many of his contemporaries may have died; the average age of marriage at this period for men was in their mid-20s, and often only when they succeeded to the family farm; this was not the case with this particular Robert Bakewell whose parents had died in 1599 and 1606. However, this delay in marrying until he was in his 40s meant that it may well have helped him to be successful, as he did not have a growing family to feed at a time when he might be struggling to establish himself. The downside of course was that there was no wife to help or children to be taught to assist with the simpler jobs. Whatever the reason for this late marriage, it was a successful one. The marriage produced 10 children, of whom six survived to adulthood. Most of their baptisms have just the father's name, only two give the mother and there may have been other children, as the run of Bishops' transcripts from which this information is taken has many missing years. Four of those children, Thomas, Henry, John and Robert (2) are dealt with in more detail below, whilst those on whom there is little or no information are given here. All the children are listed below:

Anne (1618-1683) unmarried.
John (buried 1625).
Lawrence (probably died 1625).
Catherine (1627-1724). Catherine married Richard Burges of Swannington at Whitwick in 1654 as 'Kattren Baquell'.
Elnor was buried 23rd October 1632, daughter of Robert Bakewell.
Thomas was baptised 31st January 1631, son of Robert Bakewell. He was executor to his father in 1670 and died in 1713.
Henry was baptised 29th March 1632, son of Robert Bakewell. He died in 1718.
Dennis was baptised on Christmas Day 1637, the son of Robert Bakewell husbandman and Margaret his wife. The name Dennis comes from the

Webster family where it was a family name. He is almost certainly the son of Robert and Margaret who married in 1617. Dennis probably died as a child as there is no further mention of him.

John was baptised 7th January 1639, son of Robert Bakewell husbandman and Marie his wife. He died in 1716.

Robert (2) was born around 1643. There is no surviving transcript, but his gravestone at Dishley gives his age. He was the grandfather of Robert Bakewell the Agriculturalist.

The ages at death of Thomas, Henry and John all fit with their baptism details.

In 1626, Robert's (1) brother William died; he had been farming at Chilcote, some six miles from Normanton. His brother, Robert of Normanton, was one of the appraisers of his inventory.[27] This was unusual, as it was customary for the appraisers to be the honest neighbours of the deceased. It suggests a degree of responsibility felt by Robert for his younger brother and his family, and that Robert himself was a respected farmer. Although William did not leave a will, the wills of his son (William who died in 1649)[28] mentioned his uncle, John Bakewell of Packington, whilst his grandson, yet another William (who died in 1676)[29] mentioned his kinsman Thomas Bakewell of nearby Willesley. The Bakewell family clearly kept in touch with their relatives.

In the mid-1620s, two Chancery cases throw light on what this Robert Bakewell (1) and his fellow husbandmen of Normanton were doing to try and increase their wealth. These cases involved William Hubbocke, who was the Rector of Nailstone, the mother parish of the chapelry of Normanton, and the collection of the annual tithes that formed a part of the Rector's income.

The parson had the right to choose a tithe (tenth) of every kind of produce every year; it would be the best one, whether it was the best lamb from every 10, the best calf or sheaf of corn, whatever it might be. This lessened the profits of the farmers and was very much disliked. If tithes were paid in kind then the parson did the choosing, or he could 'farm the tithes' out to someone willing to pay the amount that the tithes were worth, or what were usually collected, and anything else that could be collected would be the 'farmer's' profit. This is where the word farmer comes from, not from the farming activity that we know today. Another way was for the parson to 'compound' the tithes for a fixed sum of money agreed with the parishioners to be paid yearly. In this instance, 'compound' means group together the individual payments of the tenants into a single sum. This had clear advantages for both tenants and freeholders, as in a good year they didn't pay more, but of course the downside was that in a bad year they still had to pay the same amount. This was as well as having to spend money on rent, food and clothing, seed corn for the next year and repairing or buying new tools. The weather, famine or a glut, could make or break a farmer. However, the parson was assured of a regular income without the inconvenience of collecting and dealing with all the different produce.

William Hubbocke, alias Hubbard, became Rector of Nailstone in 1592. He decided that compounding the tithes that were due was his preference. On 14th February 1614 he entered into an agreement with Robert Bakewell and his neighbours for them to collect the tithes of Normanton and all the glebe ground (land owned by the church) in the old and new enclosures of Normanton for 21 years in return for the sum of £55 a year.

This arrangement seems to have worked well until 1623 or 1624 when Hubbocke decided to retire to his other living, the more prestigious one of St Peter ad Vincula – St Peter in the Tower (of London). He appointed his son John the Rector in his place at Normanton. Bakewell and his friends seem to have regarded their arrangement with William finished, or perhaps sensed an opportunity to make more money, and agreed with John Hubbocke to collect the tithes for four or five pounds less. But, although John Hubbocke was Rector in name, it appears that his father still wanted John to pass the tithe money on to him. When William found out later about this arrangement he took the Normanton villagers to court for breaking their agreement and trying to defraud him. The first case was in 1626,[30] when he accused the lord of the manor, Sir Richard Fermor of Somerton in Oxfordshire, as well as Robert Bakewell and five others. Fermor's answer implied that by resigning, Hubbocke had ended the lease anyway. Hubbocke tried again in 1627; this time the case was just against Robert Bakewell and the other lessees, but the real problem was a lack of communication and disagreement between father and son, the alteration to the tithe money having caused a rift between them.[31] The defendants argued that as the tithes were an ecclesiastical matter then the case should be heard under the ecclesiastical court in Leicester, not a court of law. Hubbocke was accusing them not only of cheating him of tithe money but also all the other more petty dues that could be payable, such as smoke pennies, garden pennies, Easter offerings and the like, all burdens that would eat away at the income of everybody.

What is clear from the Chancery case is that Robert Bakewell was considered the leader of the group of Normanton men '… the s[ay]d Robert Bakewel the leader of all the rest…'; the others were Thomas and John Midleton, Thomas Smallie, Lawrence Bakewell (Robert's brother) and Thomas Green. Robert signed two of the documents and the others made their marks on one. Presumably after that, things calmed down, as by the 1630s the tithes of Normanton were leased out to someone else. This method of collecting the tithes continued until the 1780s in the parish, when a new parson wished to collect them in kind. That kept the lawyers busy for the next 20 years until the matter was settled around 1804.[32]

Nichols noted that Robert Bakewell was a freeholder in 1630,[33] together with Lawrence Farmer and Richard Middleton; this suggests that he already owned some land somewhere, but no further details were given.[34]

Robert continued to lease land with his neighbours including Richard Middleton, Thomas Green and possibly Anthoney Battell. When Thomas Green of Normanton died in 1640, Bakewell, Middleton and Battell were appraisers of his estate, which was valued at £258, a high value for the time.[35] At the foot of the inventory they list Green's leases which was unusual, but of course they also had an interest in them; the emphasis is the author's.

> the lease of the howse £20
> another lease houlden of Sir Richard Farmer £20
> another lease at Snebson *wch wee hould* of Mr Kendall £20
> another lease *which wee hold* part of from the Countyss of Dover £0 10s 0d

Clearly they were clubbing together to lease more land than any one of them could on their own. Again, this would mean that they could raise more livestock. Leases of land not in Normanton parish or adjacent would almost invariably be used either for livestock or rented out for income. It would be far too impractical to take a team of oxen with plough or harrows to work this more distant land; it was time consuming enough in their own village and one of the prime reasons for enclosing the open fields.

In 1652 another lease was entered into when Robert paid £120 for a 7-year lease on three closes at Ashby Shrubs in Desford. The lease was from Francis Muxloe of Desford who said that he owned the closes known as Ashby Shrubbs, Little High Crosse Close, part of the lane leading from Desford to Enderby between the two closes, together with the 'close or lane called the Playnes with the grounds set out for highways adjoining to the said ground called the Plaines leading from Hinckley towards Leicester and from Leicester towards Market Bosworth'. This land can be identified on earlier Ordnance Survey maps as where the B582 crosses the A47 Leicester to Hinckley Road. But the lease that Robert was given by Francis Muxloe was a fraudulent one as he did not actually own the land in question and there were previous incumbrances on it. Robert was evicted by the real owner, Thomas Muxloe, shortly after the lease was made and the money paid. That lease has not been found and our information comes from the subsequent complicated legal case to recover Robert's money which was brought by Thomas Bakewell, his executor, against Francis Muxloe's son, another Francis, after the deaths of both their fathers, probably in the early 1670s.[36] The outcome is unknown. The distance from Normanton indicates that these closes were intended to be used for raising livestock or to be rented out.

The Civil War (in England this lasted from 1642-1648) brought huge disruption to those who had backed the wrong side and whose estates were subsequently sequestered and sold to pay the fines accrued. This was the case with the late Henry Hastings of Ashby de la Zouch who had owned land at Alton Grange, just next to Normanton. In an indenture dated 3rd October 1653,[37] his widow, Dame Seymore Hastings, sold seven closes of land there to Christopher Pack, a citizen and Alderman of London, the money to go towards paying the £1,500 that her late husband owed. Three of those closes, Gossey Close, which was pasture, Ram Close and Thorny Close were in the tenure of Robert Bakewell the elder, another three closes were in the occupation of Robert Bakewell the younger and the other close was in the tenure of Thomas Ellyott, gentleman. Robert paid a rent of £75 a year for his three closes.[38]

It was at this time that Robert's daughter Catherine married Richard Burges of Swannington, as already noted. The marriage took place at Whitwick on 6th February 1654 and Catherine had a marriage settlement of £170[39] as her inheritance, together with another £50 payable on the birth of her first living child. This had been paid by the time her father made his will, as she was not mentioned in it although still living with a growing family. So not only was Robert making the money which he would have needed for his expanding rents, livestock, farming tools and seeds as well as family expenses, he was also having to lay out money on marrying off a daughter.

Robert evidently was good at making money as on 2nd April 1656 he could afford a 99-year lease on 80 acres of Swepstone Heath at a peppercorn rent for the first seven years and then at £20 a year afterwards; this was from Brian Cokayne of Swepstone, son of Viscount Cullen, another indebted Royalist. Just 18 months later, on 20th November the following year, Bakewell bought the same land outright for £742; it consisted of Heathfield Close 25 acres, Swepstone Heath 86 acres, a pasture called Swepstone Slade abutting the town of Swepstone containing 6 acres, Hey Close 34 acres, the two 'Withle or Willow Hedge' closes in Church Field Close and a piece of meadow ground adjacent. The sellers were the Right Honourable Charles Cokayne Viscount Cullen, Brian Cokayne esquire his son, Thomas Cokayne and Sir Richard Farmer, the latter being lord of the manor of Normanton. Charles Cokayne, first Viscount Cullen, had raised a Royalist troop and was hit with fines and sequestration; he was said to have lost some £50,000 in the King's cause.[40] The Cokayne family needed to sell land to recoup their fortunes.

This was a large amount of land to be bought by a husbandman and gave Robert a step up the social scale to yeoman; it was also conveniently adjacent to Normanton where the family was based.

In the hearth tax of 1664, Robert Bakewell of Normanton was assessed on two hearths.[41] There were only 19 houses listed in Normanton, most of which had one hearth but three had two hearths; there was no list of those who were too poor to pay the tax. In that year he was also assessed for tax in the Lay Subsidy; his estate at Swepstone was valued at £3 and he paid £1 1s in tax.[42]

This is in contrast with the subsidy of 1600 when his goods were valued at £3 and he paid eight shillings tax.[43] He had clearly gone up in the world.

Robert (1) finally made his will on 23rd December 1669;[44] he was buried at Normanton a few days later on 1st January 1670. As mentioned previously he described himself as 'being well stricken in years'; he must have been nearly 100 if he had been living in 1575. His wife Mary was still living – 'my dear and loving wife'; there is no suggestion that he had another wife before so, presumably, Margaret and Mary were the same person. Mary was given her bed and all the furniture belonging and was to be allowed everything necessary together with 'meate, drink and lodging up to £10 a year'.

Thomas was the eldest son and executor. He was left the 'ground called Swepston Heath estimated as four score and eight acres and all the Field or Stone pitt cloase containing 30 acres, both in my occupation for ever'.

Henry was left 'that ould Rick or Stack of Maslyn w[hi]ch standeth in Swebston Heath'.[45] This suggests that Henry was already married and had received his share. Henry's son Samuel would be baptised at Swepstone later in 1670.

John was to receive a messuage in Odstone in the tenure of Godfrey Dean as well as furniture. This too would make him a landowner.

Robert (2), the youngest son, was to receive £600;[46] £400 was to come out of his father's personal estate within six months of his death, the other £200 within 18 months to be paid by the executor. If the latter defaulted then Robert was to have the rents and profits out of the Heath Field bequeathed to the executor (his brother Thomas) as well as furniture.

No married children apart from Henry were mentioned in the will, nor were any grandchildren. However one of the witnesses to the will who signed his name was Lawrence Bakewell who would either have been his nephew or even his great-nephew (see Bakewell Family Tree, Appendix 1.6).

The inventory of Robert's goods was taken on 25th January and is very brief about the house and the goods therein. He had £40 worth of money and clothes; this was a lot when most people had just 10 shillings or a couple of pounds valued for their 'purse and apparel' in most inventories; it was probably mostly money rather than clothes. The money came no doubt from his dealings in livestock. His household goods within doors and other lumber were worth £41, some 6% of the total; that is all the description of his property that was not farming; there were no rooms named and no indication of any goods he owned at all. His husbandry goods, described as 'All Utensills belonging to husbandry' were valued at £14, just 2% of the total value. These would have included ploughs, harrows, carts, tools, all the usual farming implements but no details were given here. Next followed his crops and livestock; this was the part that was most important for the appraisers. The crops were 'corn within doors and without' worth £70, corn on the ground and clods (possibly marl) £20 and hay £25, this came to 17% of his estate. The rest was comprised of livestock and this shows that Robert was making all his money by raising livestock: 16 young and old horses and mares £110, eight score ewes £96, six score wethers £44, 90 hoggrills (juvenile sheep) £40, eight milch cows £29, 20 beasts £100, and four at £12, another 11 at £16 with eight calves £8. This section came to a huge 68% of his total wealth. Like his father and grandfather he was clearly involved in raising sheep for their wool at this period of time, as well as the horses and cattle. The appraisers were none too good at adding up, with a total value given as £739 rather than the £666 that the inventory actually totalled.[47] The above percentages have been calculated from the corrected total, now worth £128,600 (2023 values).

This information makes it clear that Robert Bakewell (4) came from a well-established family of farmers and graziers, although there is no indication that they were keeping cattle other than for milk and possibly for pulling work around the farm.

Before going through the descendants of Robert (1) in detail, this seems to be the place to mention that two of his sons, namely Thomas Bakewell (1631-1713) of Normanton and his brother John Bakewell (1639-1716) of Upton were mentioned in the Disclaimers at the Herald's Visitations because they either had no right to bear arms or they did not attend the Herald's Visitations.[48] The Heralds only summoned those of substantial means in the Hundred, from a list prepared by the High Court to substantiate their claims to bear arms. This suggests that either Thomas and John did not bother to attend or knew that their family didn't have the right to bear Coat Armour anyway, knowing that only one or two generations back the family were merely tenant farmers.

Thomas Bakewell 1631-1713, great-uncle of Robert the Agriculturalist (see Bakewell Family Tree, Appendix 1.2)

Thomas Bakewell was the eldest son of Robert Bakewell of Normanton; he was baptised on 17th January 1631. In 1669 when his father Robert made his will, Thomas was aged nearly 40 and evidently unmarried; he received a substantial bequest of land in Swepstone which he would otherwise have received on his marriage as was customary at this time. It was also probable that in view of his father's advanced age, he and his brother John were running the Normanton farm or farms by then as well. As his father's executor he also had to initiate the lawsuit in Chancery to try to obtain redress for his father's money spent on the fraudulent lease of Ashby Shrubbs in the 1650s.[49]

Thomas married Mary Allsopp sometime in the early 1670s; no marriage has been found but this was not a time when parish registers were well kept.[50]

Thomas appears to have taken on the role of head of the family after his father's (Robert 1's) death in 1669.

Thomas and his wife Mary had five known children between the early 1670s and 1690s. Thomas was their eldest son and possibly the Thomas baptised at Normanton on 15th July 1675, but no parents were named; John was the second son, born around 1676; then came Robert, baptised in 1678, followed by Joseph in 1680; another son, Samuel, was buried in 1704.

Thomas's wife Mary was buried on 27th September 1709 at Normanton.

When Thomas died in 1713 aged 83, his will showed he had land on Swepstone Heath,the Stonepit close and further land at Newton Burgoland;[51] all were rented out and the tenants named. His farm at Normanton is not mentioned in his will; it was held on a 99-year lease for three lives,[52] so would have automatically passed to his eldest son Thomas or, since his whereabouts were unknown, to Thomas's brother John, on payment of the entry fine.

But it is his probate inventory which shows that he had extensive livestock farming interests, the inventory total coming to over £800. He was living at Normanton le Heath in a large property for the time; perhaps this had been his father's. Thomas's inventory is more detailed than his father's but it is typical of this area in the 17th and 18th centuries with sparse detail on the rooms and the household goods and a much greater emphasis on livestock and crops which formed the bulk of the farmer's wealth in this period. So, like his brother, Robert (1), Thomas was building on the family tradition of farming and grazing livestock.

The Descendants of Thomas and Mary Bakewell

Thomas Bakewell was probably the eldest son. His father left him half of his lands in Swepstone in his will, 'if he or his heires come again', although he was 'long since gone away'. The family had clearly lost touch with him, not knowing if he was still alive. This Thomas, junior, evidently made no will, or if he did it mentioned nothing that would link him with his father's family. Another son was **Samuel** who was buried in 1704 at Normanton about whom nothing further is known; he was probably a young child. The other three surviving brothers were, **Robert (6)** who died in1707, whilst **John** and **Joseph** both died in 1739.

Robert Bakewell (7) 1705-1737 was the eldest child of Robert and Catherine and aged only 31 when he died. Like his brother-in-law Thomas Herrick, he was a lawyer working at Loughborough. Robert married Jane Allsopp on 20th October 1730 at Ab Kettleby. The Allsopp family were mainly based in Loughborough at this time rather than Ab Kettleby, so perhaps Jane was staying with relations. They had just one son, Robert (8) who was probably born in February 1732.

Robert Bakewell (8) 1732-1793

Robert (8) was baptised at Loughborough on 2nd March 1732, the 'son of Mr Robert Bakewell'. As an infant he inherited land in Swepstone from his father as well as from his uncle John in 1739; Uncle John had bought his land from Thomas Muxloe of Swepstone. Like his father, Robert also became a lawyer. He was probably the Robert Bakewell who was articled to Samuel Harris of Castle Yard, London, in January 1747 for a premium of £210.[53] He was appointed Recorder of Leicester between 1763 and 1765. With the approval of the borough he stood as an MP in the 1766 election for Leicester but withdrew. He fell out with the corporation and was dismissed from his post as Recorder on 11th April 1765 for insulting the mayor and two other Aldermen.[54] After this he served as a JP for the rest of his career. He leased and farmed land at Gresley Old Hall in south Derbyshire in the 1770s, but after a disagreement with the Gresley family, who owned that property, he returned to his grandfather's property at Swepstone in the early 1780s; this he named Tempe Farm. Here he built a large three-storey Georgian extension to the 1660s farmhouse. Presumably whilst he farmed at Gresley Old Hall, which is another three-storied building standing on a hill, he had enjoyed the extensive views over the landscape and wanted to replicate that at Swepstone. A portrait of him in his lawyer's bands was painted by Joseph Wright of Derby, for which he paid 21 guineas.[55] So, despite a career in the law, yet another Bakewell was still wedded to the land.

His grandmother Catherine (Muxloe) left him silver salvers and a silver snuff box as well as

> … my mourning ring enammeld with black which was given me at my Brother-in-Law Mr Joseph Bakewell's funeral, my Wallnutt card table and also all my books wherein the Name Robert Bakewell is wrote.

Through his lawyer uncle of Loughborough, Thomas Allsopp, Robert had important social connections. Uncle Thomas had married Ann Hare as his first wife; their daughter Elizabeth married Christopher Heath, the well-known Derby banker in 1761. Thomas's second wife was Elizabeth Glen whose sister Mary was the wife of the Revd Caleb Lowdham of Diseworth and mother of Caleb Lowdham junior, another lawyer of Loughborough. Their daughter Mary married Samuel Phillipps of Garendon in 1766, and after his death inherited the estate for life, which she then made the most of financially. Her second husband was Sir William Gordon MP.[56] who then held the Garendon Estate for his wife's life. So Mary became Lady Gordon; she died in 1796. This was definitely a jump up the social scale for the family. These two ladies were first cousins to Robert (8) of Swepstone.

It was this Robert (8) 'The Lawyer' who was the Robert Bakewell who advised his third cousin Robert Bakewell (4), the agriculturalist of Dishley, after his 1775 bankruptcy. Robert (8) bought the Swepstone property, which lay adjacent to his own land from his third cousin Robert (4) of Dishley, out of the latter's bankruptcy.[57]

It was in 1780 that coal was found on the Swepstone land, although no evidence has been found to show that Robert (8) exploited this.

> We are informed from Leicester, that Coal, of an exceeding good Sort, has lately been discovered in Swepston, in that County, upon the Estate of Robert Bakewell Esq; that the Delve is likely to prove an extensive one; and the Bed five Feet and a half thick.[58]

This part of north-west Leicestershire was subsequently heavily mined for coal.

Robert Bakewell (2) 1643-1716, grandfather to Robert (4) the Agriculturalist, and his wife Mary Paget (see Bakewell Family Tree 5)

Robert was born around 1643, the youngest child of his long-lived father, from whom he inherited £600. He didn't marry until around 1680 so he was nearly 40, late for that date but in keeping with his father and brothers Thomas and John. He married Mary Paget (Patchet)[59] at Heather on 13th April 1681. Their marriage contract, drawn up on 10th March 1680 (1681 New Style Calendar) and signed with a signature and his father's signet ring, shows that he was a yeoman of 'Huckelscoat';[60] Mary was the daughter of Thomas Paget, also a yeoman, of Overton alias Ibstock.[61] Mary's marriage portion was £200 and Robert was to purchase 'Landes or Tenem[en]ts of the full value of six Hundred pounds'. This was a substantial settlement and provided the basis for the next generation of Bakewell farming success. The Paget family were Presbyterians, farmers and later bankers; Mary's brother Thomas was also a successful farmer and his descendants were associates of Robert the Agriculturalist. Her father Thomas, 'Tho: Padgett' was noted as one of the 'most considerable Conventiclers'[62] of those who held Protestant meetings at their houses in Ibstock in the 1669 return of Nonconformists in Leicestershire.[63]

The land that Robert bought was in Swepstone, adjacent to that of his brother Thomas. At the time he bought it there were tenants whose rent would have provided income. Maybe they had moved on, as by 1690 Robert and Mary were living in the parish where they had their youngest three children baptised. Their elder children included **Robert (3)** born around 1685 and **John** around 1687; there may also have been other children whose baptisms went unrecorded. This is no surprise as this branch of the Bakewell family became staunch Nonconformists and Robert's birth year of 1685 can only be calculated from his gravestone at Dishley. The baptisms of their three younger children though were recorded at Swepstone, **Mary** in 1690, **Thomas** in 1692 and **Samuel,** born and died in December 1694. Robert's wife Mary was also buried there on 5th July 1695. Robert and his family continued to farm at Swepstone until the move to Dishley. Whilst living at Swepstone, Robert was named as overseer to his kinsman Joseph Starkey of Packington when Joseph made his will on 26th March 1692.[64]

Robert's(2) move to Dishley is mentioned later but perhaps here is the place to note that a 'Robert Bakewell' was tenant of a share of Long Meadow Close (3 acres 13 roods 6 perches) on the Garendon Estate from at least 1684-86. The survival of the rentals is very patchy prior to 1797 but there is a small number for the 1680s.[65] The rentals cover Garendon, Dishley and Shepshed without any distinction as to who is where; however, tenants, field names and acreages are given. This Robert Bakewell paid a rental of £1 7s a year for this land. If this is Robert (2) then it looks as though he was renting land where he could, just like his father did. It would also explain how he knew about the land at Dishley, what it was like and its possibilities. He may well have been alerted to this land by brother Thomas's father-in-law, Thomas Allsopp, the rector of Hathern, who died in 1682. It is not known how long he rented the land for, as the records no longer exist.

John Bakewell of London c1687-1755

One of Robert and Mary's children that did not move to Dishley was **John**, who must have been born before 1690, probably around 1687. He was apprenticed around 1702[66] to the apothecary Josiah Coleman[67] of Leicester; John became a freeman of Leicester in 1709.[68] He had entered one of the few professions that Nonconformists could practice. There were at least four apothecaries in the Bakewell family. The other profession that was also open to Nonconformists was the law, and whilst other Bakewells were lawyers, those relatives were Anglican and their careers were probably influenced by their relatives, the Allsopp family of lawyers. It was the descendants of Robert (2) and Mary (Paget) who were Presbyterians firstly and then Unitarians by the time of Roberts (3) and (4).

John Bakewell moved to London and in 1712 he became a Freeman and Citizen of London by Redemption on 22nd October 1712.[69] On 6th May 1714 he married Elizabeth Palmer at St Mary the Virgin Aldermanbury, London. She was the daughter of Archdale Palmer (1661-1732) of Wanlip, Leicestershire, and his first wife. Archdale's grandfather was one of the leading Parliamentarians in Leicestershire, sheltering ejected Nonconformist clergy. Archdale's sympathies were of a similar persuasion; his will of 1733 left 40 shillings each to six dissenting ministers in Leicestershire. He was also a Member of Parliament 1695-98.[70] This was another step up the social ladder for the Bakewell family.

John and Elizabeth had a son **John** who was baptised on 16th October 1718 at Old Jewry Presbyterian Church, London,[71] but nothing further is known of him and he probably died young. John and Elizabeth also had three daughters, **Elizabeth**[72] who married John Summers in 1737; **Mary** who was baptised in 1720 at the Old Jewry, was married in 1740 to John Barker; their third daughter **Ann**, was baptised in 1722 at the same church as John and Mary. John lived on Cheapside in the parish of St Mary-le-Bow; he took on at least six apprentices between 1718 and 1748. His final apprentice in 1748 was his nephew William Bakewell, son of his brother Robert of Dishley. As well as being a prosperous London apothecary, he was in touch with, and no doubt visited, his late sister Mary's Wildbore family at Nottingham. In 1739 he provided a mortgage for his nephew Cornelius Wildbore for his dyehouse and croft near the River Leen in Nottingham, and John's two married daughters appear on the 1759 assignment after his death.[73]

John made his will in 1741, however he did not die until 1755, when he was buried at St Mary-Le-Bow in Cheapside, London, on 26th April 1755; his will was proved shortly after.[74] John had done very well for himself; he left £1,000 to his daughter Ann, to make her equal with the same amount as he had already advanced to his other two married daughters. His brother Robert of Dishley and his wife Rebecca, together with his nephew Cornelius Wildbore were left £10 each for mourning rings. The residue was left to his wife who was to share it equally amongst their daughters at her death. By the time the will was proved on 4th June 1755, his widow Elizabeth was still living in the parish of St Mary-Le-Bow, when she renounced the proving of John's will in favour of her daughters Mary Barker,[75] who was by now widowed, and Ann Bakewell who was still unmarried. In 1758 Ann married John Saffory who was a jeweller and a widower. Their daughter Ann was born in 1761; she married John Roger Teschemaker in 1786. He was a Hanoverian who went into partnership in the hosiery business with **Robert Bakewell (5) of Nottingham** who is mentioned later in this section. Robert (5) was the son of John's apprentice William Bakewell and nephew of Robert (4) the Agriculturalist.

The family mentioned in the following section are relevant to our story, as some of the later generation moved to Dishley Grange during the time that Robert (4) lived there and helped with the farm or the household.

Mary Bakewell and the Wildbore family

Mary Bakewell, born around 1690, would have moved to Dishley with her father but this would have been only for a couple of years, so her subsequent family are being dealt with here. In 1711 she married Edmund Wildbore of

Nottingham, by licence; this gives her age as 21 and Edmund as 23; Edmund was a dyer. Their children were registered in the Presbyterian Castle Gate Meeting at Nottingham.[76] Their only son Cornelius was baptised privately 'being very weak' on 25th March 1712; their next two children were both called Mary, the first being baptised on 24th January 1714, was noted as very weak. Presumably she did not survive long as her sister Mary was baptised on 4th January 1716. A further child, Elizabeth, was baptised on 20th June 1717; she was said to have been born six weeks prematurely and was also very weak. It would have been after this birth, and probably from the effects of childbirth, that Mary died, although no burial record has been found.

In 1718 Edmund remarried by licence to Elizabeth Huthwaite or Huthaway at Nottingham St Mary on 12th September. The marriage licence described Edmund as a 'tinctor' (dyer),[77] aged 30, and Elizabeth was aged 27.[78] The marriage would have to have taken place reasonably soon after Mary's death as Edmund had young children, some not in good health, to be looked after, and a speedy second marriage was imperative and very common at this period after the death of a wife. Edmund and Elizabeth went on to have another two children, Hannah in 1719 and Susanna in 1722; after that the chapel register has many large gaps, so there may have been more children. Edmund appears to have died in 1730.

Cornelius became a dyer like his father. He also married twice; firstly in 1738 to Mary Curtis, again by licence;[79] she died in 1744 after having two or probably three children. Maria and Elizabeth were born in 1739 and 1740 respectively; there was also James Bakewell Wildbore born c1742 whose baptism has not been found; he became a Unitarian minister at Falmouth in Cornwall and he was mentioned in his sister Ann's will of 1817.[80] In 1751 Cornelius married Jane Smellie as his second wife. Cornelius and Jane had a further eight children: Hannah (1751), Thomas (1753), Dorothy (1754), John (1755), Rebecca (1756), Ann (1758), Robert (1759) and Jane (1761) all baptised at the High Pavement Presbyterian chapel in Nottingham. There was also Mary born around 1751;[81] she married Joseph Heath in 1772 aged 21. Jane's brother John was the John Smellie of Nottingham who was one of Robert Bakewell's (4) creditors in 1780; he died in 1817.[82]

Cornelius mentioned several of his Wildbore nephews and nieces in his will. He had a dyehouse on the River Leen known as Lealand's Dyehouse; the deeds for this show Cornelius mortgaging the property to his cousin John Bakewell of London in 1739; by 1784, Thomas, his son, was involved in the business and a deed dated 2nd and 3rd August 1784 seems to be the last mention of the Wildbores, father and son's involvement, as later deeds did not mention them.[83] Thomas and some of his siblings turn up at Dishley and were probably visiting from the 1760s, if not before. Their father Cornelius seems to have left Nottingham after 1784 to perhaps be nearer his son James, who was the Unitarian Minister at Falmouth in Cornwall. The reason for this assertion is that Cornelius died at Ottery St Mary in Devon where he was buried on 19th April 1789 at the Independent church there.[84]

The move to Dishley Grange

The move of the Bakewells to Dishley Grange came about when the farm fell vacant through the sudden death of Benjamin Clarke in 1707; Clarke was buried at Dishley but his burial was also noted in the register at Barrow on Trent in Derbyshire.

> Memorandum that Mr Benjamin Clarke Steward to S[i]r John Harpur of Caulke dyed suddenly and was found dead either at Quarendon or at Mountsorell in Leicestershire (coming from Leicester fair May the first about 9 or 10 of the Clock at night being Thursday) he was buried on Saturday May 3d at Dishley in Leicestershire.[85]

Somehow Robert (2) found out about Dishley Grange needing a new tenant. It is quite possible that he was already familiar with the immediate area, both from possibly renting in the 1680s as mentioned before or perhaps Robert heard about the farm at Dishley being vacant through his family connections and friends. As already mentioned, his eldest brother Thomas had married Mary Allsopp, daughter of the Rector of Hathern, Thomas Allsopp. The Rector of Hathern also had the cure of Dishley parish. Although his father-in-law had died in 1682, it is likely that Thomas still had friends and acquaintances in the parish who may have brought him up to date with the latest news. Thomas Bakewell had been on sufficiently good enough terms with his father-in-law to be one of the appraisers of his inventory in 1682.[86]

Although Robert (2) had lost his wife Mary by the time he moved to Dishley, he would have had the help of his daughter Mary for a short time, as well as his sons Robert (3) and Thomas. The main attraction of Dishley must have been the fact that the farm was comparatively large, comprising of about 350 acres,[87] with the fields all lying together. The medieval village of Dishley had been cleared by the Cistercian monks of Garendon Abbey, given over to sheep farming, and a large medieval barn built; after the Dissolution the land was enclosed and let out by the new owners of Garendon. At Dishley it would be easier to concentrate the farming, looking after and raising the stock etc. where no travelling was necessary. Land already acquired elsewhere could be rented out for income. The farm was also to be let on a long 99-year repairing lease from 1709, which meant that the Bakewell family would reap the benefits of any improvements or alterations to the land or buildings that were made. There was another very significant attraction; although Robert's lease does not survive, one made to John Sillcock in 1662 stated that he held Dishley Grange free of Tithes.[88] The payment of tithes to the parson of the Anglican church was very much resented by the nonconformists of all persuasions as they did not see that they should pay for the upkeep of a church that they did not attend, nor to the parson who would report them to the Archdeaconry Court for non-payment of their tithes, or the churchwardens for non-attendance at the parish church. His grandson Robert (4) would surely not have done as well as he did if he had had to pay one tenth of his yearly produce to the parson.

Robert (2) moved to Dishley with his daughter Mary and two sons Robert and Thomas, but as Robert (2) was heavily involved in raising livestock, there must have been several other men and women employed as needed on the farm and dairy as well; however, nothing has been left in the record about them.

Robert (2) described himself as 'aged and infirme' when he made his will on 16th April 1716,[89] dying on 6th July that year. His gravestone recorded him as being in his 74th year, old for that time, but nothing like as old as his father was when he died. This would make his birth year around 1643. His will did not deal with the tenancy of Dishley, which would fall to his sons Robert and Thomas under the lease. Robert was to have all the Swepstone land, whilst Thomas was given a close of land called 'Shorte Trees' at Narborough which was then in his possession. In addition, Thomas was to receive the £200 that was owed to his father on a mortgage by his brother-in-law Thomas Paget and another £200 as well, and 'the Blew Bead' or bed and furniture in the further chamber. Daughter Mary Wildbore and son John Bakewell were left five shillings as the final part of their inheritance, the rest having long since been paid. His son Robert (3) was the executor, and probate was granted at Leicester on 17th August 1716.

Robert's inventory was taken on 12th July, six days after his death. It is the only inventory of the property prior to 1840 and gives an idea of the size of the farmhouse of the medieval grange. There was the hall with a clock, chest, table and chairs (£4); the kitchen, where cooking was done, had the grate, a dresser with the pewter on it and other goods (£3); the parlour with two beds in it, still being used for sleeping; a 'Lined Room', perhaps lined with wallpaper or maybe where the linen was kept, as it had a bed and chest with linen (£7); the best chamber had a bed and table (£4) and perhaps had the 34 tod of wool (£30) that was stored in it; the room over the kitchen with another bed (£2); the dairy had the brewing equipment and the copper (£5), the room over it had another bed and lumber (£2). There was a cheese chamber with 'forty hundred of cheese'[90] (£37); this would be 400 cheeses, a considerable amount; there was also a bread room and 'mault milne' (£1 10s). Unlike his father's inventory we do at least have the names of the rooms and some of the household goods, although it is quite brief.

However, the livestock listed is very much like that of his father and brothers; it was by far the most valuable part of his goods. He was evidently breeding cattle; his bull and 25 cows were worth £105 as well as 14 beasts £77, nine yearling calves at £18, another 11 heifers and a bull worth £40, together with another three cows at £10 and 14 young calves worth £14. This alone was 42% of his value. When looked at with the rest of his livestock below, the total is £365 giving 58% of the total value.

His nine swine were valued at £14, which is quite high as pigs were not a valuable animal in inventories, often only being worth ten shillings or so in total. His horses comprised three colts worth £24, seven mares and three foals worth £70; again he was breeding them.

The sheep, all 90 barren ones (£63) and the 'one hundred and lambs' (sic, the word 'ewes' appears to be missing) were worth another £100. Their produce was the wool stored in the house; 34 tod of wool was a substantial amount – there were 28lb of wool to the tod[91] and 14lb to the stone, which would approximate to 68 stone of wool. This would have contributed a serious amount of money to his income; at this stage the sheep were being kept for their wool. Apart from the use of wool for clothing etc., the Burial in Woollen Acts 1666-80 meant all bodies should have a shroud of wool when they were buried and a fine of £5 was payable for non-compliance; this Act lasted until well into the 1770s and was not repealed until 1814. In the meantime, it ensured that wool found a ready market.

The inventory totalled £793 1s[92] and was on a par with his brothers Thomas and John.

Robert Bakewell (3) and his wife Rebecca Brodhurst, parents of Robert the Agriculturalist

Robert (3) would have been in his mid-20s when his father moved to Dishley; there would have been plenty of work for him and his brother Thomas. On 29th March 1723 Robert married Rebecca Brodhurst by licence at North Wingfield, Derbyshire. Rebecca was the daughter of Thomas and Hannah Brodhurst who farmed at Williamthorpe in that parish, some 30 miles to the north. It is likely that they met either through their various farming connections or through attending Nonconformist meetings, perhaps in Nottingham where the Wildbores were living.

Rebecca's father Thomas Brodhurst had moved to Williamthorpe in the 1690s from Kirk Langley, Derbyshire, where his father William Brodhurst had died in 1692. William left his son Thomas his lands and houses with the appurtenances in Over Haddon and Duffield, also in Derbyshire.[93] At the time of writing his will in 1689, William also mentioned that Thomas was shortly to marry Hannah Redgate, the daughter of Thomas Redgate or Ridgate of Langley Lodge; no other reference to the marriage survives.

Thomas's will left his wife Hannah all his houses and lands in Derby, Duffield and Mugginton; he had also been left property in Upper Haddon and

Hognaston, Derbyshire, by his aunt, Dorothy Brodhurst.[94] After his wife's death, the property left to her was to be divided equally between William, George and Rebecca. Thomas's inventory totalled £307 19s; he too was raising livestock, although not on the scale of the Bakewell family. There was also a son, Thomas, who was presumably the eldest and who had already received his share; he was left a token five shillings.

Before her marriage, Rebecca sold two houses in Derby; this suggests that her mother had passed them on to her as her share of the inheritance from her father Thomas, and she sold them to contribute to her marriage settlement. Only the lease survives or has come to light, so the amount she received is unknown.[95] However, as the ultimate survivor of her family, any land left in the family from Thomas's will would have ended up with her or rather her husband Robert (3), and after her death with her son Robert Bakewell (4).

After the marriage of Robert and Rebecca, her mother Hannah moved to Dishley, as well as Rebecca's aunt, Elizabeth Redgate and Rebecca's brother, William. George, the youngest son, does not seem to have lived at Dishley with them or if he did it was not for long; his father's will stated that he was to be kept at school until he was 14. George was the main legatee and executor to his aunt Elizabeth Redgate's will; by then he was living in Reading.[96]

Extra helpers on the farm in the form of the Brodhursts would have been a great advantage during the 1720s and 30s, although there would still be reliance on local labourers and skilled farmworkers as well as their wives, to run the farm. The period 1725-45 was the time when Robert and Rebecca Bakewell were raising their young family. Their eldest son **Robert (4)** was born in 1725, **Thomas** was born circa 1726, **John** around 1728, **Hannah** was born around 1732, **William** about 1734, **Rebecca** around 1738 and **Mary**, their youngest child, was probably born in 1739. Robert (4) is considered after his siblings, whose families follow here.

Hannah was the eldest daughter and she was to inherit £100 at the age of 21 from her aunt, Elizabeth Redgate, in her 1733 will. Her aunt no doubt recognised that as the eldest girl, young Hannah would spend much of her early life looking after the younger members of the family and probably would be heavily involved in the domestic side of running the farm if she didn't marry. It is possible that Aunt Redgate may have been in this position herself. In the event, Hannah never did marry and acted as housekeeper for her brother with his many visitors, thereby playing a most valuable role running that aspect of the farm. She died in March 1793, aged 62. There are gravestones for her and her three young siblings at Dishley.

The 1730s were a sad decade for the Dishley family. Robert and Rebecca lost three young children; Thomas on 15th September 1730 in his fifth year, John on 22nd September 1734 in his sixth year and Mary on 2nd January 1742 in her third year; there were also losses of the adults.

Rebecca's brother William Brodhurst died on 10th March 1731 aged 34 and a year later her mother Hannah died on 30th July 1732 aged 62; Rebecca was the administrator of her goods.[97] Robert's brother, Thomas, the son of Robert (2) and Mary who was born in 1692 at Swepstone, and who came to Dishley with his brother Robert (3) and family, died there in 1735. His brother Robert was the administrator of his goods.[98] The administration bond was for the sum of £400, suggesting that Thomas's estate was worth around £200 and that Thomas evidently played a major part in the farming of Dishley with his father and his brother Robert (3). Thomas was buried back at Swepstone where he had been born and where his mother was buried.

As well as running the farm, Robert (3) was kept busy with undertaking the probate of the will of his cousin Richard Burgis of Melton Mowbray,[99] by then a wealthy grazier, who died in 1712 and who had requested Robert to be his executor and joint guardian of his children with their mother Elizabeth until they came to the age of 21. As Richard's eldest child was only born in 1699 and the youngest in 1705, this would be a responsibility for a number of years. Some 11 years later in 1723, the land in Swannington left by their father had been mortgaged, and the mortgage was then assigned to Sir John Harpur of Calke for the payment of £1,550 to Richard's executors – James Adams of Ravenstone, gentleman (a cousin of Richard), Robert Bakewell of Dishley, yeoman and Elizabeth Burgis, widow of Richard Burgis of Melton Mowbray.[100] Robert would have been relieved to have wound that up.

He was also an appraiser for the inventory of Joseph Throne of Thorpe Acre in 1719.[101] Throne was a substantial farmer whose inventory totalled just over £303, most of the value being in the livestock. As 'Master Bakewell of Dishley' he had also entered into a bond, together with Thomas Garrat of Thorpe (Acre), with Thomas Throne of Hathern prior to 1737 for £40 for Throne's kinswoman Rachel Brocas; this was to be given up to Throne's executor to be cancelled so the money could be distributed according to his will.[102] Another responsibility was as one of the Trustees in the will of Nathaniel Harryman of Hathern in 1729; Harryman was a well-off yeoman and Robert (3) was also one of the appraisers of his inventory.[103]

The Bakewells and religion

Religion played an important part in the life of the Dishley Bakewell family, staunchly Presbyterian at first with the strong influence of the Paget family who in turn were most likely to have attended the meetings at Ibstock and neighbouring villages (including Normanton) held by the ejected minister of Packington, William Smith.[104] This evolved into Unitarianism and Robert (3) must have known Samuel Statham, who was the Unitarian minister at Loughborough from 1729 to 1760, very well.

Statham was the son of a clergyman from Banbury of the same name who had property in Loughborough where his mother was still living. In his will dated 4th May 1682,[105] the father left several of the cottages and land to his son Samuel who was under 16 at the time and perhaps very much younger. His cousins were left cottages as well. Samuel junior had been apprenticed to the apothecary Joseph or Josiah Coleman and became a freeman of Leicester in 1702.[106] It is likely that he knew Robert's brother John the apothecary as he had been apprenticed to Coleman at around that year, and just two years later in 1704 Coleman died. Samuel Statham was one of the appraisers of his inventory.[107] Samuel Statham was named with Robert (3) on the bond for the administration of the goods of Rebecca Bakewell's mother, Hannah Brodhurst, in 1732.[108]

As noted previously, in 1733 Samuel Statham senior was one of the six Leicestershire Presbyterian ministers left 40s in the will of Archdale Palmer of Wanlip.[109] In 1734 Statham was noted as living at Dishley[110] when their friend Thomas Hall, a maltster of Loughborough, made his will on 13th March 1734, again with Robert (3).[111] Together with Alexander Dixwell, they were trustees for Hall to receive his rents and deliver them to his father and to eventually sell his property. This was wound up in 1741 and their receipt for the same survives.[112] Further information on Statham and the Loughborough Chapel is to be found in the section on the Bakewells' Unitarian Network (Chapter 9).

Books were part of the household, but it is not surprising that the only one we know about is a religious one. In 1753 Robert (3) subscribed to a volume of John Holland's sermons that were published that year.[113] Holland was a young Unitarian minister who had delivered a sermon on 'The Folly and Guilt of Intemperance' preached to a society of Protestant dissenters at the High Pavement chapel in Nottingham in July 1750. Perhaps the Bakewell family had been there to hear it, although it is more likely that their Wildbore cousins would have been. There would have no doubt been other books, but little is known about who bought books unless people either subscribed to them, as above, or they were listed in probate inventories.

Early farming at Dishley

There are just a few references to farming at Dishley during this period in the Garendon Account book kept by Thomas Allsopp, the Loughborough lawyer who was a trustee for Ambrose Phillips (died 1737), and who continued the accounts up to 1747.[114] As well as the rent for the extra land Robert (3) was renting from the Phillips family, there are one or two other entries. He was called upon to provide men to open the 'drayns' in Dishley Lane in March 1732 and paid 6s 4d for that work; in January 1734 he was paid 4s 3d for two of his men for working for six days in Dishley Lane, and in November of that year Robert (3) provided two hammers for breaking stone in Dishley Lane. In February 1735 he was paid 1s 2d for his charges at Ashbourne Fair; perhaps he had been on business there for his landlord, as the amount suggests this was paid for his sustenance. In August 1737 he provided oats for Garendon: 'pd Mr Bakewell for oats &c delivered at Garenton £15 12 shillings and eight pence'. In December of the same year Robert (3) bought a 'pad' for two guineas; a pad is an ambling, quiet and reliable sort of horse, suitable for ladies riding side saddle, perhaps for Rebecca or other family members? He bought an unknown quantity of wood in December 1740, paying £9. In 1742 he paid 14 guineas, and in 1745 he paid £29 8s; both payments were for the hire of a stallion to cover his 14 mares, which suggests that although he was breeding horses, he did not have a stallion of his own. Another amount of £20 0s 6d was paid in 1744 'on ac[coun]t of the horse' which appears to be for the same although the mares and stallion were not specified, and again £13 7s in 1746 which may have been for the same. Robert was also paid two guineas in 1744 for looking after a horse 'for the horse standing'; in that same year he was paid another 2 guineas 'for L[or]d Huntingdon's Mare being covered' presumably at Dishley. In 1745, 'All[owe]d Mr Bakewell for the horse £1 11s 6d' appears to be paying Robert for the cost of looking after a horse.

Robert (3) was noted for being 'an ingeneous farmer',[115] and it was he who began to dig the drainage ditches to both drain and irrigate the fields. His son young Robert (4) would have found this of great interest and seeing the difference it made would probably have inspired him to take it much further, travelling to see other farms where experiments like this had been tried. He evidently had a receptive and creative mind, as is shown by the numerous ideas of improving both the land and the animals that are the substance of most of this volume. By the 1750s he was off on his travels to see how things were done elsewhere, both in this country and Europe. His father is said to have given him a free hand with his ideas for improving the farm.

Robert (4) and his siblings will have been brought up in the more liberal Christian way of thinking of the Unitarians rather than the comparatively straitlaced Anglicans. This shows in his creative ideas about breeding 'in and in' to fix the characteristics of whatever type of animal he was working on at the time. This attitude would have caused an outcry amongst strict Anglicans.

Besides Robert (4) and Hannah, there were two other Bakewell children that survived to adulthood, **William** and **Rebecca;** they and their children can be followed up in the Family History Appendices (see Appendix 1.5).

Dishley Grange under Robert Bakewell (4)

By the time the two Robert Bakewells, father (3) and son (4), took out a mortgage for £250 with Henworth Newton in 1756, the younger Robert would have been brimming with ideas. Pawson stated that he had been thinking about how to improve livestock from the age of about 22,[116] so Robert (4) is likely to have been the main driver for this 25-year mortgage; he was aged 31, but his father was 71 and probably doubted that he personally would pay it off. Robert (4) is reputed to have taken over the running of the farm by 1760. Robert (4) no doubt used the mortgage money wisely, as he had farm buildings to improve or build for his experiments and the house probably needed more rooms – the house in 1843 seems to have had quite a few more rooms than that of his grandfather's inventory in 1716. This is probably because more livestock would have needed more farm servants to look after them. The skilled live-in farm servants would have needed somewhere to sleep, the maid servants as well as the men; they were employed by the year. Additional farm labourers would be employed by the month, week or day as needed, and those who fitted in well with his ideas and work ethic were employed for longer than the minimum, often staying four or more years before moving on. There was also plenty of travelling to be done still and visitors to accommodate – Dishley Grange was the main house in the immediate area; there was no public house for travellers, only Dishley Mill, the toll house once the turnpike was opened in 1738, an old cottage nearby and Dishley Cottage. Married farmworkers such as John Bredon and William Peet would have lived locally, perhaps in Thorpe Acre, just a mile away, or Hathern a short walk across the fields.

Visitors were many, as were pupils. The latter came to learn from Bakewell and included the better-known ones such as George Culley and his friends the Colling brothers, Robert and Charles, all from Fenton and Ketton in Northumberland.[117] John Ashworth came from Daventry where his father ran a Dissenting Academy. He later ran the Wheatsheaf in Daventry and was involved in Bakewell's bankruptcy, both as a creditor and as an assignee in 1776. Others came from as far afield as Russia and Germany. John Holland has already been mentioned; he was taken on as an apprentice at a fee of £105 for four years from 21st February 1771.[118]

Another pupil was Thomas Baker, who had been travelling for two months in Ireland for Bakewell in 1789, when Bakewell wrote to Culley in November of that year.[119] He too wrote at least one letter to Culley on Bakewell's behalf in 1788.[120] Also included in that number was Benjamin Mather who was said to be a grazier of Dishley, Leicestershire, in his marriage bond of 1793.[121] He was the son of a baker of Nottingham and had family there. He may have been the Benjamin Mather concerned in a deed for land at Wysall, Nottinghamshire, for the erection of a Wesleyan Methodist Meeting House in 1826.[122]

Robert Honeyborn, who was Robert's nephew and subsequent heir, was at Dishley Grange from at least 1783, when he was mentioned as treasurer for the Dishley Society which was founded that year.[123] He too was writing to Culley, and letters survive from 1789 to 1795.[124] Apart from John Holland, these men were not ordinary apprentices, but were already established farmers or graziers and they were often described as gentlemen. They seem to have stayed at Dishley for several months at a time and undertook whatever jobs were available.

Although Robert's bankruptcy from 1776 is dealt with elsewhere (see Chapter 17) and must have taken up much of his time as well as being a great worry, it should be noted that he was still seen as a reliable neighbour; like his father, he would help the neighbours. In 1775, when he was clearly in trouble, John Pick asked Robert, together with Joseph Halford of Hathern, yeoman, to be trustees of all his goods, cattle and chattels and to pay the interest to Pick and his wife Elizabeth and then to the longest lived of them. John Pick was his nearest neighbour, being the miller at Dishley Mill. This indenture was mentioned in Pick's will dated 26th August 1778 when he detailed the recipients of his legacies.[125]

As well as keeping track of all that was going on at the farm, Robert (4) also found time to attend at least some of the regular quarterly meetings of the recently founded Leicestershire Agricultural Society. In 1791 Robert was on the Committee; they had the Annual Meeting on 21st December at the Three Cranes in Leicester. The meeting was chaired by William Pochin Esq., MP, and attended by notable farmers Richard Astley Esq., Robert Bakewell Esq. (8), Robert Bakewell (4), Mr Robert Burgess, Mr Valentine Green, Mr Thomas Paget, Mr John Parnham Stone, Mr Samuel Stone, Mr Richard Warner and nine others. They discussed amongst other things, the premiums to be awarded by the parish for bringing up children without assistance, with payments ranging from £5 5s to Humphrey Heywood of Donington for 16 children, to Thomas Chambers of Castle Donington £4 4s for 10 children. There were 28 claimants for servants in husbandry for long and faithful service and other matters. The next meeting was given as 11th April 1792 at the same place.[126]

Hannah Bakewell, his sister, oversaw the visitors' welfare and meals, as well as all the other household duties that fell to her, which must have been an extensive and responsible job requiring good organisational skills. In 1769, Rebecca Bakewell, their mother, died just before brother Robert's monetary problems began. His father Robert (3) died in 1773 and both parents have gravestones at Dishley.

By the late 1780s, Hannah was probably feeling her age and Robert (4) was not in good health, no doubt exacerbated by his drawn-out bankruptcy; yet the farm and his ideas continued to evolve. It is around this time that their Wildbore cousins appear to be living at Dishley. The fact that several of them are noted in different documents does suggest that as Robert and Hannah got into their late 60s, the Wildbores moved in to help run the farm (see the Family Tree at Appendix 1.7); the women would have been helping Hannah who died in 1793 aged 60. After she died, somebody would have had to cope with the visitors and the work in the house and dairy. Thomas Wildbore, their brother wrote a letter to Culley on Robert's behalf in 1787[127] and he seems to have been helping with the paperwork and maybe more. Thomas also appears on a deed of 1791, as of Dishley, gentleman, as a trustee to one of the parties involved;[128] and with his sister Ann Wildbore he witnessed the marriage of Richard Hitherly and Rachel Throne at Hathern on 8th September 1794, both of whom may well have been employees at Dishley.

Ann Wildbore was 'late of Sheffield' when, together with her sister Rebecca Kirkpatrick (née Wildbore) and Robert (4), they were parties to the assignment of a mortgage in 1793;[129] Rebecca, who was then of York, had married John Kirkpatrick at Sheffield Cathedral in 1792. John Kirkpatrick was a Unitarian minister.

Dorothy, another Wildbore sister, was a witness to the marriage, again at Hathern, of William Peet and Ann Wells 'both of Dixley' on 25th December 1770.[130] She later married Thomas Vennor,[131] a merchant at Sheffield Cathedral on 16th October 1780. William Peet was another of Bakewell's trusted farmhands; his wife probably also worked at the farm. Thomas, Rebecca and Ann Wildbore were all witnesses to Dorothy's marriage. Their sister Mary Wildbore must have at least visited Dishley when she married Joseph Heath, a widower and gentleman of Nottingham, as her residence was given as Loughborough in January 1772. Her father Cornelius appeared on the marriage bond.[132]

Dishley Employees

Bakewell is likely to have had many employees, but records of their employment are limited. Because of all the travelling that Robert (4) did, he must have relied heavily on his main employees to keep the farm running. The chief of these was **John Bredon** who had been his senior herdsman for 32 years,[133] and when he made his will on 8th January 1795, he described himself as 'Ground Keeper' (i.e., farm manager) and of Dishley.[134] He had married twice, firstly to Jane Locker at Hathern in 1765; their daughter Jane was buried on 25th December 1769 and his wife Jane just three weeks later on 14th January. His second marriage was to Ann Day of Castle Donington in 1773, and a son John was buried in 1782. However, Bredon had a son William and daughter Betsey living at his death in 1796; no baptism has been found for any of his children, suggesting nonconformity. He was a witness to the will of Robert Hutchinson of Hathern in 1782[135] and he was 'of Dishley Mill in the liberty of Thorpe, yeoman' in the will of his friend Joseph Halford of Hathern in 1790, who appointed him a trustee.[136] His daughter Betsey was also living at Dishley Mill with her husband George Smith, when their first child was baptised at Loughborough Unitarian Church.

William Peet was another long-serving employee, the superintendent of horses for nearly 40 years.[137] He may have lived at Knight Thorpe where William, the son of William and Mary Peet of Knight Thorpe, was buried at Dishley on 31st October 1760. William's wife Mary was buried on 30th January 1794 and William was buried at the age of 81 on 30th November 1809. He must have spent nearly all his working life at Dishley. Again, there are no baptisms for any children.

William Arnold had been junior herdsman for 20 years and was another long-serving employee.[138] Although he was a witness to the marriage of John Darby and Elizabeth Wilmott in 1776 at Dishley, there is no other record of him.

Edward Porter was the stallion master who must have led a travelling life. He would have constantly been taking the Dishley stallions to cover mares. His parents may have been Edward Porter and Ellen Halley, both of whom were of 'Dixley' when they married in 1758 at Hathern. Edward Porter junior was buried at Loughborough on 28th May 1818, aged 62.

There was also **Martin White** who was 'of Dishley Grange' when he was buried at Dishley on 9th March 1803; he may have been there in Bakewell's time, but this is the only reference to him that has been found. **William Beston** was a labourer of Dishley when he died in late 1806;[139] he too may have worked at the farm. The administration of his goods was taken out by his brothers Matthew Beston, a cordwainer of Woodhouse, Leicestershire, and Joseph Beston, a farmer of the same.[140] The value of William's estate was under £200.

Other potential employees have tentatively been identified where their marriages at Dishley were witnessed by some of the other Dishley Grange employees or family. So, the named employees are only the tip of the iceberg, as Bakewell was known to employ far more workers than was usually thought necessary. If the men were employed on the farm, then their wives would have also been employed in the dairy or kitchen; their children would also have been helping as soon as they were able to do so, perhaps with bird scaring, which was often the employment of a child as young as four. Children would also have become familiar with the farm animals and learnt how to look after them from a young age, hence the ease with which the boys handling the bulls impressed Arthur Young on his 1771 visit and William Marshall in 1790.[141]

The following have been identified as potential employees: **John Darby** and **Elizabeth Wilmott** were married in 1776, witnessed by William Arnold and Mary Burton. **William Ward** and **Ann Harrison,** both of Garendon, married in 1775, witnessed by Robert Bakewell (4) and Edward Burington or Berrington. **Jane Locker**'s marriage to John Bredon in 1765, was witnessed by **John Ashworth** and **Ann Kitchen**. **William Dexter** of Loughborough married **Ann Porter** of Thorpe Acre in the parish of Dishley and the marriage was witnessed by Edward Porter and **Martha Hopwell**. Ann could be Edward's sister. William Dexter may have been the nephew of John Pick, the miller at Dishley Mill.[142] **Richard Hitherley** and **Rachel Throne,** who married in 1794, were witnessed by Thomas Wildbore and his sister Ann Wildbore. Thomas Wildbore was also a witness to the marriage of his cousin George Smith with Betsey Bredon in 1797, as was George's brother-in-law Thomas Ryland.

As well as all the work around the farm, there were other local responsibilities. Virtually all parishes had the parochial offices of overseer of the poor, churchwarden, surveyor of the highways and parish constable. At Dishley, which also included the hamlet of Thorpe Acre, there are no surviving accounts for any of these officers, save for a churchwarden's account book well after the time of the Bakewells at Dishley. However, records for the collection of the land tax do survive; the earliest is 1781 to 1784, then a gap 1785-87, then the period 1787 to 1832 is complete.[143] The amount to be paid was fixed from 1776 onwards and collected by one or two Assessors appointed yearly for each parish; they did not get paid for this duty. The documents record the owner, occupier and amount of tax paid, signed by the Assessors. The information is not necessarily complete.

From 1781 Robert Bakewell (4) was assessed at £33 7s 6d for his land tax. In 1783 he was one of the Assessors; in 1784 he continued as Assessor and was joined by John Bredon. In 1789 Thomas Wildbore and William Arnold assessed the tax. Thomas Wildbore was Assessor thereafter in 1787, 1791 and in 1792 with Benjamin Mather, and in 1793 with Bakewell's nephew Robert Honeyborn; he also did the job in 1794, 1795 and 1796. Although a few other men were also listed, it seems to be the Dishley Grange people including pupils and employees who were most often tasked with this job, which was probably time consuming, visiting every house and persuading people that they needed to pay, and returning, perhaps more than once, for people who were either away or who hadn't the necessary cash to pay when he called.

Towards the end of his life Robert (4) subscribed to volume 1 of John Monk's An Agricultural Dictionary, consisting of Extracts of the most celebrated authors and papers, published in 1794.[144]

Robert Bakewell's (4) Last Actions

There had been a tendency in the family for the menfolk to marry late, often because they were so busy developing their farming businesses. With his extensive farming, family and religious networks, not to mention his travelling, Robert would have met many women in the course of his earlier life, but he clearly had other priorities; then his bankruptcy, coming in mid-life, may have taken up too much of his attention to look for a companion; then again, he had his sister, Hannah, to look after all those matters that, in those days, a wife would be expected to undertake. So, in the end, Robert never married.

Having been ill for some years[145] Robert (4) finally made his will on 30th September 1795. He was generous in his bequests and had clearly recovered from his bankruptcy. His nephew Robert Bakewell (5) of Leicester was left £1,000. Nephew George Smith was left £3,000. His nieces Rebecca Kenrick, Mary Smith and Hannah Smith were left £500 each. The remainder of his estate was to be inherited by his nephew Robert Honeyborn who was also his executor and who took over the running of Dishley Grange after Robert's death. The will was proved on 29th October 1795.[146]

Like all the Bakewell wills, the provisions made in his will show that Robert the Agriculturalist was financially astute. It hardly suggests the financial incompetence that has been suggested elsewhere as a contributing reason for his bankruptcy.

He had had a remarkable life and was clearly well ahead of his time. He was not only a farmer but a businessman, constantly thinking of how to improve his livestock and farm and ultimately make money. The infecting of his sheep for the butcher with liver fluke so that they would be worthless for breeding should they fall into more unscrupulous hands, his canny methods; all suggest a business-like approach to his work. Furthermore, he had inherited much of his characteristic searching for constant improvement from his great-grandfather Robert (1). Robert the Agriculturalist had taken this to a whole new level with his innovations and experimentation.

The Character of Robert Bakewell (4)

A century later, a summary of the character of Robert Bakewell included the following:

> From his father Bakewell had an excellent training for practical & experimental farming, besides many of those special mental qualities, possibly inherited either immediately from him or through him, which were manifested in his advance beyond the traditional notions & practice of the old English farmer. 'His father,' says the writer of the obituary notice in The Gentleman's Magazine, 'had always the reputation of being one of the most ingenious & able farmers in the neighbourhood.' According to Arthur Young, who inspected the operations at Dishley on two occasions, with the space of fifteen years between, the irrigation, which was one of the most prominent features of the Dishley husbandry, had been begun by Bakewell's father.
>
> In Necrology, (Necrology: London, 1805. Article on Bakewell by "Benda"), John Lawrence, in his General Treatise on Cattle, says: 'I

formerly gave the best sketch in my power of the life & character of Bakewell in a volume entitled Necrology'(memoirs of eminent men who died between 1756-1798, edited by Lawrence). Bakewell is described as 'tall, broad in the chest & shoulders, with a benevolent countenance combining intelligence & sagacity. His manners had a rustic yet polite & pleasing frankness. He spoke neatly in few words, always to the purpose, & had a store of anecdotes & stories.' In politics he does not appear to have allied himself decisively with any party, or to have classed himself under any name; but the same writer tells us that 'he lived & died one of the warmest supporters & staunchest defenders of liberty'. He adds that one of the principal blemishes of Bakewell's character was a certain degree of acquired cunning, 'the vice of his profession'. The writer first quoted (in the Gentleman's Magazine) describes him as tall, broad set, & in later years rather inclined to corpulency, his countenance bespeaking intelligence, activity, & a high degree of benevolence; his manners frank & pleasing; well calculated to maintain the popularity he had acquired. Sir John Sinclair described him as 'a person of strong natural sagacity'; & another authority as a man of unimpeachable morals, whose conversation was never disgraced with expletives. In appearance (says Mr Prothero in his Pioneers & Progress) he resembled the typical yeoman who figures in Staffordshire pottery, 'a tall, broad shouldered, stout man, of brown-red complexion, clad in a loose brown coat & a scarlet waistcoat, leather breeches, & top boots'. In his kitchen he entertained Russian princes, French & German royal dukes, British peers, & sightseers of every degree. He never altered the routine of his daily life. 'Breakfast at eight; dinner at one; supper at nine; bed at eleven o'clock; at half past ten, let who would be there, he knocked out his last pipe'.

The benevolence of countenance, remarked by more than one of those who have described his personal appearance, was the true index of a characteristic mentioned by many who have recorded, from personal acquaintance, their estimates & impressions of him as a man. These all concur in showing that he was eminently kind-hearted, & that his natural kindliness made him the friend of man & beast. He was surrounded by old & attached servants, & so much disliked losing sight of a familiar face that he would not engage a farm man for a shorter term of service than four years. After spending four years with such a master, the man was seldom found who desired to change his place. But the happy relations of the master & servants at Dishley seem to have belonged to the traditions of the house of Bakewell, for several of the people who remained with the last Robert Bakewell to about the close of his life had lived for some years in his father's service. 'How long have you been here?' a visitor asked, in July, 1793, turning to William Arnold, the junior herdsman. 'About 20 years' was the reply. 'And you?' addressing the senior, John Breedon, who promptly answered: 'Since the King was crowned, sir' ---fully thirty two years. William Peet, superintendent of the horses, said he had served the family for nearly forty years, but had been away a few years & returned to Dishley. Others had been ten or twelve years in service.

Mr Bakewell's kindness to brute animals was proverbial, & being in constant practice at Dishley was rewarded with extreme docility in the farm animals. Powerful bulls of terrible presence, looking the more formidable for the immense horns distinguishing their breed, were led about by mere children. One writer says he saw an animal of elephantine bulk led about with a pack thread by a boy of seven; another, that a lad with a switch could single a bull out from his companions & guide him to any part of the farm by holding the switch to one side or the other to indicate the way; & a third had been greatly amused by a little boy, five years old, mounted upon one of the big bulls, & so guiding him with the point of his switch. Similar instances of docility, resulting from unvarying kind treatment, were noticed in the stallions; & throughout the live-stock departments of the Dishley farm confiding gentleness, as an effect, afforded the surest evidence of considerate & compassionate gentleness as the cause. On this subject Mr Bakewell was far in advance of his day, for his generous anger was kindled instantly by the sight or report of cruelties so often practiced in the times when the sufferings of the inferior animals, however discreditable & degrading to man who inflicted them, were thought beneath the notice of the law.[147]

The Dishley Grange Farm

The information we have about the Dishley Grange Farm at the time of the Bakewells is limited. Ironically, the best period of survival for the Garendon Estate records starts just after Bakewell's (4) death. A survey, which must have been taken shortly after Thomas March Phillipps inherited in 1796, gives full details of the tenants and their holdings.[148] It is undated, but clearly made between 1796 and 1800, as it was followed by a second survey that was dated in the latter year.[149] 'Mr Honeyborn late Robt. Bakewell' was given as the tenant. It gives a description of the Dishley farm with field names, acreages, value per acre and the yearly rent at that time. The acreage of The Homestead was given as 4 acres 1 rood 3 perches, followed by the fields, a total of 486 acres 1 rood 23 perches, with a rental of £711 3s 5d. It is the description of the farm as a whole which is of most interest though.

A Particular of the Estates in the County of Leicester belonging to Thomas March Phillipps Esqr.

> Dishley Grange – NB This Farm is in general a fine deep loam with a small mixture of Sand and in some places bordering on a Clay. – it is fenced with White thorn Hedges in fine preservation and well protected from the bite and other injury by Cattle and the Ditches and Drains kept in the utmost good condition, the Lands – remarkably clean and well cultivated, the Fields were much divided by the last Tenant Mr Bakewell, who planted on the borders of the Farm a large quantity of Withy which serves for shelter and furnishes poles to protect the Hedges and many other uses.
>
> The Drill Husbandry is here practiced with great success. Wheat but little sown, Barley Oats Vetches Clover and Turnips are much Cultivated of which these Lands produce the greatest abundance.
>
> The House in good repair, the Outbuildings, which are many pretty generally so. Let to Mr Honeyborn at £735. [150]

At the end of the survey is a summary with the acreage as given above, with the previous rent of £552, the new rent of £711 3s 5d and the difference between the two, in this case £159 3s 5d. It is noticeable how many of the farms and cottages on the Garendon Estate were in poor repair, with the land in very poor shape and in need of proper drainage and management. In contrast, Dishley Grange stands out as clearly as the best managed farm. The 1800 survey lists all the tenants again, giving their holdings and the leases which all the tenants had agreed to for improving their farms; the increase in rent from 1810 is also noted; in the case of Robert Honeyborn, this went up to £1,100 per annum. This steep increase in rent would have lessened the profitability of the farm. Honeyborn had agreed to keep the farm in good order in 1800.

Robert Honeyborn was married on 10th August 1797 by licence at Chelsea to Elizabeth Bratt who was living there; she was four years older than Robert, and there would be no children.

By 1810 Honeyborn may well have been in failing health, as he made his will on 22nd February of that year, although it was several years later, in early 1816 that he died. The nearest to an obituary in the papers seems to be a brief note of his death: 'On Tuesday the 19th inst. Robert Honeyborn, Esq., of Dishley, near Loughborough, in his 64th year, successor to the great Bakewell'.[151]

The first sale of his farm stock was three months later on 10th June, at which 'a select Part of the very valuable Stock of Rams' was to be sold, with the rest of the sheep to be sold along with the cattle and horses in September.[152] Evidently, although this sale attracted interest it was not a success:

> The sale of the nineteen rams at Dishley, near Loughborough, on Monday last, attracted a great concourse of very respectable company,

> amongst whom were many of the principal breeders of the midland part of the kingdom. Mere curiosity was evidently the predominating principle of attraction with the major part of them; and as the men of business, viz. the bidders, were therefore very few, consequently the lots, with the exception of the last, were knocked down, to use a figurative expression, almost at their own prices. Had the great Mr Bakewell lived to have witnessed this defalcation in the value of the far-famed Dishley rams, (*many of which, in his day, fetched more in being let out for one season only, than the fee-simple fetches now!*) ...

The second auction took place on 16th and 17th September 1816: '... the remaining Stock of highly useful Rams, the valuable Ewes and Theaves, and the Whole of the capital Stock of long-horned neat Cattle'.[153]

However, his stock needed yet another auction to finish selling up, this one on 25th and 26th March 1817. This was described as:

> All the remaining valuable stock of Long-horned neat cattle, ewe hogs, Draught horses, pigs & the implements of husbandry ... consisting of about 60 head of neat cattle, among which are some capital bulls, and some prime fat beasts; about 70 valuable ewe hogs, 14 draught horses of different ages; a quantity of pigs, and a large variety of implements of husbandry of all descriptions'.[154] The sale of household goods was to take place on the 31st of March: 'All the excellent, genteel and useful household furniture, plate, linen, books, paintings, brewing vessels, and numerous remaining effects'.[155]

Robert Honeyborn's brief will was proved in the Prerogative Court of Canterbury on 30th December 1816, some nine months after his death; his widow was left everything.[156] She left Dishley Grange and was living in Sloane Street, Chelsea, London where she died in 1844 aged 86.[157] Elizabeth's will left all her goods to her niece Prudence Rhodes Archer.[158]

Robert Honeyborn was succeeded at Dishley by William Smith, 1817-33, and William's son Richard Smith from 1833-43. The latter left in unfortunate circumstances which eventually led to his bankruptcy. However, a sale of Richard's goods after he was evicted from the farm gives us an insight into the farmhouse of the 1840s.

The household and other goods were sold at auction on 26th, 27th and 29th March 1844, commencing the day after Lady Day (25th March) when Richard was evicted. Described as: 'all the valuable Household FURNITURE, Glass, China, Dairy and Brewing Utensils, and other Effects, the Property of Mr Richard Smith'. The goods were itemised as 'the Dairy and Brewing Utensils, Laundry, and part of the Kitchen Requisites' which were sold on the first day; the second day saw 'the remainder of the Kitchen Requisites, Passage, Parlour, Butler's Pantry, and Servants' Bed Rooms'. The final day included 'The Furniture in the Breakfast, Dining and Drawing Rooms, Staircase, Bedrooms Nos. 1, 2, 3, 4, and 5, and Dressing Rooms'.[159]

This, so far as can be ascertained, seems to be the only piece of information on the rooms at Dishley Grange since the inventory of Robert Bakewell (2) in 1716. Clearly the house had undergone considerable enlargement since then. In 1716 only the following rooms were mentioned: the Hall, Kitchen and room over, Parlour, Lined Room, the Best Chamber, Dairy and room over, Cheese chamber and Bread room.

It was considered important to give a detailed account of Robert's (4) family and upbringing so that readers would better understand the context from which he developed as a livestock breeder, details of which follow.

Endnotes

1 Bakewell, B.G., The Family Book of Bakewell, Campbell P, (1896), p4-9.

2 Barron, F.S.A., Oswald, commenting in a review published in *The Ancestor, A Quarterly Review of County and Family History, Heraldry and Antiquities.* No. IX, (April 1904), p232.

3 ROLLR, John Bakewell, 11 December 1660, of Castle Donington.

4 ROLLR, John Bakewell, 10 July 1695, of Castle Donington.

5 Defoe, D., A tour thro' the whole island of Great Britain divided into circuits or journies (1724-1727), Letter 7, Part 2: East Midlands, Vol 2 from Vision of Britain – The Travels of Daniel Defoe.

6 The National Archives (TNA), FEC 1/1194.

7 Record Office for Leicester, Leicestershire and Rutland (ROLLR), MF 154.

8 Not to be confused with the Parliamentary Enclosure of the 1760s onwards which required an Act of Parliament to be passed.

9 Nichols, J., The History of the County of Leicestershire, Vol IV, Part 2, (1811), p812.

10 TNA, E 179/134/205; ROLLR Farnham's Village Notes, Vol 5, p277.

11 TNA, E 179/134/165.

12 TNA, E 179/134/324.

13 ROLLR, Thomas Baykwell, 28 April 1575.

14 The date appears to be before the will was made which implies an error in one of those dates.

15 Now £6,092.00. Measuring Worth (2024), 'Purchasing Power of British Pounds from 1270 to Present'. URL: www.measuringworth.com/ppoweruk/

16 The christian name Lawrence is also spelt Laurence, I have used Lawrence except in quotes.

17 ROLLR, John Bakewell, 10 May 1601.

18 One mark = 13s 4d, half a mark = 6s 8d or one noble. Now £1,814. Measuring Worth (2024), 'Purchasing Power of British Pounds from 1270 to Present'. URL: www.measuringworth.com/ppoweruk/

19 Staffordshire Record Office (SRO), William Bakewell, 2 April 1649.

20 ROLLR, William Starkey, 15 September 1632.

21 ROLLR, DE 1061 and DE 5021.

22 Now £128,600.00. Measuring Worth (2024), 'Purchasing Power of British Pounds from 1270 to Present'. URL: www.measuringworth.com/ppoweruk/

23 ROLLR, Archdeaconry Court, Trymnell v Hubbocke, Box 4/XVI, 1, 1634/5, MF 609, and Box XVII 1635/6 95/98, MF 610.

24 Not an uncommon occurrence in registers of that period.

25 Margaret also seems to have been known as Mary, but the majority of the children baptised and listed in the register transcript with just the father's name do seem to be of one family, and she was Mary in Robert's will with no indication of a former wife.

26 ROLLR, John Webster, 24 April 1612.

27 SRO, William Bakewell, 17 November 1626.

28 SRO, William Bakewell, 2 April 1649.

29 SRO, William Bakewell, 24 March 1675/6.

30 TNA, C 8/40/9.

31 TNA, C2/ChasI/H61/24.

32 Nottinghamshire Archive Office (NAO,) DD/BK/3/1-12.

33 Nichols, J., The History of the County of Leicestershire, Vol IV Part II, (1811), p812.

34 Nichols, J., The History of the County of Leicestershire, Vol IV Part II, (1811), p812.

35 ROLLR, Thomas Green, 8 April 1641.

36 TNA, C6/222/7.

37 ROLLR, DE2178.

38 State Papers Domestic 1642-1660, Government papers from the Civil Wars and Interregnum, (1875); TNA SP23/161/407.

39 TNA, C10/357/7.

40 Northamptonshire Archives, Administrative/biographical background Viscounts Cullen.

41 TNA, E179/251/4, Part 8.

42 TNA, E 179/134/315.

43 TNA, E 179/134/254.

44 ROLLR, Robert Bakewell, 2 July 1670.

45 Maslin was a mixture of wheat and rye sown together.

46 Now £115,900. Measuring Worth (2024), 'Purchasing Power of British Pounds from 1270 to Present'. URL: www.measuringworth.com/ppoweruk/

47 Now £128,600.00. Measuring Worth (2024), 'Purchasing Power of British Pounds from 1270 to Present'. URL: www.measuringworth.com/ppoweruk/

48 Rylands, J.P., FSH, A list of Persons who were Disclaimed as Gentlemen of Coat Armour by the Heralds at the Visitations in the various Counties of England, (1888).

49 TNA, C 6/222/7.

50 Various spellings, I have used Allsopp which seems to be the favoured spelling of the later generations.

51 ROLLR, Thomas Bakewell, 9 September 1713.

52 TNA, FEC 1/1194.

53 UK Register of Duties Paid for Apprentices Indentures, 1710-1811 at https://www.Ancestry.com

54 Chinnery, A., Records of the Borough of Leicester 1689-1835, V, Hall Books and Papers, (1965), p210-217.

55 Nicholson, B., Joseph Wright of Derby, Vol 2, p128, plate 208.

56 Married 1776.

57 See that action under Robert Bakewell's bankruptcy.

58 BNA *Northampton Mercury*, 13 March 1780.

59 Various spellings for this name have been standardized to Paget unless quoting.

60 Hugglescote is just three miles to the east of Normanton.

61 ROLLR, DE 365/226.

62 A Conventicle was a secret, unlawful or irregular religious meeting, being different from the established (Anglican) Church, it commonly referred to dissenters meetings. These meetings were banned by the Conventicles Act of 1664 and those attending could be fined.

63 Evans, R.H., Nonconformists in Leicestershire in 1669 in the *Transactions of the Leicestershire Archaeological and History Society*, p106 & 127, volume 25, (1949).

64 ROLLR, Joseph Starkey, 3 June 1692.

65 ROLLR, MF 182, 35-6, 37-8, no numbers appear on the microfilm and the order is mixed until the end section. Also ROLLR, A Survey of Garendon Manor, (1685), p2.

66 14 was the customary age of apprenticeship, John's birth year is calculated on that basis.

67 Josiah Coleman died in 1704 but there is no mention of whom John was turned over to for the rest of his apprenticeship.

68 Hartopp, H., Register of the Freemen of Leicester 1196-1770, (1927), p203.

69 London Metropolitan Archives (LMA) Freedom of the City Admission Papers, 22 October 1712.

70 History of Parliament online; TNA, PROB 11/657/190.

71 TNA, RG4/4408.

72 The Old Jewry registers are deficient for a few years after 1722.

73 NAO, CA/TC/2/Box/64/2.

74 TNA, PROB 22/816/ 98, John Bakewell, 4 June 1755.
75 Mary Bakewell married John Barker of the Close of Salisbury 17 July 1740 at St Mary-Le-Bow.
76 TNA, RG 4/1586.
77 Tinctor is Latin for dyer.
78 Blagg, T.M., and Wadsworth F.A., (eds.) Nottinghamshire Marriage Licences Vol II, British Record Society LX, (1935), p.152.
79 Blagg, T.M., and Wadsworth F.A., (eds.) Nottinghamshire Marriage Licences Vol II, British Record Society LX, (1935), p.378.
80 TNA, PROB 11/1602/136, Ann Wildbore, 9 March 1818.
81 Mary Wildbore born c1751 may have been the Hannah baptised that year and later known as Mary.
82 Jane and John Smellie were the children of John Smellie, a Nottingham hosier who died in 1783.
83 NAO, CA/TC/2/BOX/64/2.
84 TNA, RG4/2100.
85 Derbyshire Record Office (DRO) D2534, Barrow on Trent register 1707.
86 TNA, PROB 4/2659, Thomas Allsopp, 20 June 1682.
87 Arthur Young said the farm in 1771 was 440a.
88 ROLLR, 35/28/272.
89 ROLLR, Robert Bakewell, 7 August 1716.
90 Forty hundred in this case means 400 cheeses, not a hundredweight.
91 This amount could vary.
92 Now £140,800.00. Measuring Worth (2024), 'Purchasing Power of British Pounds from 1270 to Present'. URL: www.measuringworth.com/ppoweruk/
93 SRO, William Brodhurst Kirk Langley, 18 Nov 1692.
94 SRO, Thomas Brodhurst, 26 March 1718.
95 DRO, D754/T/11.
96 ROLLR, Elizabeth Redgate, 24 May 1740.
97 ROLLR, Hannah Brodhurst, 1 February 1732.
98 ROLLR, Thomas Bakewell Admon., 1 May 1737.
99 ROLLR, Richard Burgesse, 30 January 1712.
100 ROLLR, DG32/12.
101 ROLLR, Joseph Throne, 29 August 1719.
102 ROLLR, Thomas Throne, 1737, probate date illegible.
103 ROLLR, Nathaniel Harryman, 15 July 1730.
104 Evans, R.H., 'Nonconformists in Leicestershire in1669', in the *Transactions of the Leicestershire Archaeological and History Society*, Vol 25 (1949), p106 & 127.
105 TNA, PROB 11/385/346, Samuel Statham, 3 December 1686.
106 Hartopp, H., Register of the Freemen of Leicester 1196-1770, (1927), p193.
107 ROLLR, Josiah Coleman, 14 December 1704.
108 ROLLR, Hannah Brodhurst, 1 February 1732.

109 TNA, PROB 11/657/190, Archdale Palmer of Wanlip, 23 February 1732.

110 There was only the mill, toll house, an old cottage, Dishley Cottage and the farm Dishley Grange at Dishley, there was no other accommodation.

111 ROLLR, Thomas Hall, 2 December 1738, his goods were valued at under £20 at probate.

112 ROLLR, M 703, the receipt was for £366 1s.

113 https://www.ancestry.co.uk UK and US Directories 1680-1830 under schools, directories and church histories.

114 ROLLR, DE 7454/4.

115 Pawson, H.C., Robert Bakewell Pioneer Livestock Breeder, (1957), p18, quoting a writer in The Gentleman's Magazine.

116 Pawson, H.C., Robert Bakewell Pioneer Livestock Breeder, (1957), p18, quoting Garrad writing in 1800.

117 Stanley, P., Robert Bakewell and the Longhorn Breed of Cattle, (1995), p47-8.

118 Ancestry at http://www.ancestry.co.uk under Tax Lists. UK, Register of Duties Paid for Apprentices' Indentures 1710-1811.

119 Pawson, H.C., Robert Bakewell Pioneer Livestock Breeder, (1957), p147.

120 Northumberland Archives (Nbl A) ZCU/13.

121 Nottinghamshire Marriage Bonds 1791-1800, Abstracts of the Bonds for Marriage Licenses of the Archdeaconry Court of Nottingham (University of Nottingham, 1987), p93.

122 ROLLR, N/M/207/379 (DE 2265).

123 Stanley, P., Robert Bakewell and the Longhorn Breed of Cattle, (1995), pp24-27.

124 Nbl A, ZCU4.

125 ROLLR, John Pick, 11 May 1780.

126 BNA, *Leicester Journal,* 30 March 1792.

127 Pawson, H.C., Robert Bakewell Pioneer Livestock Breeder, (1957), p119.

128 ROLLR, DE1107/210.

129 NAO, DD/TB/1/1/21.

130 ROLLR, DE731/4.

131 Thomas Vennor died in 1787, there were no children.

132 ROLLR, Leicestershire Marriage Bonds, 13 January 1772.

133 Pawson, H.C., Robert Bakewell Pioneer Livestock Breeder, (1957), p42.

134 ROLLR, John Bredon, 22 March 1797.

135 ROLLR, Robert Hutchinson, 25 September 1783.

136 ROLLR, Joseph Halford, 10 October 1793.

137 Pawson, H.C., Robert Bakewell Pioneer Livestock Breeder, (1957), p42.

138 Pawson, H.C., Robert Bakewell Pioneer Livestock Breeder, (1957), p42.

139 This was at a time when the parish clerk recorded the deceased person's abode as either Dishley or Thorpe Acre in the Burial Register.

140 ROLLR, William Beston Admon., 1 January 1807.

141 Stanley, P., Robert Bakewell and the Longhorn Breed of Cattle, (1995), p9.

142 ROLLR, John Pick, 11 May 1780.
143 ROLLR, MF 43.
144 Accessed at Ancestry https://www.ancestry.co.uk UK and US Directories 1680-1830 under schools, directories and church histories.
145 Pawson, H.C., Robert Bakewell Pioneer Livestock Breeder, (1957), p157.
146 TNA, PROB 11/1266/257, Robert Bakewell, 29 October 1795.
147 *Journal of the Royal Agricultural Society of England,* 'Robert Bakewell', (1894), p1-31.
148 ROLLR DE 7454/1.
149 ROLLR DE 7454/3.
150 ROLLR, DE 7454/1.
151 BNA, *Stamford Mercury*, 29 March 1816.
152 BNA, *Aris's Birmingham Gazette*, 27 May 1816.
153 BNA, *Aris's Birmingham Gazette*, 2 September 1816.
154 BNA, *Stamford Mercury*, 14 March 1817.
155 BNA, *Leicester Journal*, 21 March 1817.
156 TNA, PROB 11/1623/418, Robert Honeyborn, 30 December 1816.
157 St Luke, Chelsea, parish register accessed through Ancestry at https://www.ancestry.co.uk
158 TNA, PROB 11/2005/63, Elizabeth Honeyborn, 25 September 1844.
159 BNA, *Leicester Journal*, 22 March 1844.

CHAPTER THREE

Bakewell and the New Leicester Sheep

Pat Stanley

> The breed is originally *Lincolnshire*, but Mr *Bakewell* thinks, and very justly, that he has improved it.[1]
>
> (*Arthur Young, 1771*)

Robert Bakewell's greatest and most lasting achievement must be his development of the Dishley or New Leicester sheep. The origin of this animal was described by both Arthur Young and George Culley as being of a Lincolnshire variety, although others have proffered a variety of different progenitors. However, Culley, who was Bakewell's first pupil, had monitored Bakewell's breed from their beginnings and closely followed their development. According to Pawson,

> Bakewell *had in mind* a type of sheep which, whilst helping to fertilise the land and also producing wool, would make a major contribution to the food of the nation and to the prosperity of the farmer's potential as a producer of meat.[2]

Mr Ferryman, who spoke to many of Bakewell's contemporaries, said that 'Bakewell had formed in his own mind an ideal perfection, which he endeavoured to realise, and that with this view he, with unwearied perseverance, year after year, and at something more than market price, selected from the flocks around him such ewes as possessed points, which were most likely to produce the animal he wished for'. David Lowe wrote in 1842 that Bakewell himself told Mr Chaplin, an 'Old Lincolnshire' breeder, that at one time he had used Old Lincolnshire rams.[3]

From a speech given by Sir John Sinclair, president of the first Board of Agriculture, to the Society for the Improvement of British Wool at Edinburgh in January 1791, we are told that all the large breeds of English sheep were of recent introduction, within half a century before that date. If so, they were newly-imported or newly-developed breeds when Bakewell began his work of improvement. But Gervase Markham wrote in 1614 that Leicestershire sheep were 'large boned and of good shape. ... Old Leicester was large, heavy and coarse-woolled, giving a 15-in. fleece of up to 12lb in weight'. Markham also said that Lincolnshire had the largest sheep. Lisle writing in 1757 in reference to the 17th century, gave a warning to the effect that only a 'lesser size' of Lincoln tup should be used on ewes to avoid excessively large lambs.

However, in Like Engend'ring Like,[4] Russell expresses the view that large sheep may indeed have been introduced to this country not long before Bakewell's time, as suggested by Sinclair. Any previous progress to that of Bakewell must be accredited to Joseph Allom, reported by William Marshall as being of Clifton, Nottinghamshire. After starting in farming as a ploughboy, Allom, through sheer hard work, became one of the best-known sheep breeders prior to Bakewell. Two conflicting theories of the origin of Allom's sheep say that he bought them at a Lincolnshire market or, the idea more favourably looked upon, that he purchased them mostly from a Mr Stone of 'Godeby' near Melton Mowbray. In Allom's day it was certainly customary for the superior farmers to purchase their ram lambs from him for between 2 and 3 guineas.[5]

Although the first few steps down the long road to improvement may have been taken by Allom, it was Bakewell's ability and perseverance which won the day. The way in which this achievement came about caused great debate at the time amongst the very top people of the agricultural profession. Different suggestions were made regarding the breeds of sheep Bakewell was thought to have used, but Marshall implies that there was no other ingredient in their making than his total belief in the use of breeding in-and-in, using the best specimens of one breed. The term 'in-and-in' means to repeatedly breed from animals of the same stock that are closely related; in other words, breeding from a male and female of the same parentage; this is also called linebreeding.

It is interesting to study and compare the descriptions of the Dishley sheep given by two highly competent men, namely Marshall and Culley.

Marshall, writing a description of Bakewell's superior ewes and wethers in top condition, said that their heads were long but small and hornless, with the ears somewhat long and standing backwards, and the nose shooting forward. The neck was thin and clean towards the head, but taking a conical form, standing low and enlarging every way at the base. The fore-end was altogether short, the bosom broad and the shoulder, ribs and chine exceedingly full. The loin was broad and the back level. The haunches were comparatively full towards the hips but light downwards and altogether small in proportion to the fore-parts. Legs were of moderate length with extremely fine and remarkably light bone throughout. The pelt was thin and the tail small. The wool was shorter than longwools in general but much longer than the middle wools. The ordinary length of the staple was 5-7 inches varying in fineness and weight. 'The carcase, when fully fat, takes on a remarkable form: much wider than it is deep; and almost as broad as it is long. Full on the shoulder, widest on the ribs, narrowing with a regular curve towards the tail; approaching the form of the TURTLE, nearer than any other animal I can call to mind'.[6]

Culley explained that the variety of Lincolnshire first selected by Bakewell was, 'with singular discernment and great attention', changed to that of a sheep unknown in any former period, and which surpassed all other breeds in their ability to fatten and the return given for the amount of food consumed. He described the distinct differences from other longwool breeds: 'The fine lively eyes, clean head, straight broad flat back, the barrel-like form of the body, fine small bones, thin pelt and inclination to early maturity. The mutton fine-grained and of superior flavour. Wool averaging 8lb to a fleece and in length from 6 inches to 14 inches. Wethers killed for best profit at 2 years when making between 20-30lb a quarter'.

When this is compared to the 'Old Leicestershire' breed of sheep we can fully understand Bakewell's great achievement. Marshall described a ram of the 'true old Leicestershire sort' which was shown at Leicester Ram Show in 1789. 'This creature might be said to be in the lowest state of degeneracy. A naturalist would have found some difficulty in classing him; and, seeing him on a mountain, might have deemed him a nondescript; a something between a sheep and a goat'.[7] Marshall had described a ram of the old Warwickshire breed as the worst he had ever seen, portraying it as a truly ugly animal, of large frame and heavy bone, with legs that were long and thick, culminating in large, splayed feet, the body being extremely sharp and angular; it was not even improved by good pasture, and animals generally did not go for slaughter before three or four years of age. However, having seen the Old Leicestershire ram, he decided it was worse than the Old Warwickshire.

Marshall thought that the new breed surpassed any other that he had seen and spoke of their 'positive beauty'. It appears that the most distinguishing characteristic of the sheep was its fullness and weight in the forequarters. It was usual for this breed to possess a projection of fat on the ribs immediately behind the shoulder. This gained the technical term of *foreflank*. This will seem to us most strange when in today's world an improved sheep would be heavier in the legs and saddle and not in the shoulders and breast, but Bakewell was catering for the market of the second half of the 18th century when fat mutton was the diet of the poor man. Bakewell was thus providing for the increased masses of the poor working class, his main aim of improvement being to supply their needs. Fat mutton at this time could be purchased for the equivalent of 5d per lb as opposed to 8d per lb for bacon. When salted and pickled for some time the mutton took on a striking resemblance in taste to bacon.

Marshall had the opportunity to study two carcases side by side, one of a Dishley, the other a Norfolk sheep. The differences were striking, the bone of the Norfolk being almost twice that of the Dishley and the meat of the Dishley three times thicker than that of the Norfolk. This was evidence enough for

Leicester wethers, the property of G.L. Foljambe.

Marshall to say that the Dishley was of more value to the consumer, for 'while mankind eats flesh, he throws away bone'.

There were two reasons why the wethers (castrated males) were killed at two years of age; first, this was when the most profit was to be had, and second, if kept any longer they became too fat for the more refined palate.

The following measurements were recorded by Arthur Young at Dishley, on 17th March 1770:

> I have this day measured Mr Bakewell's three year old ram, and find him as follows:

	ft	*inches*
His girt(h)	5	10
His height	2	5
His collar broad at ear tips	1	4
Broad over his shoulders	1	11½
Broad over his ribs	1	10½
Broad over his hips	1	9½
This day measured a two year old barren ewe:		
Height	1	11
Girt(h)	5	9

> Breast from the ground, the breadth of four fingers.
>
> N.B. I would have measured her breadth but for a fall of snow.[8]

Marshall tells us that 'The *degree of fatness* to which the individuals of this breed are capable of being raised, will, I am afraid, appear incredible to those who have not had an opportunity of being convinced by their own observation'. He had seen two- to three-year-old wethers so loaded with fat they were hardly able to run 'and whose fat lay so much about the bone, it seemed ready to be shook from the ribs on the smallest agitation'. In 1786 he saw wethers barely a year old selling for 27 to 28 shillings a head as opposed to the common sheep which sold for 18 shillings each. In February 1785 at Lichfield, Marshall was shown a forequarter of mutton by Mr Prinsep of Croxall which had 4 inches of fat on the ribs. Another extraordinary point in this breed was that the rams and some of the females were termed as 'cracked in the back', that is they had a channel running the length of the chine. Any animal showing this condition was said to be of the very best blood.[9] In October 1783 James Bolton of Alnwick is on record that when his three-year-old wether of the Dishley breed was killed it measured 7⅛ inches of fat upon the ribs and had a back like the fattest bacon.

The Dishley was not only remarkable for its mutton being fat but also for its fine grain and excellent flavour. The grain in the meat of other breeds tended to be coarse and spongelike.

Bakewell attracted a great many followers with his Dishley breed of sheep and numbered amongst them were many leading men of the Midland counties.

Endnotes

1 Young. A., The Farmer's Tour through the East of England, Vol 1, (1771), p117.

2 Pawson, H.C., Robert Bakewell Pioneer Livestock Breeder, (1957), p58.

3 Lowe, D., On the Domesticated Animals of the British Isles, (1842).

4 Russell, N., Like Engend'ring Like, (1986).

5 Marshall, W., The Rural Economy of the Midlands, Vol 1, (1790), p381-2.

6 Marshall, W., *ibid.*, Vol 1, (1790), p338-389.

7 Marshall, W., *ibid.*, p379.

8 Young, A., *ibid.*, p114-5 & notes.

9 Marshall, W., *ibid.*, p396-9.

CHAPTER FOUR

Ram Letting and the Dishley Society

Pat Stanley, Janet Spavold & Hilary Matthews

Although Robert Bakewell is famous mainly for the fact that he bred a new type of sheep, he in effect devised a considerable number of new techniques of sheep management. These had two main effects; the better identification of sheep worth bringing into his breeding programme, and the faster growth of the sheep, subject to some deliberate management restraints, which were in that programme. Although ram letting originated in Lincolnshire, it was Bakewell who took up the idea and popularised it. Marshall thought that the way Bakewell operated, letting male stock by the season, 'is a department of rural affairs not known to the kingdom at large, forming a new subject in the rural science.... Mr Bakewell, though he cannot be deemed the projector, has certainly been the principal promoter of this branch of rural business'.[1]

Every year, private shows were held by the principal breeders for the purpose of letting rams. An example of this can be seen at the Tate Gallery in London, in an 1810 painting by Thomas Weaver where it is entitled 'Mr Morris's Sheep Show at Barton-on-Humber, Lincolnshire'. He was in fact Thomas Marris not Morris. As well as the owner's name being misspelled, the location has previously been misidentified as Dishley Grange but is in fact the barn at Thomas Marris's Hill Farm. Furthermore, the picture indicates tups being shown, that is rams that were old enough to mate.[2]

Such shows were regularly advertised in the local papers, thus:

> DISHLEY, LEICESTERSHIRE.
> MR SMITH respectfully informs his Friends
> and the Public, that he intends Shewing his RAMS
> in their Wool on Monday the 25th, Tuesday the 26th, and
> Thursday the 28th days of May instant; – and that on Monday
> the 8th day of June, he purposes to Shew them for
> Letting.
> Dishley, May 4th, 1818.[3]

Another advertisement stated that a group of Leicestershire sheep breeders 'had agreed to make a public shew of their Rams, at Leicester' in June and at the September Fair in 1795. Dixon, writing in the Royal Agricultural Society's *Journal*, commented that 'The private shews of rams, in the neighbourhood of Leicester, with the public exhibition of them at that town, in October, are calculated to afford considerable amusement and information to an amateur'.[4]

According to Rule 18 of the Dishley Society, tup shows were held from 8th June until 8th July, at which point the tups were turned away until 8th September, when viewings resumed for any tups not let.[5] During the show period every ram breeder kept an open house. Parties of breeders would travel from one show to another. These private exhibitions closed with a *public* show at Leicester on 10th October when mostly only inferior rams would be left.

The rams were brought to Leicester in waggons to be sold or let; Marshall described their travelling arrangements.

> SENDING OUT LET RAMS. The usual time of beginning to send out is the middle of September. The means of conveyance, carriages of two wheels, with springs, or hung in slings, some of them being large enough to hold four rams. In these, they travel from twenty to thirty miles a day; being sent this way, sometimes, two or three hundred miles.[6]

Leicester Ram Show had been a fixture from time immemorial, although before Bakewell's time it had been purely a means of selling rams. It was Bakewell who introduced letting to the Midlands with a ram which he let for 16 shillings in 1760. This single event was to lead to great wealth being brought to the Midland counties. Bakewell adopted the method of showing his rams separately so that they could not be judged one against another, and Monk described one such viewing.

> Every thing at Dishley is conducted with the greatest order and regularity, and, I may add, with every politeness and attention a stranger can wish for. The sheep are shewn one by one in a place built for the purpose. They are brought in at one door; and, after you have satisfied your curiosity, they are returned back by another door, and then another is brought in; and soon I observed, they brought them regularly better and better, so that the best came last. On my asking the reason for this, I was informed, if they had brought the last first, the great superiority would have made the others much worse than they really were, which would be a great injury to them. This too I found was the reason they never shew their rams after they are engaged, it being natural to human nature to covet that which it cannot possess. I by no means profess to be a judge of cattle; but I think there were some of Mr Bakewell's sheep far superior to any I viewed in the county (or in any other county I ever saw in my life).[7]

This method, to a good judge of sheep, did not matter, but by others it was greatly disliked as it gave no guide of comparison. Bakewell and his followers adopted the custom of letting their customers value the rams as they thought, and then bid a price. 'If Mr Bakewell approves of the offer, he accepts it; if not, he rejects it'.[8]

In this way, rams could be let to the highest bidder; it also allowed the owner to regulate the price to the customer without giving open offence. Bakewell very rarely sold his rams because of the high profits from letting, but La Rochefoucauld, one of his French visitors, was told that 'he will sell to foreigners, being less afraid that their offspring will destroy the price of his own'.[9]

La Rochefoucauld saw Bakewell primarily as an astute entrepreneur who had seen a market opportunity and worked to exploit it. Most of Bakewell's visitors and admirers concentrated on the output of his business – the cattle, sheep and horses – rather than the motivation behind it. La Rochefoucauld was blunt in his assessment:

> His object is to sell [the cattle]. That is the basis of his principal business, and he sells at a very big price, so generally are his breeds admired. He told us that last year he sold a cow for fully sixty guineas, and he wouldn't have sold her for less. He sold for a hundred or a hundred and fifty a fine bull, but not for anything in the world would he sell its calves: he would profit more from leasing them out. There are a great many people who hire them during a campaign to couple them with their cows and so produce a breed much better than their own – but never as good as that of Mr Bakewell who has both the male and the female. ... It seems to me that his breed of sheep is even more refined than that of his cows. He won't sell rams and lets them out at a mad price. ...
>
> I think he has reached the point he was aiming at and no longer seeks to improve them, only to lease them out.[10]

For some, Bakewell's methods seemed very risky. Young warned that such very high prices had 'opened a field of speculation in sheep-breeding' which 'whether it shall prove durable or not, whether much money shall or shall not be made in it in future, is not what any prudent man beginning business will adventure in but with great caution'.[11] Certainly, Bakewell speculated successfully by relying on his own skills (despite his bankruptcy, dealt with in Chapter 17). Copland wrote, 'Bakewell's success was far beyond his own expectation in a pecuniary point of view. ... [he] died a wealthy man in 1795'.[12]

Bakewell's business model did however gain ground among the farming community; in 1809 Potts wrote that,

> The object every intelligent farmer ought to have in view, who breeds and maintains domestic animals, is *profit*; consequently, he ought to find out, as Bakewell happily expressed it, '*the best machine for converting herbage, and other food for animals, into money*'.[13]

There is an implication of protectionism in some of Bakewell's farming practices too. La Rochefoucauld noted the precautions Bakewell took to make sure that even his cull sheep could not be kept for breeding by the butchers, saying that Bakewell 'kills them at home and sells the butcher only dead meat'.[14] Potts described a more dubious method, whereby Bakewell deliberately infected his cull sheep with liver fluke (then called sheep rot).

Liver fluke is a type of flatworm that is ingested by sheep and lambs who have eaten grass contaminated by the worm's larvae. In the sheep's intestine, the larvae hatch and migrate to the liver, where they can cause considerable tissue damage. The adult worms then migrate to the bile duct, where they lay thousands of eggs. These eggs end up in the small intestine where they are excreted by the sheep. These flatworm eggs will hatch on the ground if the sheep are grazing on moist, warm grass. As snails favour wet ground, the early-stage larvae form of the worm, 'penetrate' the snails, using them as a host to develop further. Eventually, the more mature larvae emerge from the snail onto the grass, where they are ingested by the sheep, thus perpetuating the cycle. It follows that you need an infected sheep or contaminated grass to allow the larvae to flourish and complete its life cycle. Clean sheep grazing dry ground in cooler spring weather will be unlikely to be infected by liver fluke.[15]

> When particular parcels of his best sheep were past service, he fatted them for the butcher, and to be sure they should be killed and not go into other hands, he rotted them before he sold; which from long experience he could do at pleasure.[16]

Bakewell, being Bakewell, had discovered his own method of ensuring that ewes sold for a slaughter price were not used for breeding. At this time liver fluke was causing considerable problems for sheep farmers, but it had come to the attention of the ever-watchful Bakewell that, before the middle of May, sheep could be safely grazed on land that had been flooded, but that after that time the sheep became fluked.

Any sheep that were sold for slaughter were therefore grazed on pastures in the autumn which had been artificially flooded during the summer. Consequently, when he sold them, he knew that they were certain to be suffering from the effects of 'rot' of liver fluke. This caused no loss of fitness but, on the contrary, caused the sheep in the early stages of the condition to fatten all the quicker. By these means, Bakewell gained up to six weeks in the accumulation of condition in his own sheep over that of his neighbours' sheep and into the bargain he was assured that their fate was sealed.

There was great competition between breeders, and it was better for an owner to let his best ram for less money to a man with poor stock than to let him to a competitor with ewes nearly as good as his own, for by that method the competitor would soon overtake him. Also, if two good breeders formed a partnership to hire a superior ram, thus dispersing his blood widely, the price for letting would fit the number of partners who would gain by the use of him. No legal transaction ever took place at the letting of a ram, it was simply a gentleman's agreement. If no pregnancy resulted, no money would be paid. If the ram died while away, almost regardless of the circumstances, the loss fell on the owner. The hirer could only use the ram on a stipulated number of ewes, usually 'about eighty' according to Monk.[17] If the hirer was merely a grazier, he was not allowed to rear any ram lambs.

Despite ram letting being a competitive business, in the early 1780s Robert Bakewell met up with other significant breeders to discuss their activities and it is generally believed that by 1783, an informal grouping had become established. We now move towards the end of that decade for proof of the creation of what came to be known as the Dishley Society with Nottingham University holding in its archives an invaluable set of notes; there are tears with missing text and other parts which are difficult to read, so square brackets surround suggested text. The first page of notes has clearly been written at two separate dates, the first part simply recording who agreed to participate and the cost.

George Inn

13th [March or May 1789]

> We whose names are under written agree to pay into the hands of Mr Henry Cousin (Treasurer) the sum of ten Guineas each, in such sums, at such times, and for such purposes, as shall hereafter be agreed upon by the majority of the subscribers.

Mr Walker	J.V. Stone	Jn. Bennett	Jn. Manning
Jos. Robinson	Nath. Stubbins		Nich. Buckley
R. Bakewell	J. or F. White	Jn. Breedon	Sam. Knowles[18]

This is all that consists of the agreement at this first date and we can see why as on 23rd June 1789, Bakewell wrote from Spilsby in Lincolnshire to his friend George Culley; he mentioned some of his friends who had recently agreed jointly to hire a particular ram for a season, but as far as Bakewell was concerned,

> I do not chuse [sic] to make any further engagements at present but have desired time to concoct a Plan in which some ten more or less of my Friends may be concerned as a Company and give me such a Sum as shall be agreed upon for the use of all my Rams they taking what they chuse for their own use and having what I can let all the others for ...[19]

In other words, this group of breeders would hire whichever of Bakewell's rams they chose, leaving him to let the rams that remained for however much he could get. To put it another way, this company would have first refusal of Bakewell's rams. In that letter he then sought Culley's opinion on the idea and whether the agreement should last for a specific term of years or 'for my life'. This, then, is the date by which Bakewell had decided that he could move beyond any recent, ad hoc, arrangements to a formalised one which would come to be known as the Dishley Society.

Pawson has his own account of this meeting, but it is stated to have occurred 13th November, that Bakewell was made president and Mr Bennett vice president, that meetings would be held on the first Tuesday of every other month. There are various other additional resolutions reported, some of which were certainly agreed at another meeting on 5th January the following year (see below).[20]

Bakewell's letters are not always easy to understand as he wrote following his stream of consciousness and with little punctuation. Nevertheless, in another letter to Culley written 18th November 1789, he reports that he had not fixed a price for his rams for the previous 12 months and was convinced more than ever that it was better to let those seeking his rams for hire to compete to set a price, whether privately at home or at the market. He had recently sent a servant to the Loughborough Fair with an old lame horse and allowed bidders to compete and set the highest price. He then says that this is the same method used for the sale of timber on estates and for the sale of goods at India House, so why should this mode of setting a price not also apply to livestock. He then mentions his idea of the company that he had been working towards; meetings had been held at different places with different chairmen, but with little emerging in the way of unanimity. He had frequently been urged to take the chair but refused because of 'the state in which my affairs have been', in other words his bankruptcy; but with this now moving towards a positive conclusion, he had finally agreed to be chairman. Furthermore, at the previous meeting held 13th November, much had been agreed, despite 'the great variety of opinions' and the resolutions had been forwarded to Culley for his perusal. Various names had also been suggested as suitable for the posts of president and vice-president of the new organisation. Bakewell then asked Culley to join the group to add his abilities to 'the mutuall [sic] approbation of all concerned'.

There was a further meeting of the nascent society in the New Year, its resolutions, recorded on the same page mentioned above, are as follows, including the strikethroughs:

Three Crowns Leicester
5 Jan 1790

No. 1 Resolved.	That Mr ~~Bakewell~~ Paget be president of the society ~~That Mr Bakewell be vice president~~
No. 3	That at every meeting the members pay their share of the expenses of the day first deducting the fines on those absent
No. 4	That secrecy be kept by all the members [respecting the] business of these meetings except to the [members] absent, and that any member quitting the society keep secret upon his honor [sic] the past action before he left it. (Reference 17)

What happened to resolution No. 2 would appear to be a mystery unless it was to be the alternative role for Bakewell, which seems the sensible answer; in the event, as explained in detail elsewhere, he felt that his financial situation should exclude him from this role, presumably so as not to compromise the society in its formative stages.

This seem to be the first mention of a set of rules having been set for the Dishley Society, but what is evident is that the society agreed to new rules at different times, as its purposes became clearer or as circumstances required. This may explain why various authors suggest different numbers of rules. What most authors agree is that the society's aims became to preserve the purity of the Dishley breed of sheep, and to protect and advance the interests of the breeders who, between them, had established a breed which was capable of improving other breeds by astonishing amounts.

Much ill-will was unleashed against its principal men in the press for reportedly overcharging gullible young men of means for whom, at the time, agriculture was very much in vogue. But our improvers had invested much time, money and effort into raising the standards of the country's flocks. They felt they had a just cause for recompense for their outlay. Realising that they would be the losers if they did not combine to protect their assets, by charging high prices they ensured the quality of their product was not only maintained but continually improved, for people would be very selective in choosing the best ewes to put to an expensive ram.

The Dishley Society proved to be a great success. Prices, as bid by buyers and leasers of rams, rose sharply and so did the envy of those who did not belong to this elite band of men. Bakewell, being the founder, was singled out and held mostly responsible for alleged acts of misconduct. When word of the society began to spread it met with suspicion. Monk, writing in 1794, had heard of a Tup Society with rules and regulations, 'which are not much approved of by the generality of other breeders. I understand those rules were a *secret'* (they are listed below).[21] Wedge, writing in the same year, had heard,

> ... a report of a combination amongst some of the breeders of sheep of that kind, who are said to have agreed not to take rams to markets or fairs, nor to let one to any person who does so; not to let more than a limited number of rams, and at not less than a certain price; not to set a value on their own rams, but to receive offers from those who choose to hire them; not to shew any number of them together, in order to prevent comparison; not to sell any culling ram or ewe of the true breed, except slaughtered; and, by select meetings at markets, fairs, &c. and by every other means in their power, to keep the ram business in their own hands. There is in such a combination, something so repugnant to public spirit and utility, that it must in the end defeat its own sordid purposes.[22]

This is politically loaded language, with the mention of a combination – a term then used to outlaw any grouping agitating for reform. Wedge's language becomes more emotive when he belittles Bakewell's work in establishing the New Leicester sheep.

> As it is the business of a quack to conceal the composition of the nostrum by which he fills his pockets, so also has it been the practice of those ram-breeders, to conceal the manner in which their *new kind of sheep* have been bred; on that account I shall endeavour to explain what their owners have, with so much care, endeavoured to keep from the public. I have called them a *new breed of sheep*, because the fact they (the breeders) have laboured hard to establish; and, taking that for granted, it establishes, beyond contradiction, another fact, viz. that this *new breed of sheep* must have been a cross from some others.[23]

Housman, writing in 1894, refuted the allegations. He suggested that men of such strong morals as Arthur Young and William Marshall, to name but two, would never have vouched for Bakewell if there had been one grain of truth in them. Bakewell was not a ready communicator to men who did not share his vision for improvement. It can well be imagined that he would give short shrift to the likes of jobbing journalists who in turn would fan the flames with allegations of mystery and secret practices. Marshall probably knew more about Bakewell's breeding methods than anyone because of their close friendship but in writing about Bakewell, said that it was not his 'intention to deal out Mr Bakewell's private opinions, or even to attempt a recital of his particular practice',[24] thus perhaps encouraging the allegations.

In 1786, by letting 20 rams, Bakewell made 1,000 guineas. By 1789 he earned 1,200 guineas for letting three brothers and 2,000 guineas for seven other rams. It was estimated that the total amount made from ram letting in that season by Bakewell and six or seven other prominent Midland breeders was the incredible sum of £10,000. The famous ram *Two Pounder* '(so called because in shape he was like the barrel of the two-pound cannon)',[25] was let during one season for 800 guineas cash, and Bakewell reserved the ram's use for himself on one-third of the total permissible ewes.

Two-pounder.

It was rumoured at the time that these prices were totally false and that, although a high sum was paid out, a great proportion of it was handed back in what is termed 'luck money'. This belief was pointed out by Marshall as being totally false. Although the practice may have occurred in the early dawn of the procedures, it had long since ceased. 'Mr Bakewell, at present, has the name, at least, of being parsimonious, even to the shepherds on which his rams are employed. His highest present, I understand, is five shillings; if the price be under 50 guineas, only half-a-crown'.[26]

The high prices paid for the hire of rams was not given by the graziers, who wanted their produce for the meat market, but by *ram breeders* for the purpose of breeding ram lambs which were then let to the graziers. These principal breeders retained between 20 and 40 ram lambs, which were selected more on their breed lines than for their appearance, weaned them in July and August and then gave them the best possible feed regime to prepare them for showing. They also agreed to show 40 rams of one-shear to five-shear ages. The highest prices were given by the top or principal breeders for the purpose of getting rams which were then let to inferior tupmen, as ram-getters, the progeny of which were then let to the graziers. The price a grazier gave on average for hiring a ram was 5 or 6 guineas. The improved Leicester ram produced for him a sheep which would fatten by two years instead of the usual three or four years and make 10-15 shillings more when sold.

Rams were brought into high condition for the shows, but as soon as a let had been agreed they would be put onto poor pasture to bring them back into working condition. Rams usually remained serviceable until the age of five years, at which point they would have become too fat for use and with that condition came a multitude of health problems.

As well as the initial rules of establishment given above, the Dishley society had a number of rules, the principal of which concerned with breeding were:

- No member shall hire or use a ram not belonging either to Mr Bakewell or to one of the members of the society.
- No member of the society shall give his rams, at any season of the year, any other food than green vegetables, hay and straw.
- No member shall let more than 30 rams in any one season.
- No member shall let a ram for less than 10 guineas to any person, nor for less than 40 guineas to any person who lets rams.
- No one ram shall be let to serve the flocks of more than two persons.
- No member shall let a ram to anyone who lets or sells his rams at fairs or markets.
- No member shall take in ewes to be served by more than one ram, at his own residence, in any one season, unless they belong to members of the society, nor to be served by any ram he uses for his own flock, with the same exception.
- Mr Bakewell engages not to let any ram for less than 50 guineas to any person residing within 100 miles of Dishley.
- No member shall let a ram to any person residing within 30 miles from Leicester, and not being a member of the society, who shall have hired a ram of Mr Bakewell during the preceding season.
- No member shall sell any ewes or rams of his own breed, to breed from, unless he sells his whole flock of sheep, except to members of the society.
- From the 1st to the 8th of June the members shall not show their rams, except to one another. They shall begin their general show on the 8th of June, and continue to show their rams till the 8th of July:

from that day until the 8th of September they shall not show them to anyone, but shall open their show again, and continue it until the end of the season.

- On the 8th and 9th of June, although the rams may be shown, no ram shall be let or engaged to be let, nor shall the price which will be required for him be mentioned by any one.
- Every member, refusing or neglecting to abide by the rules of the society, or withdrawing himself from it, shall no longer be considered a member. From what time he shall not be permitted to hire any ram or share of a ram from any of its members, until readmitted into the society at a general meeting.[27]

This particular list comes from Pawson, with Youatt commenting that,

> From the inspection of these rules, it will be evident that the tendency of the first four was to preserve the purity of this new breed of sheep, and the peculiar excellencies of the sort. ... The remaining nine rules have for their object evidently the establishment of as close a monopoly as could be effected among the members of the Society.[28]

Prominent breeders who were later subsidiary members of the Dishley Society included the Duke of Bedford, the Marquess of Buckingham, Lord Egremont of Petworth, Sussex, Wm. Coke (later Earl of Leicester) of Holkham, Norfolk, and Messrs. Martin of Clifton, Nottingham, Weston of Brackley, Northants, Creek of Rousham, Oxfordshire, Prinsep of Croxall, Staffordshire, Collins of Darlington, Co. Durham, Culley of East Field, Fenton, Northumberland, and Sale of Wentbridge, Yorkshire. Being a subsidiary member of the society meant that these men could hire tups from society members but at not less than 100 guineas. However, this 'perk' was removed if a subsidiary member 'refuse to inform a member of the society that enquires what business he has done and with whom, he shall not be dealt with at less than one hundred Guineas'. Using a wether or wedder-getter meant the person hiring the tup was not permitted to keep entire sons from him. Any male lambs produced by a wether-getter had to be wethered (castrated) – Rule 34, UN MS9/24A.

Another introduction of Bakewell's was the use of teaser rams. Today, a teaser ram is a vasectomised male which is still capable of indicating when a ewe is in oestrus but is incapable of fertilisation. When introduced to a batch of ewes, he stimulates them to begin cycling, after which he is removed and an 'entire ram' replaces him.

Bakewell's teasers were rams of little value whose genitals were covered by a cloth to prevent any mating occurring. As soon as a ewe showed in season to the teaser, she was removed and put to one of the best rams, which was kept in a pen on his own. Bakewell changed the way tupping proceeded.

> Instead of turning the ram loose among the ewes, at large, as heretofore, and, agreeably to the universal practice of the island, – he is kept apart, in a separate paddock or small inclosure, with a couple of ewes only, to make him rest quietly; having the ewes of the flock brought to him singly; and serving each no more than once.
>
> By this judicious and accurate regulation, a ram is enabled to serve near twice the number of ewes he would do, if turned loose among them; especially a young ram.
>
> In the old practice, sixty or eighty ewes were esteemed the full number for a ram: in the new, from a hundred to a hundred and twenty are allowed: seven score [140] have been served by one ram, in a season.[29]

In what may have been an evolution of this approach there is an unreferenced suggestion that the ewe's head was placed in stocks for mating, to prevent fatiguing the ram. By this method a ram could cover many more females. The normal method of putting the ram to the complete batch of ewes to be served was gradually replaced by Bakewell's system.

The rams were returned to their owner by the hirer at the beginning of December. The owners then set to with a will in order to have the rams ready for the next show season. In general, the principal ram breeders kept flocks of 100-200 ewes while graziers would have 500-600 ewes.

No cost or labour was spared in the attentions lavished on the ewes at lambing by the principal breeders, as their offspring were of extreme value as potential rams for letting. The ewes were pushed to produce a good flush of milk to give the lambs the best possible start.

Cull ewes were sold by the graziers after they became four shear ewes (that is, four annual crops of wool), but the ram breeders kept their best ewes until they could no longer breed. The principal breeders did not let such valuable assets off their property alive but slaughtered them at home, so that they could not be used for further breeding by their new owners who were not in the breeding circle (eventually the Dishley Society).

As can be seen, Bakewell left no stone unturned in his quest for perfection. Every aspect of husbandry was covered. He was a great showman and his stock were also produced for exhibition to the very best of his, and his men's abilities. It is reported that his sheep were kept as clean as racehorses, with his best animals even wearing body clothes.

Bakewell's Experiments

As well as sheep of his own creation, Bakewell also kept other breeds at Dishley for use in comparison trials and also for crossing with his own rams, to see if improvement could be made with the other breed. It is probably because visitors saw sheep of other breeds, and also Leicester crosses, that it was believed that many types made up Bakewell's recipe for success. Another ingredient was his persistence in learning from experience all his life. He may not have expected success from many of his experiments but he did expect to gain useful knowledge and further his understanding of farming.

One visitor recalled seeing three sheep which were produced by one ewe but by different rams. One was by a Ryeland, one by a Spanish and one by a Dishley ram. The difference between them was said to be very great, with the offspring of the Dishley being far superior to the others. This was obviously an experiment to test the merits of the various crosses. As it was Bakewell's custom to show visitors his animals one-by-one in the yard or the buildings but not in the flock in the fields, the differences were more obvious.

Details are given in Arthur Young's *Annals of Agriculture*, Vol VI, page 11, of his second visit to Dishley in 1785. 'I was witness at Dishley to a very interesting experiment on different breeds of sheep. He has a sheep house in which six rams were tied up and stalled on turnips: one of his own breed, one Norfolk, one Teeswater, one Wiltshire, one from Ross in Herefordshire and one from Charnwood Forest. They were weighed alive when put to turnips, their food weighed to them and weighed alive again at the end of the trial'.

The trial was conducted entirely by a young Russian, John Saphonkove, resident at Dishley at the time, to whom Mr Bakewell entrusted the work, as the trial was not on a scale to which he could attend himself. It was therefore conducted in a very fair manner. Young attended the sheep-weighing and vouched for the accuracy of the figures, although he gives only the final weights for the sheep:

	Weight of the sheep	*Turnips from 19th March to 2nd April 1786*
Durham	290lb	498lb
Wiltshire	173	313
Norfolk	162	298
Dishley	158	174
Charnwood Forest	131	304
Herefordshire	115	202

Modern trialists would, of course, heavily criticise the figures on a number of grounds. There was only one sheep of each breed; they had been chosen by

Bakewell and probably not all were representative of the best of their breed; no increases in liveweight gains were given; and, most important of all, no carcase weights were attempted. They would nevertheless provide 'impressions' of the relative performance of the different breeds. Conclusions would doubtless be drawn.

On this visit he also heard about an experiment, where Dishley ewes were turned out for the summer to range on the highway verges; 'He has an experiment to prove the hardiness of his breed which deserves notice. He has 5 or 6 ewes, that have gone constantly in the highways since *May-day*, and have never been in his fields: the roads are narrow, and the food very bare; they are in excellent order, and nearly fat; which proves in the strongest manner, the excellence of the breed'.[30]

They had no supplementary feed, but they returned in excellent condition. It supported Bakewell's theory that 'The poorer the land the more it demands a well-made sheep'.[31]

Bakewell also told Young about another experiment when he kept four lambs together on the same pasture, then 'with a view to see which of them fed upon the least food', took them indoors for a fortnight. They were weighed on entry and exit, and their food was recorded. They were given 'as much food as they chuse [sic] to eat' during that time. They were a Hereford, a Hereford/Merino cross, a Hereford/Leicester cross, and a Leicester. The result was that, at the end, the Hereford weighed 149lbs and had eaten 145lbs of food; the Hereford/Merino weighed 163lbs and had eaten 179lbs of food; the Hereford/Leicester weighed 170lbs and had eaten 150lbs of food, and the Leicester weighed 174lbs and had eaten only 95lbs of food.[32] The Leicester was the most efficient in converting its feed into meat.

Arthur Young related the following information in Vol VII of his Annals of Agriculture, in 1786:

> Mr Wyn Baker, who was employed by the Dublin Society of Agriculture to make experiments, says it would require about 20 weeks to fat a sheep completely on turnips and that he would consume, during that time, one fourth of his mutton weight a day, that is, the weight of his four quarters when killed (which is reckoned about half his liveweight just before killing). Therefore, sheep, whose mutton weight would be 80lb would consume 20lb of turnips a day, or 140lb, or 10 stone a week; and as 160 stone is a ton weight, a ton of turnips would maintain 16 such sheep a week.
>
> Mr Wyn Baker remarks that the sheep to be fatted will not require above a quarter of their mutton weight a day, yet oxen will require half their beef weight per day, although beef and mutton, at market, sell at nearly the same price.

These figures from Mr Wyn Baker, allowing a quarter of the mutton weight of the sheep per day in turnips to fatten them, seem to confirm the experiment made at Bakewell's farm on six stalled rams which consumed about one-eighth of their liveweight a day. That was with the exception of Bakewell's own Dishley breed, which did not consume above one-twelfth of his liveweight per day.

Reference is made in 1828 by Mr Valentine Barford that, according to Bakewell's friend Richard Astley of Odstone, a black ram had been used at Dishley in the mid-1780s on the New Leicester flock. Barford was himself a New Leicester breeder (of Foscote, Northants), whose flock was directly descended from Bakewell's. In a letter to the Revd Henry Berry, he records that no flock of New Leicesters had produced more black rams than his, although he retained none for breeding. Barford's flock had been carefully recorded from its beginnings in a private pedigree register. It was closely interbred with the flock of Joseph Robinson, a founder member of the Dishley Society, who, in turn, had closely interbred his own flock from Bakewell's. It would appear that Black Leicester longwools have continuously cropped up through the ages, although up until the recent introduction of their own register, they were ineligible for registration.

See the illustration in the colour section: *Portrait of Valentine Barford,* by Henry Barraud, showing him amid his flock of New Leicester sheep. This portrait was presented to Barford by his friends as a token of respect for his upright conduct and his perseverance and ability in keeping up the true character and form of the pure-bred Bakewell Leicester sheep in 1850. The book Barford is holding in his right hand in the painting is his flock book, though it is sometimes misidentified as a Bible as he looks rather like a member of the Victorian clergy.

The story of the black ram is related as a point of interest but not as conclusive evidence that such an animal was indeed used. We know that Bakewell kept all manner of animals on his farm and it may be that Astley, having seen a black ram at Dishley, assumed that it was being made use of in the New Leicester flock. Another story is that Astley fell foul of the Dishley Society rules, incurring heavy fines, so his supposed observation could have been his way of getting back at Bakewell; however, as there is no written evidence that Astley had been a member of the Dishley Society, this story begins to look apocryphal.[33] Once again, it is Pawson who claims that Astley was a member of the society who, having broken the rules, was cut off.[34]

An incident revealing Bakewell's sense of humour and explaining the presence of two black-faced rams at Dishley, is related by the Revd Henry Berry. Bakewell, with the assistance of the shepherd at Holkham in Norfolk (on the famous estate of Coke, later 1st Earl of Leicester) secretly removed two Norfolk rams from Coke's flock during his absence. Coke regularly exchanged visits with Bakewell and, on his next trip to Dishley, he was treated to the usual parade of magnificent Dishley rams. But then came a sight for which he was totally unprepared – his own unmistakeable Norfolk rams were led past, each wearing a neck collar, with their characteristic thick spiralling horns, black faces and black legs. 'At a given signal, away they bolted at top speed, each clearing the hurdles in high style and then, returning, accomplished the same feat'. Who was it that told the Revd Berry this story? No other than Coke himself, who clearly enjoyed a prank, even at his own expense.[35]

Endnotes

1 Marshall, W., Rural Economy of the Midland Counties, in The Complete Farmer, By a Society of Gentlemen, (1789), no page numbers.
2 For example in Moncrieff E., S., & I., Joseph, Farm Animal Portraits, (1996).
3 The British Newspaper Archive (BNA), *Derby Mercury*, 7 May 1818.
4 Dixon, H.H., in Vol 4 (New Series) of the *Royal Agricultural Society's Journal*, p331.
5 University of Nottingham, 'Minutes of the Dishley Sheep Society', UN MS9/24A.
6 Marshall, W., The Rural Economy of the Midland Counties, (1790), Vol 1, p431.
7 Monk, J., General View of the Agriculture of the County of Leicester, (1794), p27-8.
8 Monk, J., General View of the Agriculture of the County of Leicester, (1794), p32.
9 Scarfe, N., Innocent Espionage the La Rochefoucauld Brothers' Tour of England in 1785, (1995), p32.
10 Scarfe, N., Innocent Espionage the La Rochefoucauld Brothers' Tour of England in 1785, (1995), p31-2.
11 Young, A., Calendar of Husbandry, in Dickson, R.W., Practical Agriculture,Vol 1, (1805), p1152.
12 Copland, S., Agriculture, Ancient and Modern, by the "Old Norfolk Farmer", Part II, (c1861), p110.
13 Potts, T., The British Farmers Cyclopaedia, (1809), pages not numbered.
14 Scarfe, N., Innocent Espionage the La Rochefoucauld Brothers' Tour of England in 1785, (1995), p32.
15 Sustainable Control of Parasites (SCOPS), 'Liver Fluke', https://www.scops.org.uk/internal-parasites/liver-fluke/lifecycle/
16 Potts, T., The British Farmers Cyclopaedia, (1809), pages not numbered.
17 Monk, J., General View of the Agriculture of the County of Leicester, (1794), p33.
18 Blogs.nottingham.ac.uk/manuscripts/2020/07/25agricultural-societies/ms-9-1/
19 Pawson, H.C., Robert Bakewell, Pioneer Livestock Breeder, (1957), p145.
20 Pawson, H.C., Robert Bakewell, Pioneer Livestock Breeder, (1957), p148-9.
21 Monk, J., General View of the Agriculture of the County of Leicester, (1794), p32-3.
22 Wedge, J., General View of the Agriculture of the County of Warwick, (1794), p31.

23 Wedge, J., General View of the Agriculture of the County of Warwick, (1794), p31-2.

24 Copland, S., Agriculture, Ancient and Modern, by the "Old Norfolk Farmer", Part II, (c1861), p374.

25 Pawson, H.C., Robert Bakewell, Pioneer Livestock Breeder, (1957), p69.

26 Marshall, W., The Rural Economy of the Midland Counties, (1790), Vol 1, p428.

27 Pawson, H.C., Robert Bakewell, Pioneer Livestock Breeder, (1957), p73-4, quoting Youatt-see below.

28 Youatt, W., Sheep, Their Breeds, Management and Diseases, (1837).

29 Marshall, W., The Rural Economy of the Midland Counties, (1790), Vol 1, p432.

30 Young, A., The Farmer's Tour through the East of England, Vol I (London: W. Strahan & W. Nicholl, 1771), pp116-7.

31 William Housman, 'Robert Bakewell', *The Journal of the Royal Agricultural Society of England*, Third Series, Vol V (London: John Murray, 1784), p19-20.

32 Young, A., The Annals of Agriculture, Vol 6, (1785), p20.

33 Information provided by Dr Hilary Matthews.

34 Pawson, H.C., Robert Bakewell, Pioneer Livestock Breeder, (1957), p74.

35 Housman, W., Robert Bakewell. Story reprinted in Otago Witness, Issue 2710, 21 February 1906, p5.

CHAPTER FIVE

New Leicester versus Old Lincoln and Other Breed Crosses

Pat Stanley & Janet Spavold

> I have not used any Lincolnshire Rams for 20 years past – why have you at different times from the year 1773 to 1786, hired from this county?
>
> (*Robert Bakewell,* 1788 – letter to Mr Chaplin – see below)

It has been mentioned before that the Dishley or New Leicester sheep were used to improve many other breeds. Among them was the breed to which they are reputed to owe their derivation, the Old Lincolnshire. This breed was crossed so extensively with the New Leicester that they were eventually to disappear and illustrations of them are hard to come by. For over a quarter of a century great ill-feeling was generated between the purist breeders of Lincoln and the new improvers.

In 1788, Bakewell fell out most bitterly with Mr Charles Chaplin, a breeder of the Old Lincoln. Chaplin strongly held that, in his opinion, the Old Lincoln was a far better animal than the new improved type, which had been produced by the introduction of New Leicester blood.

Arthur Young's *Annals of Agriculture,* Vol X, gives us access to the angry exchanges between these two men. A proposal had been made in 1788 to hold a ram show at Partney, one of the ancient Lincolnshire sheep fairs, so that the Old Lincoln could be compared with the 'New Improved Lincoln'. Bakewell declined to allow his sheep to be viewed before the event, when they would be in top show condition. But, being in his rival's neighbourhood, he took advantage of Chaplin's absence to have a sneak preview of the competition, in their natural condition. Chaplin greatly resented this act and wrote Bakewell the following letter:

> The extraordinary art made use of in the exhibition of your stock at Dishley, points out, in the strongest manner, the impropriety of showing it in a disorderly state; and, after refusal on the 21st instant to let you see my sheep before they were collected and sorted at home, I did not expect to hear of your meanly sneaking into my pastures, at

> Wrangle, on the 24th, with two other people, driving my sheep into the fold, and examining them. Such unwarrantable conduct can only be accounted for by your great anxiety about the show of rams at Partney, near Spilsby, on the 18th of September, which was proposed for the purpose of making the comparison between those bred from your sheep and the original breed of this county. The small sheep that have no cross of the Durham kind, which you have had the address to impose upon the world, without size, without length and without wool, I have always held to be unprofitable animals; but that I may not appear to be too tenacious of my own opinion, I hope you will produce them at Partney, on 18th September next, to meet the Lincolnshire sheep, where there will be many better judges than ourselves to decide on their merits.

Bakewell replied:

> On my return home on Tuesday last, I saw your letter addressed to me of the 26th August. In the Leicester paper of the 6th instant, in which you are pleased to notice the extraordinary art made use of in the exhibition of the stock at Dishley; which you have seen at several different times. Surely you cannot say you have observed any unfair practices, or that you was [were] ever denied seeing what was not engaged for the season, on account of their not being sorted. Access to anything that was not out on loan, because it was not sorted or in a disorderly state? At Horncastle, on Thursday, the 21st August, I asked you if I might see your rams near Saltfleet. You did not say I should not, but that they were not sorted, and that when they were you would be glad to see me at Tathwell. I did not go to Saltfleet, but into the marshes, near Skegness; and from thence, on the Saturday afternoon following, to Wrangle; the next day, Sunday the 24th, to Freeston, where I met with two graziers, with whom I had not any acquaintance till that day. They proposed on Monday to go to Skegness, and asked me if I thought they could see your rams. I told them I was informed on my way to and at Wrangle, that they might. We set forward together, and called at the inn at Wrangle, which I came from the day before, and there passed what you are pleased to term my 'meanly sneaking into your pastures on the 24th'. We asked a young man if you had any rams there, he informed us you had. 'Where are they?' 'In the close next the house'. 'May we see them?' 'Yes'. 'Who would show them?' 'I will'. From which we supposed he had frequently shown them to others. We then alighted and went into the close; he opened the pen-gate, and we assisted him in driving them in, about fourteen in number. The age or breed of any of them I do not know. From thence we went to

the person who has the care of your rams, about a mile and a half nearer Skegness, and asked if we could see them; he refused us, saying he had received orders by a letter from you, not to show them to anyone. He was then asked if they had been shown before. He answered they had. 'When did he receive the order not to show them?' 'On Saturday night last'. Had we known this before, we should not have been guilty of what you term 'such unwarrantable conduct'. I have long made it a rule not to find fault with another person's stock. Why should you be so severe on mine? And I now take the liberty of requesting you to explain what you mean 'by sheep without size, without length, and without wool,' which you say I have had the address to impose upon the world; and of informing you that I am fully persuaded there are ten rams without a cross of the Durham, or any other kind, let for 1,000 guineas more this season than the same number of the 'true Old Lincolnshire breed, of the long staple', some of these at the highest prices, into the counties of Lincoln and Nottingham; and to breeders, many of whom have used the Dishley sort of sheep for upwards of twenty years, and who have agreed for some, and offer higher prices for others, for future seasons, than they have yet given, and may surely be supposed capable of knowing the value of what 'you have always held to be unprofitable animals,' Did they not find their interest in so doing, would they persevere? The address must be extraordinary, indeed, that could impose upon them against their interest and so long experience.

Give me leave to ask, Sir, had you such sheep, would you at this time of the year incur the expense and run the hazard of sending them 80 miles from home, to gratify the curiosity of breeders in this neighbourhood? If you are still desirous of a public exhibition, please to say if you would choose to send on the 5th July next, to Lincoln and to Leicester (as there is a fair at both places on that day) 2 rams of the 'true Lincolnshire breed of the long staple' to each place, to be shown against the same number of the Dishley sort, weighed alive, killed, and an exact amount given of the carcasses and the offals, for the information of the public.

I am, Sir, your humble servant.
Robert Bakewell Sept 12, 1788.[1]

Bakewell's gasping indignation at Chaplin's remarks is quite tangible. His anger is conveyed in his lengthy sentences, even to the reader after all these years.

Chaplin had not finished. Continuing with Pawson's quote, from Tathwell on 19th September 1788 Chaplin wrote:

TO THE GENTLEMEN who attended Partney Fair on the 18th of this month:

GENTLEMEN,

It is unnecessary for me to say anything about the show of sheep yesterday, but as Mr Bakewell and myself, after having attempted in your presence to agree upon an experiment, likely to give to the public a fair insight into the comparative merit of our respective breeds of sheep, parted yesterday, under an agreement to meet this morning at Sir Joseph Bank[s]'s and there finally to settle the business. It is proper that I should give you an account of the result of that meeting.

Mr Coke of Norfolk, Sir Joseph Banks, and several gentlemen whom Mr Bakewell had brought with him, were present, when after some previous conversation, I made him the following offer, as a fair modification of the proposal made by him to me in the General Evening Post of Tuesday the 16th instant.

OFFER
To show at Midsummer next, two rams of my own breed, and my own property, with their wool on. To have them shorn, and then killed, and to have their fleeces, mutton, tallow, and offal weighed separately, on condition that I should name the sheep now – and that Mr Bakewell would also now name two rams of his own breed, and his own property, with which he would meet mine, and which he would subject to the same experiment at the same time.

To this proposal Mr Bakewell declined giving any answer, but he seemed inclined to close with a similar one, provided he should be allowed to produce such rams, as between now and Midsummer next, he might be able to procure, and he said, that I had the same opportunity as he had, to search all England for the best that could be got.

In which, knowing that your enquiries chiefly pointed at a comparison between the Lincolnshire and Leicestershire breeds, and unwilling to show against all England, as in that case Rams of my own sort might be brought in competition with me, I positively refused my acquiescence, a refusal, Gentlemen, I hope you will think me justified in.

And now I take leave to say that, as on several occasions in which I have had an opportunity of observing Mr Bakewell's conduct, it has not, in my opinion, been bounded by that rigid line of propriety with which I have ever earnestly endeavoured to circumscribe my own, I think myself at liberty to decline all further controversy with him, and unless he, or some of his people, are again caught sneaking into my pastures, to examine my unsorted stock, or doing some other thing of similar impropriety, he may rest secure from being again either publicly addressed, or answered by.

Your most obedient servant

CHARLES CHAPLIN.

P.S. If anyone wishes to be satisfied of the comparative merit of the Lincolnshire and Leicestershire wool, let him enquire at Leeds and Wakefield, and at Smithfield he may learn from which county the best sheep for mutton are sent.

Bakewell eventually saw this letter of Chaplin's, as was doubtless intended, and he replied:

To CHARLES CHAPLIN Esq.

Sir,

Your letter of the 19th Sept., addressed to the gentlemen who attended at Partney, I should have noticed sooner, but that I waited for some information relative to what Rams you have had from this county, a business you always seemed to transact in the most private manner. On the 25th of August, it is true, I went to a piece of rich grazing land, next to the public house in the town of Wrangle, and there saw fourteen Rams, which I understood had been shown before to anyone who asked to see them, therefore, I could not consider them as unsorted or in a disorderly state. This did not arise, as you are pleased to state it, from any anxiety of mine about the Show of Rams at Partney, with which I never intended to have any concern, though probably it might cause your anger, lest I might have discovered how those were bred, which were intended for that exhibition.

For this *heinous offence,* in your first letter, you severely condemn my sheep. In your second, hint at my character, and compliment your own. Your manner of noticing my conduct and stock, I consider as a very fortunate circumstance, as it gives me an opportunity of making some remarks on yours, for which I should otherwise have mounted an apology. I shall take the liberty of asking you the following questions, to which, when I have your answers, I will lay before you and a candid public, the conditions on which I will meet you at *Lincoln, Leicester,* and as many other county towns as you please. I conceive that you and I showing two rams each, without saying how they were bred, and which on this account might not do any business this season, and might be highly fed nine months for the showing at *Lincoln* only, can throw but little light on this important subject, and be of small consequence towards ascertaining the merits of the true old Lincolnshire, and the new Leicestershire sort of sheep.

Why do you give your Rams anything but grass, hay or turnips in winter, or anything but grass in summer? Why do you refuse to show them, except at particular times? Mine that are not engaged, you may see any day in the year, Sundays, and the first seven days in June, excepted. When may I be favoured with a sight of yours?

Why, at a time when shearlings are let at such high prices, and lambs are letting at still higher, for the next season, do you show or let so few of that age?

Why, when Rams are let so high, and sixty Ewes descended from this stock were sold at auction the 14th of Sept., 1787, at more than 5 gns. each, and some of them into the county of Lincoln, and many cull Ewes have since been sold at much more, do so many of yours of both sorts go to market?

Why are they not rather bought to breed from, when they may be had at butchers' price?

Why do so many of the Lincolnshire breeders, who will give the best prices, go into other counties for their Rams?

Why do you, who have formerly sent so many Rams into this county, now send so few – if any?

Can you inform me of any of the Leicestershire Breeders, who will show Rams at this time, and say they are of your kind?

Since the time when 5 of your Rams were advertised to serve Ewes, in as many different places, at half-a-crown each – a Ram, bred in your county, son of a Leicestershire Ram, has been advertised in your neighbourhood, to serve Ewes at a guinea each! Have the prices of yours advanced in the same proportion?

I have not used any Lincolnshire Rams for 20 years past – why have you at different times from the year 1785 to 1786, hired from this county?

Why did you in the year 1783, *buy four*, besides what you hired? If you had not found an advantage in this, who could have address enough to impose such upon you, and induce you to continue the practice for so many years? If such as you have had from this county have improved the good old Lincolnshire Sheep, some think you had not a cross before it was wanted, and that some of the Rams you showed at Partney, were descended from the Leicestershire kind – Can you prove the contrary?

Can anyone after this be at a loss to know what the Leicestershire Breeders think of the Lincolnshire sheep, and *you*, of those which *you* say, you 'have always held to be unprofitable animals'. If you have not more of this sort, do not take more pains in handling those you choose, and pay a little more attention to the *carcass* than you have hitherto done. Beware, that (notwithstanding all the art you can make use of, *feeding*, in *shearing* and *dressing*) – some of your spirited and unprejudiced neighbours, do not cause you to go out of the RAM BUSINESS – *in the same manner as you have represented me going into your pastures at* WRANGLE.

I am, Sir, your humble servant

ROBERT BAKEWELL

Dishley, Nov 19, 1788

This exchange of letters and many others between Bakewell and his correspondents can usefully be found in Pawson's Robert Bakewell Pioneer Livestock Breeder, which gives a better idea of the normally very civil manner in which Bakewell addressed his friends and acquaintances.

OTHER BREED CROSSES

Twenty one years later, this public exchange of letters was still remembered by Lawrence. Writing about the Lincolnshire, he remarked that by 1809 pure-breds hardly survived in the county 'where the Dishley cross has been almost universal'. The resulting animal produced both more mutton and more wool per acre but with the loss of some of the 'high flavour of the genuine Lincoln mutton'. The wool was finer, shorter and with better colour but less tough for combing. Chaplin, by then dead, had strongly preferred size and bone in his sheep but clearly that was not the choice of consumers. Lawrence's judgement on this dispute was that 'Probably a more rational system would have enabled him to improve his favourite stock to the utmost, and to equal that of his rival Bakewell, in number, at any rate, in weight per acre, retaining the natural superiority of quality of the Lincolns'.[2]

Disagreements such as this must have been frequent, for they were acknowledged in Bakewell's obituary:

> ... his sheep continue as universally unrivalled, notwithstanding the competitions excited at various times by motives of interest and envy.[3]

While the bitterness of this exchange had its origins in the arguments over Bakewell's source choices of rams in the early stages of his improvement experiments, there was clearly also a financial struggle for supremacy as the prices for the ram lettings increased. Once Bakewell had created his ideal type of sheep, breeders from all over the country began to experiment with the use of New Leicester rams on their own local breed. It was generally accepted that breeders sought to produce either mutton or wool for the market, and that the choice depended on where they were – on the local geology. Rudge summed it up neatly when considering Gloucestershire's Cotswold sheep.

> ... those soils which produce the finest wool, are not adapted to raise the greatest weight of mutton; and, on the other hand, those soils which fatten most, in equal proportion deteriorate the wool, it being a fact ascertained by experience, that wherever fine woolled sheep have been introduced on more luxuriant soils than where they were bred, or accustomed to, the wool has increased in quantity, but become coarser in quality.[4]

The quality of the wool itself was a point of contention in the dispute with the Old Lincolns, 'the only other really strong wooled breed'.[5] A modern breeder describes the Leicester as having 'long, lustrous strong wool with its defined crimp and style [which] makes the breed what it is ... the wool type that has been valued so much in the past and is one of the English Leicester's most recognisable traits'.[6] The trade-off between mutton and wool is often mentioned by contemporary breeders, who usually point to the New Leicester giving better mutton than wool. Young's judgement was that the large, long-woolled breeds were more profitable on rich pasture, but with poorer grass, the small, short-woolled breeds were preferred (both types being improved).[7] He further commented that a close dense coat threw off rain, whereas the long-woolled fleeces tended to absorb it.[8]

Although Culley saw the New Leicester as 'unquestionably a valuable breed', his principal objections to it were that they were...

> fattening too much, and the mutton in consequence becoming less delicate in its flavour than in that of other breeds; on the deficiency in the quantity of wool which they produce; and on their not being calculated for the fold.[9]

He described Bakewell's reasons for abandoning folding: that keeping large flocks together resulted in the strongest sheep taking most of the food; that small, dispersed flocks allowed for folding them according to their age, type and vitality and reduced the need for movement round the farm, and that large folds distorted the management of pastures.

The Border Leicesters proved to be an enduringly successful cross (see black & white illustration). They were developed by the Culleys when they moved from Durham to the Borders in 1767. When Usher was writing about this cross-breed more than a century later in 1877 he commented on Bakewell's 'good knowledge of animal physiology' and thought that...

> the flocks tracing the closest lineal descent from the Dishley, untainted by any other strain of blood, selected and crossed with taste and judgement, tended with care, and 'all appliances and means to boot', are still the best in the Border district.[10]

Border Leicester sheep.

Monk noticed that Bakewell himself experimented with crossing, using a Ryland ewe with a Ryland, a Spanish and a Dishley ram; the latter proved 'far superior' to the others.[11]

Joseph Wilkes of Measham, Leicestershire, is best remembered now for his industrial developments there – cotton and bleach mills, coal mines and brickworks. But he came from a farming family including graziers and maintained that interest. He supplied cheese to the navy. He was part of Bakewell's social circle, present when Young met Bakewell, Wilkes, Paget, Cresswell, Green, Lee and other breeders for dinner at the Queen's Head in Ashby de la Zouch. The following day, Young went to Measham to see this industrial empire, which included 'a large and handsome inn, which is to be the sign of the tup'.[12] Influenced by Bakewell, he developed his own Measham breed of sheep which survived long enough to be remembered in the 1950s by an old sheep judge as 'bred for wool not meat, a squarish animal'.[13] Wilkes started the Smithfield Club in December 1798 and was its president until his death.

This group of breeders were Bakewell's contemporaries, friends and business associates. The next group discussed took on Bakewell's mantle after his death, some having known him in later life, but all still able to access the pure Dishley breed until 1831, while Honeyborn (died in 1816) and then

Smith (left in 1831) were at Dishley Grange. The importance of this link can be seen in the frequent newspaper advertisements for the sale of farm stock, where the lineage of the sheep is given, as here:

> TO BE SOLD BY AUCTION.
> *At Hetton-le-Hill, on Thursday the 14th of September next,*
> 26 BROOD EWES AND 10 GIMMERS. The Ewes have all had Lambs this Season, and are by a Son of Darmpton, bought at Mr R. Colling's Sale, and Mr Mason's No. 20, and by No. 10 Shearling, 1824.
> Also 26 TUP LAMBS, & 20 GIMMER LAMBS. The Origin of this Stock was bought upwards of 30 years ago, in Leicestershire, and tupped at Dishley, and has been crossed from Mr Robert Colling, Mr I. and Thomas Stone, and Mr Mason's, and are worth the Notice of the Public.[14]

By 1805, the introduction of New Leicester blood had been widely accepted as improving other breeds. Dickson listed six such breeds. New Leicester blood had been 'most advantageous' to the South Downs. The Somerset Polled sheep had displaced the Dorsets on the Somerset Levels because they were 'capable of improvement' by Leicester crosses. The 'greatly neglected' Teeswater had been improved via Leicester rams in Northumberland, and similar improvements were recommended for the West Riding of Yorkshire. The Dartmoor or Devonshire Natt 'is said to have received considerable improvement by being crossed with the New Leicesters'. In Herefordshire, the Ryland had been crossed with the Dishley and 'an useful kind of sheep has been produced' with increased wool and carcase weight, but the wool was not as fine. Dickson passed on Young's advice on improving the flock of the less well-off graziers who could not afford high-priced rams: to select the best five or six per cent of the ewes each year, to be covered by a 'ram better than any of your own', either from a top ram-letter or one who charged a reasonable fee.[15]

In Gloucestershire, the pure Cotswold had become rarer by 1807 following its improvement by New Leicester blood. Its wool became shorter and finer. Its carcase weight was reduced but made compact with finer bones and better meat and fat. Cotswolds crossed with South Downs produced finer wool and there were disagreements as to which was better. The balance lay with the Leicesters for mutton, because 'from their aptitude to fatten, they will require less food for their support, and consequently bear a larger stock on any given quantity of land or food', but with the South Down for wool.[16]

Rudge noted that the Ryland in Gloucestershire had been crossed both with the New Leicester and the black-faced Shropshire, which had increased the carcase weight and wool quantity, but decreased the wool quality. Worse, the Leicester cross had introduced shab, 'a disease not known before on the Ryelands',[17] so the breeders were trying to cross back to the original sheep.

By 1809, Potts recorded that, if kept on rich marsh and grazing, the Lincolnshire 'may be kept with more profit than the new Leicester, to three years old'.[18]

Before discussing the individual crosses, Lawrence summarised the New Leicester's sweeping success by 1809: 'The Dishley cross has made its way into every part of this island; to the Land's End, to the bottoms of the Welsh mountains, and of the Scottish Highlands, to Ireland, and even to Russia'.[19] Even so, some attempts failed, as with the Bampton or Western long-woolled sheep. The recently popular Ryland crosses, he said, had resulted in the sheep, once very hardy, having worse constitutions. He singled out the Cotswolds where a single cross enhanced the forequarters but further crosses were detrimental. He said that trials with Spanish rams had failed.

He described Culley's success in crossing the Northern Heaths and Cheviots, followed by a Dishley cross. The Northumbrian/Dishley cross pioneered by the Culleys was widely found in lowland Scotland. Cheviot, Dishley and Northumberland tups were successfully used in the West Riding of Yorkshire on Wensleydale (also known as Craven or Penistone) sheep. Staffordshire's original polled long-woolled sheep was improved with Dishley rams. English Merino tups had been continually improving so that 'a fair field now offers to the ambition of those enterprizing tup-breeders, who are emulous of rivalling Bakewell in a fine woolled cross, in producing *Dishley* Merino tups'.[20]

Mr Hose of Melton Mowbray, Leicestershire, was an important grazier in 1809. He had crossed some of his Dishley ewes with a Merino ram, 'with decisive success'. Lawrence took samples of the resulting fleeces to Hawksley's Arnold Works, a hosiery factory on the outskirts of Nottingham. Hawksley offered immediate contracts for quantities of this wool at almost twice the normal price for Dishley fleece. The difference was that the Merino blood shortened the staple but thickened the pile, with little carcase loss.[21]

Later writers and breeders had the advantage of perspective on the development and influence of the Dishley and its various crosses. Professor Low's evaluation of the New Leicester sheep traces changes in breeding, from Bakewell's own method of using only his own stock to the choice of New Leicester ram families available by the 1840s. This meant that some of the faults of the early New Leicesters could be avoided and contemporary sheep could be larger and more robust than Bakewell's. The Cotswolds, the lowland Gloucestershire, the Devonshire, the Lincolnshire and the northern crosses were by the 1840s all heavier than the original Dishley sheep. Low's opinion was that,

> no doubt can be entertained of the vast benefits conferred on the breeders of the country by the formation and diffusion of the beautiful breed of Bakewell. Its superiority over all the older races of the long-woolled districts, is attested by the degree in which it supplanted them, and the eagerness with which it was everywhere received. In less than fifty years from the first establishment of the shows at Dishley, it had either superseded all the older long-woolled sheep of the country, or been so mingled with them in blood, as to have effaced their former distinctions. Not only did it supplant, or become mixed, with the older races of heavy sheep, but, after a time, it effected an important change in a great part of the lighter sheep of the country.[22]

In Low's opinion, the long view that can be taken shows that the common objections to the New Leicester were more than balanced by its advantages, accepting that it is 'a breed of artificial formation' which has been removed from 'the natural state'. Its defects by then were only damaging if 'refinement of breeding is carried to an excess, which every breeder has now the power to avoid'.[23]

Caird noticed another variant crossing in Yorkshire, between a Shropshire ram and a Leicester ewe, which gave a better balance of fat and lean meat.[24]

Andrews claimed that by 1853 Bakewell's Leicesters 'were celebrated all over the world' and that they 'have had more effect in improving other breeds than any other class'. The Teeswaters, while 'a very similar class of animals', only owed 'their celebrity to Mr Culley's exertions in crossing local animals with the Dishley breeds'.[25]

Copland listed the Leicester crosses which, in his opinion, were successful and well established by the 1860s. They were the Cheviot cross which 'has been very successful'; and the Southdown and Leicester, or Leicester and Southdown, which 'are now established breeds'.[26]

The publication of Darwin's On the Origin of Species in 1859 led to a reassessment of the in-and-in breeding method which had often been condemned in Bakewell's day. An unexpected scientific comment came from the West Country in 1862. An anonymous journalist writing about the need to increase food supply during the Industrial Revolution may be the first person who linked Bakewell's thinking to Darwin's evidence, writing that Bakewell 'took the matter up, and adopting that great principle of 'selection' which Mr Darwin tells us is at the bottom of nature's operations, founded a valuable breed of Leicester sheep'.[27]

Coleman surveyed the leading breeds of sheep in 1877. He placed the Leicester first because Bakewell created a 'type of animal that eventually impressed its qualities more or less upon every other variety of long-woolled sheep', even though at that time, the Shropshire was making 'great strides in public favour'. The main breeders buying at the 1831 sales were: Mr Stubbins of Stone Barford, whose purchase was the origin of the Holmepierrepont sheep, by then giving way to the Shropshires; Paget, of Eleman (a mishearing for Lubenham), Leicestershire; and Philip Skipworth the elder's Aylesbury flock, which in 1877 was one of the most valuable in the country.

Coleman wrote at length about the influence of the Leicesters on the northern breeds: the Cheviots, the Black-faced, and the 'Barmshires' or Border Leicesters, in Roxburghshire, Berwickshire, and Northumberland. He discussed whether the early Border Leicesters had been further improved by crossing in with the Cheviots, or by selecting pure-bred Borders specifically for high legs, white faces and clean bone, or whether the soil and climate had influenced change. He inclined to the first of these factors but acknowledged that these questions 'can never be satisfactorily answered'. Nevertheless, it was clear by then that the Yorkshire and the Border Leicesters had diverged, even though they originated with the same stock. The Yorkshire-bred sheep had 'blueness in their faces and a tuftiness in their legs' while the Border Leicesters 'are white and clean in both, and, more, what are generally called *upstanding* sheep'. The then recent tendency to try further improvement by crossing with sheep 'of a looser frame, and wool of an opener and stronger staple' had, he felt, generally failed. He saw the best crosses then as with the Cheviot and the Black-faced, both mature at two years old with good quality mutton, and the Southdown. Pure-bred Leicesters by then had spread beyond the Borders even to the far north of Scotland.[28]

As late as 1919 Bakewell's lasting influence was acknowledged even though some breeders claimed that it was 'played out'. Wrightson paid well-deserved tribute to Bakewell when he wrote:

> Good stock, although general, is by no means universal, and, besides, the Leicester cross produces wonderful carcases. What, for example, can excel the Southdown or Hampshire Down and Leicester cross? Again, what can be better for certain situations than the cross between Leicesters and Cheviots, or Leicesters and Blackfaces. Going further afield, the Leicester-Merino is an improvement upon the Merino, as it produces a butcher's sheep, as well as a mere wool-carrier.
>
> There is scarcely a breed which has not felt the influence of the Leicester. No race has been so largely employed as a means of improvement, and Southdowns, Cotswolds, Lincolns, Shropshires, Hampshire Downs, and probably every sort of sheep, directly or indirectly has benefited by it.[29]

Endnotes

1 Quoted partly from Pawson, H.C., Robert Bakewell Pioneer Livestock Breeder, (1957), p188-190.

2 Lawrence, J., A General Treatise on Cattle, The Ox, The Sheep, and The Swine, (1809), p553-4.

3 Anon., 'The Obituary of Robert Bakewell', in *The Gentleman's Magazine*, November 1795, pp969-71.

4 Rudge, T., General View of the Agriculture of the County of Gloucester, (1807), p306.

5 Heazlewood, B., *Should the English Leicester have Modernised?* n.d., p2.

6 Heazlewood, B., *Should the English Leicester have Modernised?* n.d., p2.

7 Young, A., The Annals of Agriculture, in Potts, T., The British Farmers Cyclopaedia, (1809), no page numbers.

8 Young, A., General View of the Agriculture of the County of Sussex, (1793), pp57-8.

9 Culley, G., Observations on Livestock (1786), in Potts, T., The British Farmers Cyclopaedia, (1809), no page numbers.

10 Usher, J., *Border Leicesters* in Coleman, J., (ed.) The Sheep and Pigs of Great Britain, (1877), p35.

11 Monk, J., General View of the Agriculture of the County of Leicester, (1794), p27-8.

12 Young, A., Tours in England and Wales, selected from the Annals of Agriculture, in *No. 14 in a series of reprints of scarce tracts in Economic and Political Science*, (published by the L.S.E. in 1932), p274.

13 Information from Mr K. Elliott of Measham Museum.

14 The British Newspaper Archive (BNA), *Newcastle Courant,* 19 August 1826.

15 Dickson, R.W., Practical Agriculture, or, A Complete System of Modern Husbandry, Vol 2, (1805), pp1029, 1137-8, 1140, 1153.

16 Rudge, T., General View of the Agriculture of the County of Gloucester, (1807), p305-6.

17 Rudge, T., General View of the Agriculture of the County of Gloucester, (1807), p309-10.

18 Potts, T., The British Farmers Cyclopaedia, (1809), pages not numbered.

19 Lawrence, J., A General Treatise on Cattle, The Ox, The Sheep, and The Swine, (1809), p330.

20 Lawrence, J., A General Treatise on Cattle, The Ox, The Sheep, and The Swine, (1809), p331-33, 336-7, 352-54, 401, 406.

21 Lawrence, J., A General Treatise on Cattle, The Ox, The Sheep, and The Swine, (1809), p591.

22 Low, Prof., 'An Evaluation of Robert Bakewell's New Leicester Sheep', from Illustrations of the Breeds of Domestic Animals of the British Islands No. XII, in *The Sporting Review* Vol VII, January-June 1842, p59-61.

23 Low, Prof., 'An Evaluation of Robert Bakewell's New Leicester Sheep', from Illustrations of the Breeds of Domestic Animals of the British Islands No. XII, in *The Sporting Review* Vol VII, January-June 1842, p59-61.

24 Caird, J., English Agriculture in 1850-51, (1852), p294-5.

25 Andrews, G.H., Modern Husbandry, (1853), p180.

26 Copland, S., Agriculture, Ancient and Modern, by The "Old Norfolk Farmer", (c.1861), p397.

27 BNA, *Trewman's Exeter Flying Post or Plymouth & Cornish Advertiser*, 19 February 1862.

28 Coleman, J., (ed.) The Sheep and Pigs of Great Britain. Part II: The Breeds of Sheep, (1877), p30, 32, 36, 37, 38.

29 Wrightson, J., Sheep, Breeds and Management, (1919), No. 1 of Sinclair, J., (ed.) Live Stock Handbooks, p21.

CHAPTER SIX

Bakewell's Legacy: Farming Practice

Pat Stanley & Janet Spavold

> Perhaps, during the last century, no man did so much towards increasing the agricultural interests of the country as Mr Bakewell.[1]

The experiments described later in which Bakewell touched his sheep with his magician's wand are legion. His work seemed so extraordinary to the Duchess of Exeter that she described him as 'the Mr Bakewell who invented sheep'. The legacy of his work on sheep was, as will be seen, extensive into the 19th century, and an example of its enduring success can be seen in the long life of the Speeton flock, owned by the Coleman family of Speeton, Filey, in Yorkshire, which descended directly from Bakewell's stock in the 1790s.[2] But let it be said that he tested his capabilities on many subjects, first with the Longhorn cattle (see Part Two) and then including the production of a superior 'Black Leicestershire Carthorse' and improved pigs and fowls, discussed in Chapter 7. The legacy of his work on the black horse was its development into the familiar Shire horse; the lighter draught horses he developed, and coach horses and cavalry horses, eventually were replaced by mechanisation. The pigs and fowls did not develop into long-lasting types.

There was, however, a sudden total failure in the value and saleability of at least the Dishley sheep in 1816 and 1817 following Honeyborn's death, as evidenced by the newspaper sale advertisements. On 27th May 1816 the sale of 'a select part' – the best 19 – of Mr Honeyborn's rams was advertised, to be held on 10th June 1816. The rest of the rams and ewes, plus the cattle and horses, were to be sold in September 1816. The auctioneers clearly expected the 19 rams to sell well given their pedigree and past performance:

> It is sufficient to say of this Stock, that it is all directly descended from that which the celebrated Mr Bakewell (Uncle to Mr Honeyborne) left at Dishley at the Time of his Decease, and to which all the principal breeders of New Leicester Sheep and Long-horned Neat Cattle, owe the Origin of the Pedigrees of their respective Stock.[3]

But the sale did not go to expectations. A Stamford paper recorded the event. The sale had attracted 'a great concourse' including 'many of the principal breeders', but they had not come to buy. Most attended out of 'mere curiosity' and there were few bidders; 'consequently the lots, with the exception of the last, were knocked down, to use a figurative expression, almost at their own prices'. The published estimates for these rams ranged from the cheapest, No. 1, at 15 guineas, to the most expensive, No. 19, at 95 guineas. The journalist's lament reflected his shock at this outcome.

> Had the great Mr Bakewell lived to have witnessed this defalcation in the value of the far-famed Dishley rams, (*many of which, in his day, fetched more in being let out for one season only, than the fee-simple fetches now!*) he might, with Mr Pitt, have justly exclaimed, 'Oh the times – Oh my country!'[4]

Some of those attending thought this shocking fall in value was due to the 'general pressure of the times'; others were 'imagining the stock to be in some degree degenerated'. In the end, only six rams were sold and the rest, including No. 19, returned to the farm. The article ends with an account of an even more unfortunate ram sale at the same time. A breeder brought his rams 'in carriages' the 25 miles to Loughborough, thinking it to be the best place to sell them. But as only one person came to bid, the sale was aborted 'and the finely-proportioned animals of course all taken away unsold'.[5] Sales of more stock were advertised for 16th and 17th September 1816.[6] More of Honeyborn's remaining stock was advertised to be sold on 25th and 26th March 1817 together with his husbandry equipment,[7] followed by a final sale of livestock, husbandry equipment and 'All the excellent, genteel, and useful household furniture, plate, linen, books, paintings, brewing vessels, and numerous remaining effects, late the property of Mr Honeyborne, at Dishley, Loughborough, deceased'.[8] This sale was, in fact, the breaking-up of *Bakewell*'s home; it would be fascinating to know more about the books and paintings.

Honeyborn's rams certainly did not fail to sell because they were degenerated. He was unfortunate to die just as the severe depression following Waterloo and the end of the Napoleonic Wars hit the country. In Leicestershire, the worst effects were felt by the framework knitting and worsted trades; the knitters made stockings for army contracts, which immediately ceased. They depended on the long staple wool of the New Leicester sheep, so the knock-on effect was that nobody was buying the wool and nobody needed to replenish their flocks as they had been doing for the duration of the war. It was the New Leicester's economic value, not its intrinsic worth, that was hit and as the economy recovered, so did the sale values. Even in the depth of the depression the value of a ram's pedigree was still acknowledged in sale advertisements; for a Frolesworth sale in 1817, the rams were advertised as 'descended from

Rams of Mr Bennett, Mr Buckley, and Mr Smith, of Dishley, whose Stocks stand in the highest Reputation'.[9] William Smith had taken over the lease on Dishley Grange and the legacy of the New Leicester continued.

A more diffuse legacy can be seen in Bakewell's influence on farm management. The way he farmed at Dishley Grange was much admired by contemporaries and his methods were then copied. Bakewell's quest for knowledge led him down every avenue of agriculture. In order to improve his livestock he also had to improve the land from which they were fed, and in order to improve the land he utilised both manure and irrigation.

Feeding and Manure

In his early years, when stock were housed, little if any use was made of bedding or litter. Instead, his animals, which were kept tied, were fed little but often to keep their appetites keen, with all forage going through their systems and not onto the floor as bedding. This naturally produced very rich manure, the making of which Bakewell believed was one of the most important points of husbandry.

He was so convinced of the superior value of manure produced in this way, as opposed to that of loose-housed animals in bedded yards, that he would take in neighbouring stock over the winter months of November to March, providing he had room for them, his only return being the dung they produced.

It was the normal practice of many farmers to run their cattle out all winter, with the fodder stacks round about them. However, this caused the land to become badly poached. Furthermore, the stacks were destroyed, leading to great wastage, and the valuable dung was scattered thinly over the field. Alternatively, they were kept in yards open to the elements, with access to fodder round the sides.

Bakewell could, and did, winter up to 170 cattle all tied in standings which he devised and had built into purpose-made sheds or old barn conversions. This was all done at his own expense and not that of his landlord. The standings he created were to aid the easy retrieval of dung and to reduce wastage of this most valuable product. For many years, he maintained that manure increased its fertilising power with age and that it should be kept at least three years before use until it had the consistency of peat dust (however, he later changed this opinion).

The floors of the standings where the beasts stood were paved and were 6-8 inches higher than the level of the barn floor. They were just broad enough for a beast to stand on, though with some difficulty, the result being that all the dung fell beyond the standing and into the lower channel. When the beast lay down it drew itself up onto the higher standing and in this way was elevated above the dung, keeping quite clean without the need of litter. A man was employed to keep the channels clean and barrow the dung to a central clamp in the open yard which was surrounded by the cattle sheds.[10] (Bakewell's design of cow standings presages the continuing discussion of recent years on appropriate dimensions for cow cubicles for dairy animals.)

La Rochefoucauld described the efficient method Bakewell used for feeding his cattle, which were each tethered to a small manger in a great barn. They stood parallel to the walls, all facing the centre, with a passageway behind them, and their straw, hay and turnips were brought into the centre in carts. They were not let loose in the yard to stop them trampling and wasting their fodder, which was regulated so they ate it all. As a result, there was 'an abundance of muck', easily collected from the standings. Bakewell thought it was 'better to cart the dung on to the land than to let the cattle loose to manure it unevenly'. La Rochefoucauld was deeply impressed by the way the system ran: 'The layout of this farm, which is immense, gave me the greatest pleasure. All is in tremendous order, everything arranged to avoid wasting time in carting from one place to another'.[11] But some of Bakewell's ideas on the production of fertiliser did not meet with the approval of one writer, who thought that his plan 'to dry his dung to the fineness of a pinch of snuff, I have ever thought absurd and unprofitable in the extreme'.[12]

By not utilising straw for litter, but keeping it all for forage, he was able to greatly reduce his winter feed bill. Large quantities of hay were produced from his water meadows and the means for maintaining his stock were also increased by the cultivation of vetches, turnips, cabbage, carrots and potatoes. The first reference to turnips being eaten by sheep is in Houghton's Collections on Husbandry and Trade, a periodical work begun in 1681. Other crops tried were coleseed and borecole. Carrots, although expensive, were highly valued as stock feed. They were preserved in various ways through the winter, some being buried in earth, while others were stored in buildings. Tops and roots were fed when fresh, but after storage the tops were removed. Potatoes were used when the price was low, and coleseed was sown into oat stubble for use by sheep, when it had grown a sufficient bulk of leaf in the period from late autumn to early spring. It was with resources such as these that Bakewell managed to support such large numbers of animals on a limited acreage.

Near to his house Bakewell had several small plots of grassland of less than an acre in size, which were kept very well fenced. These were used for isolating small lots of cattle or sheep. The rest of the farm was divided into 10 acre closes. These divisions were carried out by Bakewell at his own expense, even though a tenant of the farm, and were mainly created by growing hawthorn hedges, back-fenced against the animals (see section below on hedge planting).

Irrigation

In 1771 Arthur Young described the irrigation methods used by Bakewell at Dishley.

> Of his hay he is also very choice; and the means he has taken to command as large a quantify as possible, are perhaps to be reckoned amongst the rarest instances of spirited husbandry ever met with among the common farmers of *England*. It is that of watering his meadows that lie along a small brook [the Blackbrook] which runs through one part of his farm. This improvement was begun by his father, now living, and carried on and finished by himself.[13]

Marshall classed him as 'standing first in the kingdom as an improver of grassland by watering' and from Monk's *Agricultural Report* we find that Bakewell was able to cut grass four times a year after growing it by means of irrigation. Young further described Bakewell's methods and the following is best considered as a collation of different sources.

> Mr Bakewell has been in the practice these several years, of watering his meadows & pasture lands, which he considers a great improvement, as superseding the necessity of manure, & producing more food by 20s. per acre; though the original expense of making canals & intersections, did not exceed that sum, & the annual expense attending keeping the drains in proper order, & all the watering operations, is not more than 5s. per acre. He has no regular or fixed period for watering, as in Wiltshire, but keeps watering all the year round.[14]

Originally, 60-80 acres at Dishley gained considerable benefit from natural flooding by the River Soar, but after a trip to the west of England to study irrigation methods, accompanied by a Mr George Boswell who had published a book on the subject entitled A Treatise on Watering of Meadows, Low Lands Etc., Bakewell proceeded to irrigate a further 100 acres of his land artificially.

On his return from this fact-finding mission, Bakewell put his genius to work by diverting, for his needs, the Blackbrook, which skirted the farthest boundary of his farm. He changed its path at the highest point he could and carried it, by means of a freshly-cut canal of one-and-a-quarter miles in length, not only around the higher points of the land available for the scheme on his farm, but also to supply the farmstead with water. As well as a multitude of other purposes to which it was put, he was able to create a drinking pool for livestock and a wash pool for his sheep.

The 100 acres of his newly-converted water meadows were originally in ridge and furrow, as was most of the county. Bakewell ploughed this up and cultivated and reseeded it, leaving grassland with a perfectly even surface. Then, by means of ditches known as floats, he conveyed water from his newly-constructed canal, causing the water to flow over the land at will. Water was returned to the canal by drains or sumps. While this operation was in progress the whole surface area to be flooded was covered with a sheet of water at least an inch in depth. In very hot weather it was allowed to lie for up to 48 hours, in moderate weather for 4-5 days and when cold for up to 14 days. If there was a possibility of frost the operation was stopped.

Always one to be sure of his facts, Bakewell devised experimental plots to establish the validity of his methods. One plot would be flooded, another not; a further plot would be manured, while yet another would be irrigated with spring water and another flooded with canal water. In this way he could compare the value of irrigation against other forms of fertiliser and the different methods of irrigation. Young's judgement on Bakewell's watering was that,

> Where the situation answers, it certainly ought to be adopted, especially in countries at a distance from lime, & other foreign manure; & perhaps is one of the greatest means of improvement, that can possibly be introduced into many parts of Scotland. Mr Bakewell thinks it would also be of considerable benefit, even to arable land, where it could be accomplished.[15]

'Our farmer', Young remarked, 'has expended large sums in these uncommon undertakings; he richly deserves the enjoyment of their profit'. The difference in the colour of the grass on the watered and unwatered land was said to be 'astonishing', and the amount of grass produced on the watered area was in the proportion of 10 to 1 against the unwatered sickly, yellow grass. Neighbouring land remained full of great quantities of rushes and coarse, sour grass, while Bakewell's land was so well tended it was almost free from weeds. Young reported that, prior to Bakewell improving his land by the creation of new water meadows, it had been in a state similar to that of the rest of the county, covered in ant hills and disfigured by irregularities of surface. His neighbours are said to have looked upon his grassland operations with horror, feeling that the ploughing was 'burying good land to bring up bad'.

Fearful that his flood water should overflow their land and 'poison it', some people actually started legal proceedings against him to stop the work. Agreement seems to have been reached by Bakewell undertaking to construct banks to prevent the water straying off his own farm.

Bakewell used the spoil thrown up by cleaning out his local Blackbrook to fill in the hollows along its course and he kept both the brook and its tributary ditches clean and repaired. He extended his original 60 acres of naturally-flooded meadow gradually, feeling his way gently and experimenting constantly to prove to himself the efficiency and economy of the system.

Another purpose to which he turned his canal was to use it as a cheap and easy means of moving his crops of turnips and cabbages from the fields to the buildings, and manure from the buildings to the fields. He designed a flat-bottomed boat for this purpose but, finding it not the easiest thing to use, he hit upon the simple idea of throwing the turnips straight into the water and allowing them to float down to the barn. In 1789 he designed a reservoir for them to fall into, with a grate at the bottom to release the water but retain the turnips. They came out beautifully clean and ready to be stored for the winter.

The canal he cut was narrow and, of course, had a minimum fall but the actual reduction in labour costs, both human and animal, in using the waterway for conveying root crops, etc. was said to be 'huge'. Remarkable though it may seem, evidence still remains of Bakewell's revolutionary irrigation system, although it is well over 200 years since it was installed. Timber liners of elm for the canal and brickwork drainage wells still survive to help us piece together this man's great achievement.

Bakewell even turned his gaze to the humble cabbage, for which he became nearly as famous as for his sheep. The cabbage he propagated was a large stock-feed variety of high nutritional value and hardiness. Great care was taken in raising the seed to keep it pure and prevent any cross-fertilisation by other brassicas. For this reason, it was grown on plots surrounded by corn. It was chiefly fed to sheep, although occasionally used for other stock. They were planted well apart; Marshall recommended four feet by two and a half, so they could be cleaned using a plough. Marshall declared Bakewell to be the 'greatest cabbage grower' he had ever seen.[16] Another visitor remarked on the experimental plots of both drilled and broad-cast sowing for cabbages, wheat and barley, 'to shew the comparative difference. The drill is preferred'.[17]

Hedge and Tree planting

He also had revolutionary ideas on hedge planting, described in detail by Young in the Annals of Agriculture.

> Mr Bakewell is very curious in his fences, and plants his quicks [quickthorn] in a different manner from what is common in various parts of the kingdom. He plants one row at a foot from set to set, and, making his ditch, lays the earth which comes out of it to form a bank on the side opposite to the quick. In the common method, the bank is made on the quick side above it.[18]

Bakewell's method encouraged better root growth and avoided the roots being covered by earth dug from the ditch. But Young did not approve of the addition of post and rail fences, one outside the bank and the other outside the hedge, which he thought a waste of land. La Rochefoucauld thought Bakewell's method was better than the old way that he had seen in Norfolk and Suffolk, because it made it more difficult for horses or cattle to break out.[19]

He also put in large plantations of white willow (*Salix alba*) which, when mature, were used for gates, hurdles, fence rails and other purposes. Elsewhere, these trees are described as being Dutch willow. His famous chair, which was held by the Royal Agricultural Society of England, was made from willow grown on his farm. It is from this chair that he is said to have held court in his kitchen at Dishley, and at night would have been found writing up the day's proceedings, settled in its welcome embrace.

Bakewell's chair.

The R.A.S.E. journal of 1896 recorded that Bakewell's chair (see page xvi) was presented to the society by one of its members, Mr Henry C. Woodcock of Rearsby, Leicestershire, and that it was accompanied by a painting representing Bakewell on horseback (see colour illustrations section).

One stand of willow was planted almost the full length of the canal which he constructed through his farm, while several plantations were situated near to the house and buildings. These were cut in a seven-year rotation, providing coppicings which could be very long and large and could be cut and split to make posts, gates and rails.[20] He reinforced all his gates with cross-footed hurdles made from these coppicings to prevent them being forced by any strong and determined males, of which he kept a large number, from reaching the females before they were ready for breeding. Coppicings were also used to make handles for rakes, pitchforks and other tools. All the trimmings from the willow, when producing these various articles, were put to good use; a visitor in 1793 saw a peeled stack propped up in the farmyard; the bark had been neatly tied in bundles ready for use as the base of straw and hay ricks if needed, though permanent brick bases were constructed.[21]

Farm roads

His ideas on road building were heavily influenced by Joseph Wilkes, for whom this was something of an obsession; he was a promoter in several turnpike trusts. Many turnpike roads were being constructed in this period and there was disagreement as to how best to build the surface. Telford and MacAdam favoured convex, to drain water off the road into the ditches, while some turnpike promoters favoured a flat surface. But,

> Mr Bakewell and Mr Wilkes are advocates for the concave. This is nearly the reverse of the common practice of making roads, by making the middle *the lowest*, though flat about one third of the width, with a small slope from the sides to the middle, where the best materials are placed, and an equal but gradual descent, sufficient to carry off the water from the middle.[22]

Monk noticed that the roads at Bakewell's Dishley and at Measham, where Joseph Wilkes was the proprietor, were built this way, and they were in good order. The reasons for this preference were that carriages could be driven on three parts of the road surface without undue pressure on the wheels, which made it easier for the horses. The surface would be worn evenly and therefore damaged less. They would not be badly affected by heavy rain and were easier and cheaper to repair. While convex surfaces were by far the most common, as late as 1805, the concave surface was still in vogue, because 'it has been contended for by some very intelligent advocates, as the late Mr Bakewell, and Mr Wilkes; and different roads in Leicestershire have been formed on the plan, it is asserted with great advantage'.[23] It does not seem to have been so popular in other counties.

Ploughing

Bakewell was credited with being the first to plough with only two horses abreast instead of the more usual four in tandem. He was a great supporter of new machinery inventions, using, amongst other things, the Rotherham plough and Cooke's improved drill, a variation by the Revd Cooke on the principle established by Jethro Tull.

He was also fortunate in having machinery makers in his local village, Hathern. Messrs. Hanford and Co. were no doubt able to turn his ideas into reality. It was certainly from them that he obtained a hopper attachment for his plough, by means of which he was able to sow beans at the same time as ploughing – now quite a common practice.

In his report on Leicestershire Agriculture to the first Board of Agriculture in 1794, John Monk noted that there was a blacksmith's shop at Dishley where two men were employed in making tools and implements for the farm, and in repairing those that were broken. This clearly gave Bakewell the capacity to develop many of his practical ideas ranging from new designs for gates to innovations for his irrigation system. The gardens are reported to have been 'sometimes irrigated', even though they were described as being of a 'swampy nature', not far from the River Soar. He was not only a practical inventor however; he was interested in the scientific theories of his time, attending a lecture on experimental physics in Loughborough one evening with the La Rochefoucauld brothers. They entertained him to supper afterwards and they could see how well, 'beneath a heavy and rough exterior', he had studied his breeding methods 'with as much care as one would put into the study of mathematics or any of the sciences'.[24]

The great agricultural improvers of the period, like Bakewell, had to finance their experiments out of their own pockets, as there was no Board (or Ministry) of Agriculture funded by the state to carry out such research. When the Board of Agriculture was established in 1793, because it was never an official government body but floating somewhere between a club and an official department, its small stipend of £3,000 a year was never enough to fund research. Both Bakewell and William Marshall were, though, strong advocates of such an institution, as was Arthur Young, who became its first secretary. Provincial banking was just beginning towards the end of the 18th century. The Leicestershire Unitarian network was significant for this development, but there is no evidence that Bakewell used a formal bank, though Leicestershire had two in the 1790s. By 1803 the son of Bakewell's friend Thomas Paget, also called Thomas, had established the Leicestershire Bank with three other partners, also Nonconformists. In 1825 Paget junior left the partnership and established another bank, Thomas Miller & Sons.[25] Banking was therefore hardly available to Bakewell's generation, more so to the next one and many in the local Nonconformist networks became shareholders and customers.

When we learn, then, that in November 1776 Bakewell was declared bankrupt, it is perhaps not altogether surprising. The causes of his plight were cumulative, with the balance being finally tipped by bad debtors owing money for the hiring of sires. His position could not have been helped by the lavish hospitality to be had at Dishley, which was said to be in the same grand style as in the homes of the landed gentry when they entertained. It must also be allowed that this was a most unfortunate period for making money by the cultivation of the land. The disastrous war with America was said to have considerably reduced the price of produce. Wool, which constituted a proportion of Bakewell's income, greatly fell in price, selling for as little as nine shillings per tod (28lb). And the personal cost which he shouldered in order to conduct his many and varied experiments was no small matter.

The day was saved by his many friends and acquaintances who felt they could not, and should not, stand idly by and witness the fall of a man to whom so much was owed by his contemporaries. A subscription list for voluntary contributions was opened and the list of subscribers reads like a *Who's Who* of the day. The Duke of Rutland himself gave 200 guineas. Bakewell's bankruptcy is dealt with in detail in Chapter 17 and Appendix 7.

Even at the time of his bankruptcy, Bakewell kept up his high position, for when his stock was valued (the unexpired term of his lease being included in the valuation), there was a balance in his favour of from £1,500 to £2,000. From documents that may be considered of the highest authority – the valuations that were placed in the hands of the trustees at that time – the following extract shows the prices at which a portion of his sheep were estimated:

		£	s
Willow Piece	26 ewes and lambs	81	18
Top Piece	19 ewes and lambs	99	15
Balk Piece	6 shear hog rams	42	10
Marl Piece	6 ewes and lambs	31	10
Thorpe Close	26 theaves	109	4
Mill Leys	10 shear hog rams	200	0
Mill Close	10 shear hog rams	260	0
Cow Close	3 rams	50	0
ditto	3 rams	300	0

Ironically, after being made secure by his friends and followers, he was repaid the money, the lack of which had forced him into difficulty in the first place.

So neither his herd nor his flock was ever dispersed in his lifetime and he was quickly able to take up his mantle once more and carry on with the task for which the nation should evermore be grateful.

Praise for, and appreciation of, Bakewell's legacy can be seen in the tributes to him published on his death. The South Devon Agricultural Society supported the call for a memorial to him at their meeting on 27th October 1796.

> It being represented at this meeting, that a monument is intended to be erected to the memory of Mr Robert Bakewell, late of Dishley, for the very great and beneficial improvements made by him in the breed of sheep and other cattle – it was unanimously voted, that this Society should contribute the sum of Ten Guineas towards the same, as a token of its sense of his merits.
>
> James Cornish, Secretary.[26]

It would be interesting to know who was proposing the memorial, and what happened to the South Devon Agricultural Society's ten guineas.

The Earl of Leicester, a great admirer of Bakewell and the founder of the Holkham sheep-shearings, instigated a standing after-dinner toast in memory of Bakewell, raising their glasses to his aim of 'small in size and great in value'.[27]

The most articulate and impassioned plea for due recognition of Bakewell's legacy came in 1919.

> That Bakewell deserves to be considered a genius there can be no doubt, and few inventors have merited higher praise than he, or have added more to the greatness and prosperity of their country. The man who led the way to the improvement of all our breeds of live stock, and showed how in a few years the animal form might be modified, and improved incalculably beyond anything which had previously existed, surely deserves a national monument, and yet how little recognition has he received! The name of Bakewell is rarely mentioned among those worthies who have made England. It is not seen inscribed on the cornices of public halls, and his bust is missing in the corridors of the national pantheon. Novelists, dramatists and painters, soldiers, engineers, and scientists may have their names recorded on the walls of fame, but Bakewell, the apostle of the art of improving the domestic animals of the world, is passed by unnoticed, and doomed to lie forgotten in his grave without a mark by which those who participate, as all do, in his grand work may know to whom they owe so much. It is not out of place to call attention to this omission, and we cannot but think that the suggestion should receive attention, seeing that agriculture has now been raised to a position of greater eminence by the formation of an Agricultural Department of the State. The master-mind from whom Culley and Charles Colling derived their inspiration should not be lost sight of or allowed to be forgotten, for if the torch and fire of genius are to be claimed or accredited to any it is to the pioneers of a new and great movement like this. A monument to Bakewell, flanked with effigies of the horses, cattle, and sheep which he ennobled, and adorned with a fitting inscription, would be as dignified in its silent teaching as that

> of a Stephenson, a Watt, or a Cartwright; and no place would be more suitable for its site than the metropolis of the country which he so well and unconsciously served. He was no seeker of or candidate for future fame, but as a farmer he worked without ostentation, until his influence spread itself over every civilised country of the globe.[28]

The civic recognition of his legacy is still missing, but another of his legacies lies in the creation of some important agricultural institutions both during his lifetime and subsequently.

The Dishley Society

There is also some lack of clarity as to when the Dishley Society formed; although it is variously reported as 1783, there is a lack of evidence for this date, with most agricultural historians agreed on 1789. Amazingly, and by great good fortune, the Leicestershire Agricultural Society held its 154th show on Dishley Grange in the very year of the bicentenary of Bakewell's death, 1995 and a strong contingent of Longhorn cattle was present.

The Leicestershire Agricultural Society

There seems to have been some uncertainty over the date of the founding of the Leicestershire Agricultural Society, so the Rutland History Society researched the matter, finding as follows: an initial meeting was held in February 1788 at the *Three Cranes* in Leicester to establish a society for the encouragement of agriculture in the area. On 2nd April of that year a list of founder members and the resolutions of the earlier meeting were published. Although initially postponed, the first AGM of the society was eventually held on 7th August 1789, following which its business was regularly reported in the *Leicester Journal*. The society was reported to have about 100 members, including as we understand, Robert Bakewell. Members were invited to attend from outside Leicestershire and eventually the organisation supposedly merged with a similar one based in Rutland. However, it seems to have gone into terminal decline thereafter. In 1833 a new organisation, the Leicestershire Agricultural Society came into existence, but a separate Rutland Agricultural Society had been established two years previously.[29]

The Smithfield Club

Mention is made elsewhere of Joseph Wilkes of Measham, Leicestershire, whose farming was influenced by his friend Bakewell. In 1798, three years after Bakewell's death, he had the idea of starting the Smithfield Cattle Club with an annual show of livestock. This operated for only two years until it was disbanded by the 5th Duke of Bedford, who formed a new club in December 1800 and became its first chairman. The new club ran with no committee but was managed by the club members, whose number was restricted to 50 and consisted of the 14 committee members from the old club, plus 22 members whom that committee proposed, including Sir Joseph Banks, and the remaining 14 places were decided by ballot.

A print engraved by Charles Green, probably published in the *Illustrated London News*, was made for the centenary of the club and it shows Wilkes, in waistcoat, breeches, boots, hat and caped coat, looking at a pen of fat pigs on the left.[30]

One wonders what the great man himself would think of the vast strides in all sectors of agriculture which have been taken since he himself set the wheels in motion, with such vision and industry, so many years ago.

Arthur Young felt he could not conclude his observations of Bakewell's many undertakings 'without expressing the satisfaction I felt at viewing them. Nowhere have I seen works that do their author greater honour. They are not the effect of a rich landlord's attempts to be a good farmer on his own land but the truly meritous [meritorious] endeavours of a tenant performing great and expensive works on the property of another. Let me exhort the farmers of this kingdom in general to take Mr Bakewell as a pattern in many points of great importance. They will find only good can come of it and the kingdom in general will be benefited not a little'.

Although no memorial has ever been raised to the memory of the man whose achievements must stand alongside those of Jethro Tull, Charles Townshend and Coke of Norfolk, there can be no more fitting tribute than that which we see paraded before us at our agricultural shows today. They corroborate the achievements of the dedicated stockbreeders who have followed Bakewell down the ages and have given our country the reputation of being the stockyard of the world.

Robert Bakewell's direct legacy

The following is an extract from the valuation of Dishley, taken in 1795, the year that Robert Bakewell died.

Mill Close	- one red cow and cow calf	£35	0s	0d
	- one cow and bull calf	50	0	0
	- one cow and cow calf	40	0	0
	- one two-year bull	70	0	0
Land Piece	- one red cow and calf	40	0	0
	- one cow and calf	60	0	0

Brook Piece	- one cow and calf	35	0	0
	- one cow and calf	30	0	0
	- one cow	30	0	0
	- one cow and cow calf	30	0	0
Cows Close	- one bull	20	0	0
Barn	- four bulls	105	0	0
	- three cows	150	0	0
	- three cows	60	0	0

After Robert Bakewell's death, the herd passed intact to his nephew, Robert Honeyborn, at valuation figures of between 20 and 70 guineas. Honeyborn lived and farmed at Dishley for a further 20 years, until his death, but sadly never measured up to his great predecessor. In his hands the great race of Dishley Longhorns was to wither and die, so that within a short period of Bakewell's death, there could scarcely be found a dozen true beasts of the kind within a 12-mile radius of Dishley itself. In 1844 Youatt likened the disappearance to that of 'some strange convulsion of nature, or some murderous pestilence' that had 'suddenly swept away all of this wonderful breed'.

During Honeyborn's time, Dishley lost all its importance and it quietly slipped back into the shadows, holding only a tenuous grip on the memory of those who quickly forgot this once venerated corner of the Leicestershire sod.

Bakewell's last few years were beset by health worries which he is said to have borne 'stoically'. He had always had a tendency to stoutness and, nearing the end of his days, 16 stone on the frame of a 5ft 9 inch man can hardly have helped.

The year before his death he was again made president of the Dishley Society but, due to ill health, had to stand down mid-term and hand over to his friend and compatriot, Thomas Paget of Ibstock, Leicestershire. Even when he was only 61, in 1787, it is to be seen in a letter he wrote to George Culley, that he felt sufficient time would not be left to him to carry out his life's work. He urged Culley, and all others of a like mind to his, to pursue his methods with 'unremitting zeal' as far as they possibly could. He reminded them always to have an eye for a superior way of thinking, should it emerge.

On 1st October 1795, at the age of 69, Robert Bakewell passed from this world to a better. His many obituaries are a great tribute to the man who was so obviously appreciated by many in his own lifetime. He provided the foundation and the inspiration for the great men who were to follow him.

Although in his lifetime he had a great aversion to selling his prize specimens, he developed an extensive letting business, with rams letting for as much as 1,000 guineas a season. He had always sought quality rather than size in an animal and founded a trend for small bone which was, after his death, taken to such an extreme that it can be cited as one of the main reasons for destroying the popularity of the Longhorn breed. J. Neville Fitt quite rightly stated: 'Moons wane, and races wither to the tomb'. The time of Bakewell and his contemporaries had finally ended.

Many of the men who followed in his wake lacked the skill of the master, and again in the words of J. Neville Fitt in 1876: 'Not content to leave well alone, in trying to gild refined gold, (they) overshot their mark, and produced misshapen animals, delicate in constitution and, worse than all, uncertain breeders'. They did not understand the principles of breeding as Bakewell and his contemporaries had and, ignoring qualities for which the Longhorn had been renowned, in their scramble to produce finer and finer bone, they sacrificed both size and constitution. The male of any species ought always to possess rough masculine properties but this most necessary qualification was neglected by the later breeders who strove only for long patchy rumps and hips, which had become the height of fashion, depth of rib and largeness of chest being considered of no or very little importance.

Towards the end of Bakewell's life, however, the great estimation in which the Longhorn was held can be judged by the following results of two sales held in September and November 1793. They were both conducted by a Mr Boot, the first for Nathaniel Pearce, to sell his Longhorn cattle and New Leicester sheep, the second for C. Cartwright to disperse his Lincolnshire sheep and Shorthorn cattle.

	Mr Pearce			Mr Cartwright			Difference		
	£	*s*	*d*	*£*	*s*	*d*	*£*	*s*	*d*
Best score of ewes	116	19	0	27	0	0	89	19	0
Best score of wethers	50	11	6	29	0	0	21	11	6
1 x bull 4 x cows	331	5	6	38	15	0	292	10	6
3 x yearling heifers	63	0	0	8	8	0	54	12	0
TOTAL	562	16	0	103	3	0	458	13	0

Bakewell the Teacher

In Bakewell's time, agricultural students were sent onto a farm to learn their trade, the cost varying with the ability of the tutor and the treatment the pupil expected to receive. The usual term was four years, and the cost ranged from £40 to £200.

Bakewell's first pupil was George Culley of Fenton, Northumberland, born in 1730 and destined to become one of his most fervent admirers and friends. He became famous for his flock of New Leicester sheep and for his chosen cattle breed, the Shorthorn, to which he applied his teacher's breeding

Robert and Charles Colling from an engraving by T. Thompson after the painting by T. Weaver.

principles. He was the author of Observations on Livestock (1786), probably the first book ever written specifically on farm animals. This was a project about which Bakewell was most enthusiastic and encouraging.

Matthew Culley sent his sons to Dishley; Matthew in 1762, followed by George in 1763. George in particular became a close friend and follower of Bakewell's methods. When staying at Dishley he also met Joseph Huskison of Osgathorpe in Leicestershire, another of Bakewell's close network and a neighbour. Huskison's son was much the same age as George; in 1765 he stayed with the family for about three weeks, as recorded in his first travel journal.[31] He visited Dishley again in 1771, then in 1784-5 when he went on a tour to East Anglia with Bakewell where they visited Arthur Young's farm. He was not impressed.[32]

The Culleys went on to create the Border Leicester by crossing a New Leicester ram with a Teeswater ewe; they hired their first ram in 1764. They consistently worked with Dishley rams for the next 20 years and they followed Bakewell's example by letting their own rams by the season in Northumberland, then expanding into Durham, Yorkshire and the Scottish Borders. As the Dishley influence spread, other northern breeders did the same. Following the success of the Dishley Society in 1789, of which Culley was a member, a Northumberland Tup Association was proposed in 1792 in an attempt to control the market, but as the northern breeders were not able to agree with their Leicestershire counterparts on their respective areas of influence the attempt failed. A later Culley development was the successful Dishley/Cheviot cross in south-east Scotland.[33]

Culley was a family friend of the Collings of Ketton, near Darlington, and was instrumental in arranging for the family's eldest son, Charles (who at that time was working for his father on the farm) to go, at the age of 31, as a pupil to Dishley. Charles, born in 1751, closely studied Bakewell's breeding system and was later joined by his brother, Robert. On returning home, the two of them applied their new-found knowledge to the popular breed in their area, the Shorthorn, and eventually eclipsed the deeds of their master. In their hands lay the doom of the Longhorn.

The foundation on which the Shorthorn empire was to be built was a bull called *Hubback*, which had caught the eye of Charles one Sunday as he made his way to church. The animal was quickly purchased but it actually did very little for the Collings, as he was sold on when his daughters entered the herd. It was to be some time before the Bakewell teachings were to sink in. In fact, it was to take an accidental mating, very much on the inbred line, with the result highly superior to the usual outcross, before the words of Bakewell began to hit home. Colling doubted no longer and from that point on his herd was as closely inbred as Bakewell's or Fowler's had been.

One of his earliest achievements was breeding the famous *Ketton Ox*, an animal regarded as faultless. It was by *Favourite* out of a common black and white cow bought at Durham Fair. Colling kept the beast for exhibition only, until, at the age of four years and weighing 3,024lb (27cwt or 1,371kg), he was sold to Mr Bulmer of Harmby, North Yorkshire, in February 1801, for £140. The animal was sold again after six weeks to Mr John Day for £250 who, after changing its name to the *Durham Ox*, exhibited him all over the country at fairs and shows, causing tremendous excitement. Day received several large offers for him but refused because the animal had become a pet of the family. The *Durham Ox* was finally slaughtered at Oxford in February 1807, after dislocating a hip.

It was to be to this animal that the writer, the Druid (H.H. Dixon), attributed the fall of the Longhorn race.

Endnotes

1 *The Farmers' Magazine*, August 1842. Second Series, No.2, Vol VI.
2 Information courtesy of Mr C. Coleman.
3 The British Newspaper Archive (BNA), *Aris's Birmingham Gazette*, 27 May 1816.
4 BNA, *Drakard's Stamford News*, 21 June 1816.
5 BNA, *Drakard's Stamford News*, 21 June 1816.
6 BNA, *Aris's Birmingham Gazette*, 2 September 1816.
7 BNA, *Stamford Mercury*, 14 March 1817.
8 BNA, *Leicester Journal*, 21 March 1817.
9 BNA, *Northampton Mercury*, 20 September 1817.
10 Young, A., The Farmer's Tour through the East of England, Vol I (London: W. Strahan & W. Nicholl, 1771), p120-3.
11 Scarfe, N., Innocent Espionage The La Rochefoucauld Brothers' Tour of England in 1785, (1995), p32.
12 A Farmer and Breeder, The New Farmer's Calendar, (1801), p162.
13 Young, A., The Farmer's Tour through the East of England, Vol I, p124.
14 Primarily from Young, A., The Annals of Agriculture, Vol XX, p20.
15 Young, A., The Annals of Agriculture, Vol XX, p20.
16 Encyclopaedia Britannica, 3rd edn., Vol VIII (1797), p762.
17 J.H. of Derby, 'A Visit to Robert Bakewell's Farm, July 9th 1793' in *The Gentleman's Magazine* for September 1793, p792-5.
18 Society for the Encouragement of Arts, Manufacture, and Commerce, The Complete Farmer: or, a General Dictionary of Husbandry (1789), no page numbers. Also quoted by Encyclopaedia Britannica, (1797), p383.
19 Scarfe, N., Innocent Espionage The La Rochefoucauld Brothers' Tour of England in 1785, (1995), p33.
20 Monk, J., General View of the Agriculture of the County of Leicester, (1794), p45
21 J.H. of Derby, 'A Visit to Robert Bakewell's Farm, July 9th 1793' in *The Gentleman's Magazine* for September 1793, p792-5.
22 Monk, J., General View of the Agriculture of the County of Leicester, (1794), p53.
23 Dickson, R.W., Practical Agriculture, Vol I (1805), p152-3.
24 Scarfe, N., Innocent Espionage The La Rochefoucauld Brothers' Tour of England in 1785, (1995), p33.
25 Temple Patterson, A., Radical Leicester (1954), p88-9, 168.
26 BNA, *Sherborne Mercury*, 31 October 1796.

27 Copland, S., Agriculture, Ancient and Modern, by The Old Norfolk Farmer, Part II, (c1861), p375.

28 Wrightson, J., Live Stock Handbooks, No. 1 Sheep, Breeds and Management, (1919), p18-19.

29 *Journal of the Rutland History Society*, No. 2, (1981).

30 Information courtesy of Keith Elliot of Measham Museum.

31 Orde, Dr A., (ed.) Matthew and George Culley Travel Journals and Letters 1765-1768, (2000), p41.

32 Orde, Dr A., (ed.) Matthew and George Culley Travel Journals and Letters 1765-1768, (2000), p221-2.

33 Orde, Dr A., (ed.) Matthew and George Culley Travel Journals and Letters 1765-1768, (2000), p7.

CHAPTER SEVEN

Bakewell's Legacy: Horses, Pigs and Poultry

Janet Spavold

Horses

Bakewell tested his capabilities on many subjects, including the production of a superior 'Black Leicestershire Carthorse'. There was already widespread interest at home and abroad in improving working horses of all kinds, from the very heavy draught horses used in London, to the growing need for stagecoach and private carriage horses in the turnpike era. People needed saddle horses of every sort for everyday mobility. There was always demand from the army for cavalry and cart horses. Lincolnshire had been the leading county for cart horses from the middle ages – Horncastle's great August horse fair was established in 1229 – but other Midlands centres developed fairs specialising in different types of horses. Easily accessible from Dishley were Derby and Nottingham, which both drew on horses like the widely used Midlands black cart horse bred in the Trent Valley area, and in Yorkshire and Lincolnshire,[1] as did Market Bosworth, Leicestershire. Hinckley, Leicester, Lutterworth and Market Harborough, all in Leicestershire, specialised in coach horses bred in the area, though the most important market for them was Northampton.[2] Given the local activity on improving working horses in northern Leicestershire and southern Derbyshire, it is hardly surprising that Ashby de la Zouch held a specialist horse market, the Ashby Black Stallion Show, through the 18th century. Bakewell's first famous black stallion, *K*, was bred in 1766.[3]

The trend to specialisation in breeding horses for specific purposes had already begun by Robert Bakewell's time. Probate records show that his great-grandfather, Robert (1), who farmed at Normanton-le-Heath, Leicestershire, and died in 1669, was breeding horses for sale; his inventory lists '16 young and old horses and mares' worth £110,[4] £20,980 in 2023 prices[5] and 17.6% of his total inventory. On average they cost £6 to £7 each. This was at a time when most local farmers only owned one or maybe two horses and many still used oxen. In 1681 for example, Thomas Carter of Church Gresley, south Derbyshire, was the only farmer in the whole of Gresley Parish with a team of five working mares valued at approximately £5 each. Out of a total of 214 inventories for the parish, 59 were still using oxen.[6]

Robert's (1) three sons continued horse breeding along with their other farming. When his eldest son Thomas died in 1713 he had '1 mare & one gelding worth £12 and four fillies valued at £12 at Normanton, six mares and two foals at Swepstone £42 and two colts at Bagworth Park £10'.[7] The value of this stock was £76, £14,430 in 2023 values,[8] and it was 9.5% of his inventory value. On average a mare was £6, a filly £3 and a colt £5. On the death of his second son John in 1716 he had 12 horses and foals worth £60, £10,640 in 2023 values,[9] and 13.6% of his inventory total.[10] His third son Robert (2) who later farmed at Dishley Grange, also died in 1716; he had seven mares and three foals worth £70, and three colts worth £24.[11] They were worth £94, £16,670 in 2023 values[12], and formed 12% of his inventory. His son Robert (3) continued at Dishley Grange but owned land in other Leicestershire parishes which was inherited by Robert (4), the subject of this book. Most of the family's land in these and other Leicestershire parishes was described as 'closes'; they were not grazing their horses on common land so they would have been well able to control their breeding stock. There was a long accumulation of skill and experience in the family behind Robert's (4) experiments. The Bakewell family clearly formed part of the Midlands horse economy described by Pitt who questioned whether anywhere else could supply the demand for

> strong black dray or road horses ... as well as some of a secondary weight for coaching and the army, except from where they now are, Leicestershire and Derbyshire principally? If this subject be properly considered, it will appear that agricultural horses are here principally a nursery for raising a supply for commercial purposes ... Horses are necessary in great numbers in a rich, commercial and luxurious country, for other purposes than those of agriculture. They can only be produced from the land, and therefore must be bred by farmers ...[13]

During this time, others were developing the black horses locally. Philip Stanhope, 4th Earl of Chesterfield (1694-1773), had his family seat at Bretby, Derbyshire, just over the north-west boundary with Leicestershire. Like many of his contemporaries, he was interested in the agricultural improvements of his time. He was a Francophile and widely travelled on the continent before his first appointment as ambassador to The Hague in 1728. He was appointed for a second time in the early 1740s before becoming Lord Lieutenant of Ireland in 1745-6. While in The Hague he became familiar with the German and Flanders horses. The German horses were heavier and slower, but 'The *Flemish and Dutch* horses are large, and strongly and beautifully formed',[14] and were preferred for improving the English cart horses.

In the considered retrospective of the Encyclopaedia Britannica's contributor in 1797,

> The horses of Germany are much better and more handsome than those of the Low Countries. They are of great use for carriages; but much more for the army, and for drawing the artillery. They have a great deal of hair, especially about the legs. They are not large, but they are well set; and yet they have tender feet. ... The horses of the Low Countries are very fit for the coach, and they are best known by the name of *Flanders-mares.*[15]

The writer noted the interest in improvement in Great Britain, in that 'the frequent introduction of foreign horses has given us a variety that no single country can boast of'; and as a result 'No country can bring a parallel to the strength and size of our horses destined for the draught; or to the activity and strength united of those that form our cavalry. ... Those we employ for that purpose [i.e., for the cavalry] or for the draught, are an offspring of the German or Flemish breed, meliorated by our soil and a judicious culture'.[16] Later this writer (William Marshall) states that 'The midland counties of England have for some time been celebrated on account of their breed of the black cart-horse' and explained its history.

> The present improvement in the breed took its rise from six Zealand mares sent over by the late Lord Chesterfield during his embassy at The Hague. These mares being lodged at his lordship's seat at Bretby in Derbyshire, the breed of horses thus became improved in that county, and for some time it took the lead for the species of these animals. As the improved breed passed into Leicestershire, however, through some unknown circumstances, it became still more improved, and Leicester has for some time taken the lead. It is now found, however, that the very large horses formerly bred in this district are much less useful than such as are of a smaller size.[17]

The 'unknown circumstances' were probably the work of Bakewell's ancestors and others in the area. The six mares were probably Frisians and likely to be a diplomatic gift.[18] Culley stated that a set of six Frisian black coach stallions was brought to Castle Donington in Leicestershire from Holland by the 10th Earl of Huntingdon who persuaded his tenants to use them, resulting in a good type of working horse.[19] What attracted Bakewell's attention was that these animals excelled in the points he thought defective in his own stock and this prompted him to go to Holland and part of Flanders with Mr G. Salisbury[20] to buy a set of West Friesland mares for his own use in improving the local black horses.[21] This was not easy; while the Frisians were willing to sell stallions, they were very reluctant to sell their mares. Bakewell did buy six, but at a high price in effort and money.[22]

Drawing on his experience with sheep and cattle, he applied the same selective methods to his horses. 'With great labour, expense and judgement he produced a breed of capital horses,' and he had the honour of exhibiting to George III 'by Royal command', according to the Dictionary of National Biography. A horse, the stallion *G*, not named in this letter and wrongly named in the following quotation, was taken for the king to see in 1788, as Bakewell described in a letter to Arthur Young.

> I left Dishley about a fortnight past, and have brought a Black Stallion to London which has been seen by the King, Prince of Wales and many other personages to whom he gives more general satisfaction than others I have yet shown. ... I had near an hours [sic] conversation with the King on the subject of breeding which he seemed much pleased with and listened to what I said with great attention.[23]

Bakewell's strain was lighter-legged and flatter-boned than the horses that came later, but of great strength, especially in the forequarters, with erect heads and fine crests. He was famed as the best breeder of horses during his lifetime. Even his mares were used in work teams but they were treated gently and the use of the whip was entirely prohibited at Dishley. (See illustration: *Robert Bakewell's Black Cart-Horse Stallion* in the colour illustrations section)

He preferred to let his stallions by the season rather than sell them, a system which became widely accepted and allowed him to study the results and thus refine future offspring.[24] The fee for a season was between 25 and 150 guineas.[25] A later writer in 1861 added that while Bakewell's results 'were found useful in improving the Leicester black breed, they were considered too heavy for carriage-horses, for which a lighter bay began to be substituted; and in the present day, with the improved roads which have been made throughout the country, a heavy horse is not required in harness'.[26]

The Earl of Chesterfield was part of the intellectual circle of Voltaire and Montesquieu which included the Duc de la Rochefoucauld, an educator and social reformer, who had a model farm at Liancourt. The Duc visited England in 1769 to study farming and may well have visited Bakewell at Dishley Grange at this time, though this is not recorded. But he sent his sons François and Alexandre, who met Bakewell when he and the brothers were both visiting Suffolk; de Rochefoucauld sent his sons back the next year with introductions and recommendations. In a letter to Arthur Young,[27] Bakewell confirmed that the Duc himself certainly visited Dishley Grange in 1783:

> On Wednesday and Thursday last the Duc de la Rochefoucuald and his friends were at Dishley, the former day I was not at home but spent the evening with them at Loughborough, they breakfasted here on Thursday and about noon set off for Derby. ...

In that same letter, he requested that Young visit him soon so that he could see some of his horses before they went out and 'compare them with the Suffolk kind. Opinion of the true form of a horse puzzles me very much and my desire is to be open to invitation and to listen to information that may be rec'd, from any part of this or any other Kingdom'.

Bakewell's success in improving cattle and sheep prompted Marshall, in his Rural Economy of Yorkshire, to see another business opportunity:

> The present prices given for hunters and coach horses, and more especially the declension of the breeding of the former, are incitements sufficient to induce men of spirit to make an attempt. Nothing appears to me to be wanting, but a Bakewell to take the lead.[28]

He did take the lead and was famed as the best breeder of horses during his lifetime. His methods of handling his stallions (indeed, all his animals) were exceptional for his time, as described in 1771. In the quote below, 'stone horse' means stallion.

> The stone horses the finest I ever saw, of the strong kind, are trained in the same manner, & though full of spirit & fire they are perfectly tame & tractable; he says there is no such a thing as a vicious horse naturally, but by improper treatment, the great error is in correcting & beating them; if they dread harm from men, they will stand on their defence; but convince them by gentle treatment & gradual means, that you mean to use them well, & intend no hurt, they become perfectly tame, & are fond of us. Thus a young horse is apt at first to be uneasy, & fling out when you lift his feet; if you beat him when he does so, his fear of ill treatment makes him worse; but the way to bring him to submit to it with calmness, is to begin with handling him where you find he is least disturbed, which is generally about the shoulders & forehead; then by gradual advances, never persisting when he is uneasy & resists, you, by degree's, gain the point.[29]

Young noted that Bakewell 'apprehends that bad drawing horses, can be owing to nothing but bad management. All his were perfectly gentle & willing workers, slow, but of great power'.[30] A 'drawing horse' is one that is in harness.

In 1771, Bakewell had over 60 strong black horses and mares in his breeding herd. He valued a few stallions at £300 and these were let out for the season at £100, to cover a maximum of 80 mares. Stallions were also let out by the leap,[31] as shown in the newspaper advertisements, with Young confirming that the charge for one of his stallions at the time was 5 guineas per 'leap'.[32] It is interesting to contemplate that in the letter written to Young 12 years later (see above), Bakewell expressed his doubt that he had the best horses possible.

Marshall was not the only writer to urge horse breeders to follow Bakewell's example in cattle and sheep because, as Richard Parkinson observed:

> If horse-breeders, possessed of good judgement, would pay the same attention to breed and shape as Mr Bakewell did with sheep, they would probably attain their wishes in an equal degree, and greatly to their advantage, whether for the collar or the road, for racing or hunting.[33]

In Northumberland, George Vardy of Fenton Farm near Wooler did just this and bred a distinctive working horse to suit the northern farmers and carriers, which 'could have been easily developed' into a valuable breed.[34] He had been a pupil of the Culleys and like them he initially visited Bakewell to buy sheep. Bakewell's horses were already familiar from the Culleys' introduction of them; they were known locally as Bakewell Blacks. They were stronger and heavier than the Clydesdales, but the local farmers preferred a working horse that was not so big, looking for 'clever clean-legged horses with good action'. Vardy bought several of Bakewell's mares, and from them and from the local hardy Fell ponies, created a breed known as the Vardy horse. It stood about 15.2 hands high, was predominantly brown or black with a bit of white on one or both hind feet and had clean legs. They were described as 'clever horses and could handle the light 7cwt block cart beautifully'. An unusual characteristic was that they trotted when working. A Vardy called *Farmer* featured as the exemplar of a draught horse in Stephens's Book of the Farm of 1940; a portrait of this horse was painted in 1838 and it was described as '16 hands high, girth below the elbow 80 inches, 9½ inches below the knee – sound – hardy – great courage', (see the black and white illustration entitled *Draught Horse*).

That stallion is a grey and is virtually identical to that identified as *Farmer*, a Vardy Horse in millfieldgreys.co.uk/3_george_sandyhouse/sibs_margvardy.html. Other greys, such as *Silvertail*, were recorded. The breed was popular in Durham, Northumberland and Scotland from the 1790s to the 1850s, but gradually fell out of favour as more powerful horses were needed for working with heavier machinery. The movement to pedigree types worked against them too when they were rejected for both the Shire and Clydesdale stud books. The Vardy did not qualify for the Cleveland Bay stud book either, being a heavier type. The 1880 English Cart Horse stud book mentions one horse called *Vardy* which claimed Bakewell's *G* as an ancestor.[35] By 1881 when the breed was in danger of being lost, Albert Grey of Hawick tried to revive it.[36] A prize of £20 was offered at the Coquetdale Agricultural Society Show for the best mare nearest to the old 'Bakewell' or 'Vardy' breed to encourage its survival. But it was too late and this specialised breed gradually died out, the last known mention of one being in 1905.[37] In Chivers's opinion, 'if only he had been kept separate, he might have had much to offer both Scotland and England at the end of the century'.[38]

Draught Horse. From Stephens's Book of the Farm, 1940, Plate VII. A Vardy Horse.

Horse breeding was a more limited market than that for sheep and cattle, so Bakewell's work here was somewhat overshadowed, as Brown observed: 'Mr Bakewell's stallions are heard of, but not half so much as the show of Leicestershire rams and bulls'.[39] Nevertheless, the extent of his breeding range is confirmed in his own statement in his petition, sometime before 1784, for support during bankruptcy which,

> SHEWETH That your Petitioner has for a Series of Years employed his Attention on a Plan for improving the Breed of Horses for Cavalry, Harness and Draught, as also of Meat Cattle and Sheep. ... That your Petitioner apprehends he has brought all the Different Kinds of Stock abovementioned to a greater Degree of Perfection than has been done by any other person and thereby rendered important Services to this Country.[40]

As part of the bankruptcy proceedings, Bakewell's farm stock was advertised in the local press for sale on 5th August, 1778, at Nottingham, though the sale did not in fact take place until March 1779 at Dishley. It illustrates the extent of his involvement with horses; he was said to carry 60 horses at Dishley.[41]

> The entire STOCK of BROOD MARES and FOALS, two and Three-year-old Fillies, of Mr Robert Bakewell, of Dishley, being of the Coach

> and Cart Kind. The above Stock were bred by Mr Bakewell, and chiefly got by Blake and George, which Horses covered several Seasons at Five Guineas a Mare.[42]

Most of the farm work was done by mares in the Midlands at this time, including brood mares. Even Bakewell used his mares in work teams. Brown noticed how widespread this practice was when writing about Derbyshire in 1794.

> The cattle [the term covered horses too] used for the plough in the fertile part of Derbyshire, are chiefly brood mares, of the strong cart kind; and in general are excellent stock for breeding waggon and dray horses, which are in such request in every part of the kingdom. The introduction of brood mares to do the work, seems to me one cause why we find no oxen employed in tilling the soil; and in brood mares, great improvements have certainly been made, and I trust will continue to go on, from the conveniency of the pastures, and the fair field they open for extending and improving the breed of draught horses.[43]

Contemporaries were aware however of drawbacks to the breeding of very heavy dray horses, which were better suited to urban than to farm work: 'The midland counties of England have for some time been celebrated on account of their breed of the *black cart-horse*; though Mr Marshall is of opinion that this kind are unprofitable as beasts of draught in husbandry'.[44] By 1790, Throsby quoted Marshall as thinking that 'The breed of horses in the midland district is allowed to be on the decline'.[45] The Encyclopaedia Britannica writer concurred: 'It is now found, however, that the very large horses formerly bred in this district are much less useful than such as are of a smaller size'.[46] As late as 1805, R.W. Dickson saw the same drawbacks to the very heavy horse, but approved the specialisation for work:

> The *Heavy Black horses* are a sort that may be employed with advantage for some purposes of draught, but are not in general well adapted to the uses of the farmer. This kind is distinguished by the colour, being constantly dark black; by being clumsy in their form, and seldom well-proportioned; slow in their motion; with rough fleshy legs; the size large.
>
> This breed of horses is found to prevail in the Midland districts, as in Leicestershire, Derbyshire, Lincolnshire, &c. where the mares are chiefly employed in farm-labour; the horses being disposed of according to their sizes; the largest for dray-horses, the next for draught in waggons or other teams, and the smallest for the purposes of war.[47]

This pattern of selection continued into the next century: 'The largest of this heavy breed of black horses are used as *dray-horses*. The next in size are sold as *waggon-horses*; and a smaller variety, and with more blood, constitutes a considerable part of our *cavalry*, and is likewise devoted to undertakers' work'.[48] Coach horses were the same type as those for cavalry.

With the perspective of over a century, by 1831 it was possible to see the great improvements made to the heavy working horse, using Bakewell's methods of breeding in-and-in from the best stock he could find. In the case of horses, this was from abroad.

> All our heavy draught horses, and some even of the lighter kind, have been lately much crossed by the Flanders breed, and with evident improvement. Little has been lost in depth and bulk of carcass; but the fore-hand has been raised, the legs have been flattened and deepened, and very much has been gained in activity. The slow heavy black, with his two miles and a half an hour, has been changed into a lighter, but yet exceedingly powerful horse, who will step four miles in the same time, with perfect ease, and has considerably more endurance. …
>
> We would wish, then, to impress it on the minds of breeders, that peculiarity of form and constitution are inherited from both parents; that the excellence of the mare is a point of quite as much importance as that of the horse … It is recognised in the midland counties in the breed of cart-horses; and the strict attention which has been paid to it, has brought our heavy horses to almost the same perfection in their way as the blood-horse. …[49]

There is however a clear recognition that, in general, the black horses had declined from the magnificence of Bakewell's own output in the years after his death.

It was reported in *Finns Leinster* Journal for 15th April 1795 that Bakewell's horse *English Farmer* was taken to Ireland to stand at Clonmel, Boytonrath, Tipperary and Cahir Abbey. This stallion was 'got by his noted Horse B, that was shewn in London to the King and Royal Family, and a great Number of the Nobility and Foreign Ministers, who allowed him to be the finest Horse of his Kind that ever was bred in England'.

In 1783 the La Rochefoucauld brothers visited Bakewell at Dishley Grange and wrote up their impressions in a journal. It has lively and pertinent details not noted elsewhere.

> He has also raised a fine breed of cart-horses. I've seen eight of his stallions which are truly magnificent. All are black, their coats are short and glistening although they are never groomed. They are almost six

> French feet high; a long and powerful neck, with a small head and lively eyes gives them a majestic appearance; they are strong enough to pull a house along. These stallions never work, so one can't tell how good they are; but Mr Bakewell has many others of the same breed, which are as good as they are handsome. He lets them out at about 60 or 80 guineas for a season. It's after a great many trials that this degree of superiority is reached, and he has the glory of being the only man in England, perhaps in the world, who has reached such great perfection in cattle, sheep and horses. I think he has reached the point he was aiming at and no longer seeks to improve them, only to lease them out.[50]

Even the horse shown to King George was not Bakewell's most impressive stallion.

> Mr Marshall describes in magnificent terms one of these large horses, a stallion belonging to Mr Bakewell named K, which, he says, was the handsomest horse he ever saw. 'He was (says he) the fancied war-horse of the German painters; who, in the luxuriance of imagination, never perhaps excelled the natural grandeur of this horse. A man of moderate size seemed to shrink behind his fore end, which rose so perfectly upright, his ears stood (as Mr Bakewell says every horse's ears ought to stand) perpendicularly over his fore feet. It may be said, with little latitude, that in grandeur and symmetry of form, viewed as a picturable object, he exceeded as far the horse which this superior breeder had the honour of showing to his Majesty, and which was afterwards shown publicly at London, as that horse does the meanest of the breed.' A more *useful* horse, bred also by Mr Bakewell, however, is described as having 'a thick carcase, his back short and straight, and his legs short and clean: as strong as an ox, yet active as a poney [sic]; equally suitable for a cart or a lighter carriage'.[51]

Although ploughing was usually done in Leicestershire and Derbyshire with six or seven horses on the coal measures clays, Bakewell himself used only two of these smaller, powerful horses, with a Rotherham plough and no driver; they easily turned one acre in the day – four times the work of his neighbours using more horses.[52] As late as 1858, a horse, said to be descended from Bakewell's stud, was shown at the RASE Chester meeting.[53] Young agreed that Bakewell's horses, like *K,* were in demand from farmers, carriers, those 'whose teams must draw great weights' and the heavy cavalry, noting that the 'great black horse is in high estimation in this kingdom'. But he claimed to be 'no friend to the horse at all. For the use of farmers, if they want great strength to be applied, a large ox I hold to be a much better animal, & if they want activity, mingled with a certain portion of strength – more nimbleness than oxen are capable of, then a lighter & smaller horse will be more applicable'.[54] He was not alone in preferring oxen for strength.

As for his other animals, Bakewell devised special equipment for the horses. A visitor in 1793 recorded that in the farmyard,

> simple, neat, useful, and movable racks, for feeding horses, call forth the attention. To prevent the horses quarrelling or injuring one another during the contest, there is only one aperture at each corner. The nearest division is appropriated for their corn, and in the centre part a rack for the hay. One pair of shafts serves to take on and off, and move the whole or any number of the racks to any part of the farm. Adjoining the house is a smith's shop, where two men are generally employed in making repairs for the farm.[55]

A small sample of the many advertisements for stallions at stud show something of the geographical range of Bakewell's breeding influence, and the cost of hiring one of his stallions. They could be leased out by the season, or sold to new owners, or led round on an advertised route so that local people could bring a mare to be covered. Bakewell also operated a subscription system with limited numbers of mares per season, for his best stallions.

The earliest advertisement seems to have been placed in the *Leicester Journal* in 1762, for two black colts for stud; in 1764 Bakewell won first prize at the Ashby de la Zouch horse show.[56] The *Stamford Mercury* for 28th March 1771 advertised an unnamed black draught horse bred by Bakewell standing at Uppingham and Oakham at one guinea a mare plus one shilling for the groom. It also advertised a black stallion, *Merryman*, standing at Oakham and Market Harborough, which had been at Oxfordshire in 1765, 1766 and 1767, at Clipstone in Northamptonshire in 1768 and 1769, and in Rutland in 1770. This horse, with a track record, charged 2 guineas a mare plus the shilling for the groom. The *Oxford Journal* for 20th April 1771 advertised Bakewell's own stallion, *David*, travelling between Shipston-on-Stour and Chipping Norton in the Cotswolds each week at 'One Guinea and a Half a Mare, or One Guinea a single Leap'. *Saunders's News-Letter* for 16th April 1773 was placed personally by Bakewell to guarantee his colt's pedigree and such information was often included in the advertisements. The colt, *Farmers Delight*, was to be at Carnew two days a week, and at Gorey the other days (both in County Wicklow, Ireland), covering at one guinea a mare and half a crown (2s 6d) for the groom. *Farmers Delight* was 'a beautiful black Draft Colt, rising 3 Years old, bred by Mr Robert Bakewell of Dishley, Leicestershire; he was out of a favourite Draft Mare of his own breeding, and got by his black Horse *BALL*'. The previous week (2nd April 1773), *Farmers Delight* had been advertised in the same paper as standing at the Roe Buck, Chequer Lane, Dublin under the same description.

The *Ipswich Journal* for 5th May 1781 advertised 'A Good Black CART or COACH-HORSE, bred by Mr Robert Bakewell, of Dishley, in Leicestershire' to stand near Stonham Pye (probably the Magpie Inn, Little Stonham, Suffolk. If it was this inn, it survived into the 20th century as a landmark on the A140 between Ipswich and Norwich, but is now gone) at half a guinea a mare and a shilling for the groom. The *Edinburgh Advertiser* for 21st April 1775 carried this: 'A remarkable strong Black Horse, proper for getting either draught or carriage horses: his sire covered three years in Leicestershire at Five Guineas. This Horse was bred by Mr Bakewell, at Dishley, in Leicestershire, and is thought to be the best of his sort that ever was in this country'. In an 1827 report, Bakewell was 'said to have let stallions for 100 guineas & upwards; another authority says from 25 to 150 guineas. At home the fee was 5 guineas'.[57]

Evidence that Bakewell had bred successful stallions as early as the 1750s comes from an advertisement in the *Derby Mercury* for 16th March 1797, when William Harrison of Little Eaton, Derby, advertised his 'noted OLD HORSE' at stud. The horse was by the 'Dishley B.L.' stallion, one of whose sons was also advertised with an exemplary pedigree: his dam was 'a daughter of the late Mr Donisthorpe's Ball of Packington & grand daughter of the Packington Blind Horse; this noble and excellent production is the fifth generation, each succession being attended to by Wm. Harrison, for a series of 40 years, beginning with Mr Hood's Blind Horse, Mr Bakewell's Old G. Mr Massey's G. Mr Bakewell's B.L. and Mr Donisthorpe's Ball'. Mr Hood bred the famous *Packington Blind Horse* (Packington is in Leicestershire). The descendants of the *Blind Horse* are set out by Chivers from the Shire Horse stud book. While two other stallions, *G* and *Blaze 184*, are listed in the stud book as true Bakewell lines, a third, *Big Ben 130*, was rejected.[58]

Whereas Bakewell's huge Longhorns were renowned for their docility, his stallions were not. A much later article on the cart horse mentions how vicious Bakewell's stallions could be and that one boy was killed by one. The article also discusses that the famous *Packington blind horse* was not only 'very celebrated for the superiority of his stock; but he also became so unmanageable and so dangerous, that his eyes were either purposely put out, or he lost the sight of them in his mischievous struggles with his grooms or attendants. Even in this state, it was highly dangerous to approach him: he lived till he was eighteen years old and continued as vicious as possible to the last'.[59] Of course, there is always the possibility that at such a late date, the original story may have become exaggerated or is even apocryphal, but as Bakewell always sent his man to accompany his stallions, we should not ignore the possibility.

Pigs

Leicestershire had a long-established trade in breeding pigs for the London market, and for salting for the navy. Markham, writing in the early 1600s, said that 'No county in England breedeth naturally better swine one than another, but if any have the preference, then I must prefer Lestershire [sic] and some parts of Northamptonshire, and clay counties bordering on Lester'.[60] This reputation persisted, though not for the 'largest and best' swine; these Midland counties and Lincolnshire in 1801 'breed large and good stock, somewhat quicker feeders, than those of Shropshire, also very handsome porkers'. This author considered the Berkshires and Oxfords as the very best but condemned any crossing with China pigs as a 'fashionable' mistake.[61] In 1809 Lawrence described Leicestershire as

> A very ancient breeding district; the original stock large, deep and flat-sided, light-spotted, with rather handsome head and ear. I am not informed whether the Bakewell variety has much honour in its own country. I have seen in the county, and in Rutlandshire, remarkably handsome and good spotted porkers.[62]

There were known deficiencies to the old breeds of native pigs; Lawrence further noted that they had huge heads and ears but little power of growth, and as a result the native pig 'tends solely to the production of blubber'. If the pigs were over-crossed with China pigs, they lost their fertility and had fewer piglets per birth.

In his The Complete Grazier, William Youatt wrote:

> The Dishley Breed, which were at one time as celebrated as the other kinds of Mr Bakewell's stock, are remarkably fine boned and delicate, and are supposed to be partly descended from Crosses of the Berkshire & Chinese.
>
> They were certainly carried to great perfection, and possessed considerable weight in a small compass.
>
> They were broad over the back and deep on the sides: thick in the shoulders, round in the ribs, full in the hams, and capable of fattening with far less food than the generality of old breeds.
>
> They answered well for pork and small hams.[63]

Quoting from the Agricultural Survey of Leicestershire, p295, Youatt described the following:

The Measurement of two pigs of this breed were as follows:

A Boar used for stock.	
Length from nose to rump	5ft 4ins
Girth round the shoulders	5ft 4ins
Thickness at do	1ft 8ins
A Hog not quite fat, but estimated at 31 stones.	
Length from nose to rump	5ft 0ins
Girth round belly	6ft 9ins

Lawrence described Bakewell's New Leicester pig as being,

> from their dish-faces, their size and colour, most probably a mixture of the China and the old Leicester breeds, [which] is liable to the above-stated objections. This breed has never attained any equal share of that high and universal repute possessed by the new Leicester sheep; yet it might be a good cross, had we not a far better in that of Berkshire.[64]

To date, a reliable illustration of the Dishley Pig has proved elusive.

From William Pitt comes this 1809 comment on Bakewell's pigs, thought to be of Berkshire origin. 'The improvement of hogs in Leicestershire has been attended to with the same care and success, as that of other livestock. At Dishley, some years back, a fine-boned sort, of small dimensions, had been carried to great perfection: I have seen there a hog of small size, when lean, fatted to 20 score weight, or more; his length, height, and thickness, being nearly equal; his belly touching the ground, the legs being enveloped in fat, and the eyes scarcely to be seen for fat, the whole appearing a solid mass of flesh. I have measured a small hog killed there, 13 inches and a half through the chine'.[65] The Dishley pig had not been forgotten by 1837 when this description was published.

> The *Dishley's*, which are supposed to be partly descended from crosses of the Berkshire and Chinese, and were at one time as celebrated as any other of Bakewell's stock, are remarkably fine-boned and delicate. They also possess considerable beauty, and are said to lay on a larger quantity of meat in proportion to bone and offal, than any other kind known – being, when fat, nearly equal in height, length, and thickness; their bellies almost touching the ground, and their eyes scarcely to be seen for fat: the whole carcase, indeed, appearing to be a solid mass of flesh. To this, however, is to be opposed the defects of being neither very prolific, nor very good nurses; slow of growth; tender constitutioned; and requiring more food in fattening than the larger breeds.

> Notwithstanding these objections, improvement has been made on several of the native breeds of the different counties by crosses with a Dishley boar; by which a roundness has been given to the frame with a proportionable depth of body. Their legs have also been shortened; a finer bone produced, with a better appearance when growing as a store [a pig being fattened for the butcher rather than breeding]; and a disposition to feed quicker and more kindly in the sty.[66]

This anonymous author refers to an unspecified Survey of Leicestershire for his first paragraph and describes two experiments in a footnote. A Mr Knight of Downton in Herefordshire compared some Dishley hogs with

> hogs of the mixed Shropshire and Berkshire species; both kinds being fed alike. The Dishley's kept themselves in better condition than the others, but grew so slowly, that at the end of twelve months they were greatly inferior in weight to their competitors.
>
> In another experiment they fattened well, but the quantity of food which they consumed, was proportionately greater than a single hog of Mr Knight's, whose weight was more than equal to that of Bakewell's two.[67]

Bakewell applied his usual methods of selection and breeding in-and-in to the local pigs, but with unreliable results. Unlike Lawrence, several contemporaries thought Bakewell used only Berkshires as crosses. He carefully reared his pigs, recording their feeding as he did for his sheep, writing to Culley in 1787: 'I am making experiments of this nature in the Pig way which I shall also communicate and I think it will throw much light on this important Subject'.[68] He was still experimenting with 'different kinds of Sheep and Pigs' when he wrote to Young in 1788.[69] He set up a purpose-built pigsty at Dishley.

> In one particular stye are conveniences for measuring and weighing the food of the hogs, and also noting down the same, which may be done by any servant with a piece of chalk already lying upon the spot, till it can be entered in a more regular manner upon paper in a proper book. This contrivance is the most simple that can be imagined, in order to prevent mistakes, and to render the experiments conclusive. There is also a weighing machine in the stye, particularly appropriated for the purpose of weighing the swine which are undergoing the experiments; nine in number, divided into classes of three in each department, are under experiment.[70]

Monk approvingly quoted part of this description the next year (1794) in his report on Leicestershire. But none of the records has survived. However, in 1783, La Rochefoucauld did comment briefly on the way the pigs were fed: 'Lastly his pigs are large, with big bellies and very short legs; they fatten very well on potatoes, which other pigs do not'.[71] Having put nose rings on them, Bakewell allowed his pigs 'to range at will because he believed in their value in consuming material which otherwise would have been wasted'.[72]

Contemporaries clearly had mixed opinions over the success of Bakewell's experiments with pigs, some claiming success but others thinking the pigs were 'All rickety' or 'all fools'.[73] Marshall claimed that Bakewell had 'a superior breed' of pig more 'improved by breeding in-and-in' than his cattle, sheep or horses had been:

> He continued to send his sows, year after year, to the same celebrated boar ... which boar is the father of the entire family; his daughters, and his daughter's daughter, having been regularly sent to him! The consequence is, the breed, so far from being worn out or weakened, has been highly improved, by this incestuous intercourse.[74]

When Youatt wrote his book, The Pig, in 1847, he neatly summarised the rise and fall of the Dishley pig.

> The old Leicestershire breed was a perfect type of the original hogs of the midland counties; large, ungainly, slab-sided animals, of a light colour and spotted with brown or black. The only good parts about them were their head and ears, which showed greater traces of breeding than any other parts. Mr Bakewell improved them, and the variety thus obtained was called after him, and was superior in value and beauty to the old stock. Within the last few years various crosses have been tried, and the original breed is now fast losing all its peculiarities and defects.[75]

Although Dishley pigs are not mentioned in this sale advertisement, the breeds that are reflect the sparse mentions contributing to (or degrading) the Dishley pig. It was placed in the *Derby Mercury* for 15th February 1832. (BNA) C.G. Mundy, Esq. of Burton-on-the-Wolds, Leicestershire sold all his livestock on 28th/29th February. It included '2 capital white boars, 5 cows [sic, sows] 2 Chinese sows, half-bred Neopolitan [sic] ditto, 2 home-bred ditto'.

While the pigs may not have attracted the lasting attention given to Bakewell's cattle, sheep or horses, the Dishley breed was recognised well into the next century. Like his other improved stock they were exported, as these advertisements from the USA âttest. In March 1821, the editor of *The Daily National Intelligencer* (Washington DC) printed a discussion on the best way to set up an American stock farm. This had been started by a letter to the paper from Sir John Sinclair, dated 8th December 1820. The Editor then laid out a

A Bedford Pig by William Youatt.

plan for improving breeding in America, and his proposed foundation stock included four breeds of pigs: the Byfield, the Bedford (see black and white illustration), the Angus and the Dishley.

A farm auction in Charlestown, Massachusetts, advertised in *The Atlas* for 28th March 1839 included '3 Dishley Sheep, and 8 Swine'. The 'Hints on Swine' printed on 27th October 1860 in the *Charleston Tri-Weekly Courier* commented on cross-breeding with Suffolk hogs, which,

> have long been in repute as a hardy and prolific species, though generally of only moderate, or rather small, size; and, when crossed with either the Berkshire or the Dishley breed, produce animals which are very general favourites. They also fatten quickly, but their shape is not considered altogether valuable for making bacon.

James Wilson, writing in 1912, noted that Bakewell correctly interpreted the economic and social changes affecting agriculture in his own day where others did not.

> Bakewell saw these movements; he saw what the increasing population required, and that, in future, stock must be bred, the horse for his strength, and cattle, sheep, and swine for the economic production of food and clothing. ... There is considerable doubt as to the kind of swine he bred.[76]

The Berkshire Pig from the Sheep and Pigs of Great Britain.

Black Suffolk Pigs from the Sheep and Pigs of Great Britain.

By 1907 it was noted that the old Midlands breeds, having been crossed with the Berkshires, had 'become nearly all black', or if crossed with the Tamworths, were 'rusty coloured, but very hardy'.[77] Leicestershire's pigs were judged the next year to 'have been so greatly altered and improved by crossing, principally with the Large and Middle Whites, as to lose much of their distinctive character'.[78] In summary, Bakewell's work with pigs did not have the lasting impact of his improved cattle, sheep and horses. (See black and white illustrations of The Berkshire Pig and Black Suffolk Pigs from Sheep and Pigs of Great Britain)

Poultry

There is very little evidence for Bakewell's improvements to poultry breeding, and he seems to have had even less success than he had with pigs. It is not a major part of his reputation and may even have detracted from it. But it was tried, as described by Copland.

> The rule adopted by Bakewell applies as well to the smaller and feathered denizens of the farm-yard as to the sheep and oxen. A spasmodic deviation from it was attempted, a few seasons back, by the introduction of those huge monstrosities the Cochin-China fowls, whose gaunt, clumsy, gigantic forms, and hideous attempts at crowing, were enough to frighten the modest occupants of an English *basse-cour* [probably meaning the barnyard/hen-house] out of their propriety. The rage for these ugly foreigners has, we believe, nearly subsided, and people are at leisure to marvel at the infatuation which, a few years since, raised their price to that of a prize ox at the Smithfield Show.[79]

See the illustration of a *Partridge Cochin Cock* by Ludlow, c1890 in the colour illustrations section.

In 1911, Fream quoted Darwin rather than Bakewell when discussing the origin of domestic fowls and did not include Bakewell's in his extensive list of varieties.[80] By then, poultry seems to have been bred for novelty rather than practicality. One anonymous writer, who must have been partly a contemporary of Bakewell's, clearly did not think they had much value.

> Poultry are an article of luxury, for which the little farmer never obtains an adequate price. He had better allow his wife a certain annual sum, for pin-money, than suffer her to keep these devourers.[81]

Endnotes

1 Edwards, P., Horse and Man in Early Modern England, (2007), p14-5.

2 Edwards, P., *ibid.*, p224-5.

3 Chivers, K., The Shire Horse. A History of the Breed, the Society, and the Men, (1976), p37.

4 Record Office of Leicester, Leicestershire & Rutland (ROLLR), Probate inventory of Robert Bakewell of Normanton-le-Heath, 1670.

5 Measuring Worth, (2024), 'Purchasing Power of British Pounds from 1270 to Present'. URL: www.measuringworth.com/ppoweruk/

6 Spavold, J., (ed.) In the Name of God Amen. Everyday Life in South Derbyshire 1535-1700, (1992), p139-40.

7 ROLLR, Probate inventory of Thomas Bakewell of Normanton-le-Heath, 1713.

8 Measuring Worth, (2024), 'Purchasing Power of British Pounds from 1270 to Present'. URL: www.measuringworth.com/ppoweruk/

9 Measuring Worth, (2024), 'Purchasing Power of British Pounds from 1270 to Present'. URL: www.measuringworth.com/ppoweruk/

10 ROLLR, Probate inventory of John Bakewell of Swepstone, 1716.

11 ROLLR, Probate inventory of Robert Bakewell of Dishley Grange, 1716.

12 Measuring Worth, (2024), 'Purchasing Power of British Pounds from 1270 to Present'. URL: www.measuringworth.com/ppoweruk/

13 Pitt, W., A General View of the County of Leicestershire, including a survey of the County of Rutland, (1809), p288-9.

14 Library of Useful Knowledge, The Horse, (1831), p20.

15 Encyclopaedia Britannica, Vol VI, (1797), p708.

16 Encyclopaedia Britannica, Vol VI, (1797), p709.

17 Encyclopaedia Britannica, Vol VIII, (1797), p773.

18 Chivers, K., The Shire Horse. A History of the Breed, the Society, and the Men, (1976), p33.

19 Chivers, K., *ibid.*, p33-4.

20 Wykes, D.L., 'Robert Bakewell (1725-1795) of Dishley: farmer and livestock improver', in *Agricultural History Review* 52, I, p41.

21 Pawson, H.C., Robert Bakewell Pioneer Livestock Breeder, (1957), p63.

22 Chivers, K., *ibid.*, p49.

23 Bakewell to Young, 10 March 1788, in Pawson, H.C., Robert Bakewell Pioneer Livestock Breeder, (1957), p172.

24 Chivers, K., *ibid.*, p51-2.

25 Young, A., The Farmer's Tour through the East of England, Vol 1, (1771), p119.

26 Copland, S., Agriculture, Ancient and Modern, by the "Old Norfolk Farmer", (1861), p330.

27 Pawson, H.C., Robert Bakewell Pioneer Livestock Breeder, (1957), p171.

28 The Complete Farmer: or, a General Dictionary of Husbandry in all its branches; By a Society of Gentlemen, members of the Society for the Encouragement of Arts, Manufacture, and Commerce, (1789), no pagination, alphabetical subjects.

29 *Annals of Agriculture*, 'Observations made at Mr Bakewell's in 1771', Vol 28, p588.

30 Young, A., The Farmer's Tour through the East of England, Vol I, (1771), p120.

31 *Annals of Agriculture*, 'Observations made at Mr Bakewell's in 1771', Vol 28, p588.

32 Young, A., The Tour ... *ibid.*, p120.

33 Parkinson on the Breeding and Management of Live Stock, Vol II, p59, quoted in Library of Useful Knowledge, The Horse, (1831), p222.

34 Chivers, K., *ibid.*, p73.

35 Chivers, K., *ibid.*, p522.

36 Chivers, K., *ibid.*, p243.

37 We are grateful to Mr J. Gall, formerly of the Beamish Open Air Museum, for much information on the Vardy horses.

38 Chivers, K., *ibid.*, p77.

39 Brown, T., General View of the Agriculture of the County of Derby, (1794), p25.

40 Pawson, H.C., *ibid.*, p181.

41 Caddy, A., Robert Bakewell and His Times, (1925), p7.

42 We are indebted to Brenton Heazelwood for this information.

43 Brown, T., General View of the Agriculture of the County of Derby, (1794), p25.

44 Encyclopaedia Britannica, Vol VIII, (1797), p773.

45 Throsby, J., Leicestershire Views, (1790), p414.

46 Encyclopaedia Britannica, Vol VIII, (1797), p773.

47 Dickson, R.W., Practical agriculture, (1805), p1171-2.

48 Library of Useful Knowledge, The Horse, (1831), p41.

49 Library of Useful Knowledge, The Horse, (1831), p41, 222.

50 Scarfe, N., Innocent Espionage, the La Rochefoucauld Brothers' Tour of England in 1785, (1995), p32.

51 Marshall. W., in Encyclopaedia Britannica, Vol VIII, (1797), p773.

52 Young, A., The Tour ... *ibid.*, p130.

53 Housman, W., 'Robert Bakewell', *Journal of the Royal Agricultural Society of England*, 3rd Series, Vol V (1894), p1-31, reprinted from *The Gentleman's Magazine*, Vol LXV, Part II, (1795).

54 Young, A., Annals of Agriculture, Vol VI, (1786), p468-469.

55 H-.J-., 'A Visit to Robert Bakewell's Farm, July 9th 1793', in *The Gentleman's Magazine*, (September 1793), p792-95.

56 Wykes, D.L., 'Robert Bakewell (1725-1795) of Dishley: farmer and livestock improver', in *Agricultural History Review*, 52, I, p40.

57 Housman, W., 'Robert Bakewell', *The Gentleman's Magazine*, Vol LXV, Part II (1795), reprinted in the *Journal of the Royal Society of England*, 3rd Series, Vol V, (1894), p1-31.

58 Chivers, K., *ibid.*, p550-51, 554-5.

59 *The Farmer's Magazine*, Vol V, (July to December 1836), p213-215.

60 Quoted in Long, J., The Book of the Pig, (1886), p3.

61 A Farmer and Breeder, The New Farmer's Calendar, or Monthly Remembrancer, (1801), p549-550.

62 Lawrence, J., A General Treatise on Cattle, The Ox, The Sheep, And The Swine, (1809), p441.

63 Youatt, W., The Complete Grazier, or Farmers' and Cattle Breeders' and Dealers Assistant, 5th edn., (1830), p287.

64 Lawrence, J., A General Treatise on Cattle, The Ox, The Sheep, And The Swine, (1809), p435.

65 Pitt, W., A General View of the County of Leicestershire, including a survey of the County of Rutland, (1809), p295.

66 Library of Useful Knowledge, British Husbandry, Vol II, (1837), p510.

67 Library of Useful Knowledge, British Husbandry, Vol II, (1837), p510 footnote.

68 Pawson, H.C., Robert Bakewell Pioneer Livestock Breeder, (1957), p118.

69 Pawson, H.C., *ibid.*, p174.

70 H-.J-., 'A Visit to Robert Bakewell's Farm, July 9th 1793', in *The Gentleman's Magazine*, (September 1793), p792-95.

71 de la Rochefoucauld, F., Marchand, J., (ed.), Roberts, S.C., (trans.), A Frenchman in England 1784. Being the Melanges sur Angleterre of François de la Rochefoucauld, (1933), quoted in Pawson, p39.

72 Pawson, H.C., *ibid.*, p83.

73 Pawson, H.C., *ibid.*, p65.

74 Pawson, H.C., *ibid.*, p66.

75 Youatt, W., The Pig, (1847), p56.

76 Wilson, J., The Principles of Stock-Breeding, (1912), p10.

77 Wallace, R., Farm Live Stock of Great Britain, 4th edn., (1907), p366.

78 Youatt, W., Fream, W., & Bear, W.E., The Complete Grazier and Farmers' and Cattle Breeders' Assistant, 15th edition, (1908), p534.

79 Copland, S., Agriculture, Ancient and Modern, by the "Old Norfolk Farmer", (1861), p308.

80 Fream, W., Elements of Agriculture, 8th edn., (1911), p584-5.

81 A Farmer and Breeder, The New Farmer's Calendar, or Monthly Remembrancer, (1801), p551-2.

CHAPTER EIGHT

Distribution of Bakewell Livestock

Janet Spavold & Brenton Heazlewood

The stock types, names and values of livestock sold in this country at major sales during and after Bakewell's life are discussed and tabulated elsewhere in this book. But his fame spread into the wider world, and some of this can be traced through various sources such as newspapers, trade directories and sale information. The Dictionary of National Biography's survey of his life summarised this, writing that Bakewell had produced his 'special breed of Leicesters' which spread into 'every part of the United Kingdom and to Europe and America in little more than fifty years'.[1] While it is acknowledged that there were earlier improvers of the Longhorn cattle, namely Sir Thomas Gresley of Drakelow Hall in Derbyshire by the 1720s, then Welby of Linton (close to Drakelow) and Webster of Canley in Warwickshire,[2] nobody else made such lasting changes over such a great geographical area. Bakewell chose some of his initial cows from Canley.

During Bakewell's lifetime, Marshall listed the 'respectable ram breeders' who were 'leading men in the midland district, favourers of the new Leicestershire or Dishley breed'. It gives a good idea of their geographical spread by 1790 in Leicestershire, Lincolnshire, Nottinghamshire and Northamptonshire. They were:

> Mr *Stubbins* of Holm; Mr *Paget* of Ibstock; Mr *Breedon* of Ruddington; Mr *Stone* of Quarndon; Mr *Walker* of Woolsthorpe; Mr *Bettison* of Holme; Mr *White* of Hoton; Mr *Knowles* of Naileston; Mr *Deverel* of Clapton; Mr *Princess* [*Prinsep*] of Croxall; Mr *Burgess* of Hugglescott; Mr *Green* of Normanton; Mr *Robinson* near Welford; Mr *Moor* of Thorp; Mr *Ashley* of Oddeston; Mr *Henton* of Hoby.[3]

Another purchaser from Bakewell and his Dishley Society members was Sir Tatton Sykes of Sledmere House, where the Rare Breeds Society now has a farm park. He came with Mr Robinson, who became well known for his Leicester sheep, especially for rams. This led to the formation of the well-known Speeton flock of Leicester Longwools in the East Riding of Yorkshire. It was one of the longest-surviving flocks which could trace its breed lines directly back to Robert Bakewell. From 1834 to the 1850s it was almost exclusively bred from Robinson's flock,[4] then continued to be a prizewinning flock, surviving into the 21st century.

Pitt, in 1813, noted the improvement in Staffordshire's sheep; he thought that Mr Miller of Dunstall had the best sheep in the county. They had been bred first with Dishley rams, then he bought rams from 'Messrs. Green, Farrow, Stone, and Buckley, at from 50 to 200 guineas each, as well as by purchasing ewes up to ten and twelve guineas each'.[5] As listed above, Green and Stone were from Leicestershire; there were two Buckley brothers, John and Nicholas, while Farrow was probably from Staffordshire. Both the Buckley brothers, John and Nicholas, who were very big Dishley breeders, were from Normanton on Soar, with Nicholas being close to Bakewell. There were other Dishley breeders in the county.

The Culley brothers, followers of Bakewell's methods, moved from County Durham to Northumberland in 1767 and introduced the sheep into Scotland both as pure-bred Leicesters and for crossing with the Cheviots and the Blackfaces in the Border areas. The Border Leicesters were especially popular in Roxburghshire, Berwickshire, and Northumberland. The Culleys, in turn, converted other sheep breeders to Bakewell's work, namely 'Messrs. Thompson, Chillingham; Jobson, Chillingham Newtown; Robertson, of Ladykirk; Smith, Learmouth; Compton, New Learmouth; Smith, Norham; Riddell, Timpendean',[6] and others.

Bakewell himself listed the places that he had sent his livestock to, especially the cattle, sheep and horses, in his appeal for support during his bankruptcy. They had been sent into:

> the Counties of Bedford, Bucks, Cambridge, Chester, Cumberland, Derby, Devon, Dorset, Durham, Essex, Gloucester, Hereford, Herts, Huntingdon, Kent, Lancaster, Leicester, Lincoln, Norfolk, Northampton, Northumberland, Nottingham, Oxon, Rutland, Salop, Somerset, Southampton, Stafford, Suffolk, Sussex, Warwick, Westmoreland, Wilts, Worcester, Yorks, into North-Britain, Wales, Ireland, Germany and Jamaica.[7]

He took his stock to London for sale and to show, most notably the black stallion shown to King George III in March 1788. His knowledge was in demand for other agricultural practices too; the Bishop of Llandaff wrote to him in 1788 asking for the loan of one of his men who could instruct Llandaff's steward on establishing water meadows on an estate he had bought in Ambleside.[8] The following year Bakewell was involved in a scheme to petition the king to use some of the Crown lands to establish experimental farms.[9] As well as eminent people visiting him at Dishley to learn from his methods, he undertook long journeys to see and study other agriculturalists' estates; there are many such places listed in his letters.

Some other places in Great Britain and Ireland not previously noted and which are known to have had Dishley sheep were as follows, though there will have been many more.

England Leicestershire: Lockington, Normanton Hill, Loughborough, Hugglescote, Snareston and Isley Walton.
Derbyshire: Hartshorne, Foremark Park, Ticknall, Idridgehay, Borrowash, Longford, Overseal, Doveridge and Hoon.[10]
Nottinghamshire: Retford.
Oxfordshire: Foscote.
Lancashire: Manchester.
Yorkshire: Winterton near Hull, Aylesbury Manor, East Riding.
Devonshire: Barton near Exeter.
Sussex: Sheffield Park, Uckfield.
Northamptonshire: Watford near Daventry.

Wales Flintshire: Mold.

Scotland Perthshire: Balgowan.
Midlothian: Edinburgh.

Ireland Londonderry: Eglinton.
Dublin: City of Dublin.
Roscommon: Township of Roscommon.

After Bakewell's death, Dishley Grange and his stock passed to his nephew Honeyborn (see two black and white illustrations by John Boultbee of *Mr Honeyborn's Ram* and *A New Leicester Ewe* painted 1802), then to William Smith. Both maintained his breeding policies and practice, so his bloodlines continued to be available to other graziers. After Smith, the stock was sold and the primary breeders who bought it were, according to Coleman, Mr Stubbins of Stone Barford, south of Warwick; Mr Paget, of Eleman,[11] and Philip Skipworth the elder of Aylesbury Manor, East Riding of Yorkshire. 'The latter's purchase of ewes laid the foundation of the Aylesbury flock, which ranks at present [1877] as one of the most valuable in the country; from Mr Stubbins' purchase descended the once celebrated Holmepierrepont sheep'.[12] Stubbins had moved from Warwickshire to Nottinghamshire.

Until breed societies with stud books were established, quality stock for sale were advertised with their pedigrees so that buyers knew their provenance. Bakewell's life coincided with the steady rise of local newspapers,

Mr Honeyborn's Ram by John Boultbee.

A New Leicester Ewe by John Boultbee.

though coverage was patchy. Where they do exist they are a valuable source of information, both for pedigrees and distribution, as a small selection of examples shows. As with the horses, we often learn the animals' names too.

In 1776, rams 'of Mr Chaplin's and Mr Bakewell's breed' were advertised for sale at Cranston, Dalkeith. In 1795 rams from Charlecote, Warwickshire were for sale, 'closely descended from the Dishley Breed'. They were 5-shear ewes by a son of *Dishley G* and a son of the *Dishley Two Pounder*, 4-shear ewes by a son of *Dishley J* and a son of *Two Pounder*, 3-shear ewes by Mr Luc'leys *Short Tail*, a son of *Dishley OH* and by a son and grandson of *Two Pounder*, and 2-shear ewes by *Short Tail*, a son of the *Dishley Young G* and by a son of *Dishley HH*. The ewes were descended from stock bought from Mr Hall of Cotes, Mr Davy of Loughborough and Mr Fowler of Rollright among others. When Honeyborn's stock of sheep, cattle and horses was sold after his death in 1816, their provenance was summarised thus: 'It is sufficient to say of this stock, that it is all directly descended from that which the celebrated Mr Bakewell (uncle to Mr Honeybourne) left at Dishley at the time of his decease, and to which all the principal breeders of New Leicester Sheep and Long-horned Neat Cattle, owe the origin of the pedigrees of their respective stock'.[13]

Thomas Hawksworth of Swarkestone, Derbyshire, sold a bull in 1819 and he listed its pedigree by place and parentage: 'the Bull was bred by Mr Moorwood, of Rowden Hall, in the Parish of Higham on the Hill, and was got by the very justly esteem'd *Cowley Bull*, out of a celebrated Cow bred by Richard Auley [Astley], Esq. of Odstone Hall, Leicestershire, and is allowed to be the best Long-horn'd Bull in the Country'.[14] Nearly 30 years after Bakewell's death his influence was still a significant factor in the standing of sheep breeders. The sheep and cattle breeder Mr Harris, whose celebrity 'renders any comment unnecessary', nevertheless wanted readers to know that 'he commenced at Dishley'.[15] Mr Rowlands of Rough Heanor, Derbyshire, sold named rams with ewes and lambs in 1823: 'The Four-Shear, by a son of Mr Smith's *Old Rumps*, and a son of Mr Farrow's *Bumper*. – The Three-Shear, by a son of Mr Smith's *Magnum Donum*, and *Bumper*. – The Two-Shear, by *Fat Back*, a son of *Old Rumps*, and a son of Mr Astley's *Pillion Rumps*. – The One-Shear, by a son of Mr Smith's *Dishley*, and *Pillion Rumps*. – The Lambs, by a son of Mr Smith's *Magnum Donum*, and a son of *Loadsman*. These Sheep have been bred with the greatest care for thirty Years from the Flocks of the most eminent Breeders of the Dishley Sheep'.

So active was Richard Astley in Ireland that he tenanted a large farm there; he was also made an honorary member of the Farming Society of Ireland. Astley, Honeyborn, Buckley and others regularly judged at Ballinasloe.

In the 1760s, Bakewell's stallions were advertised at stud in Clipstone in Northamptonshire, Rutland and at Little Rollright near Chipping Norton in Oxfordshire. In the 1770s, they stood at Dishley, Bruern in Oxfordshire, Uppingham in Rutland, Stamford in Lincolnshire, Siddall near Stockport,

Toad Hole Furnace, Heage and Allestree in Derbyshire, Chipping Norton, and several places in Northamptonshire and Buckinghamshire by arrangement. In 1781 one stood at Little Stonham, Ipswich.[16] One was sent to Dublin in 1784, as advertised in *Saunders Newsletter* on 6th April:

> To be sold, or hired out for the Season, a beautiful English Draft Stallion, remarkably fruitful; his Get far exceeds any Thing that appeared in this Kingdom yet. He was got by Mr Bakewell's famous Packington, that was leaped in England at five Guineas a Mare. Inquire at No. 48, King's street, Stephen's green, Dublin.

In 1785, Bakewell had sent a stallion to George Culley in Northumberland for the season, and the following year offered to send the stallion again, 'on the same Terms as last Year (Viz.) Eighty Guineas, you bearing all Expences [sic] except me paying the Man's Wages'.[17]

The Dishley references declined by the 1830s, as preferences moved away from the Dishley stock. The Southdowns became a major challenger for the Dishley sheep, and the Shorthorn for the Longhorn cattle. By the 1850s, the Dishley sheep mainly appeared in advertisements for crosses, including the Merino; the resulting fleece was often described as 'silky' because of its lustre, 'an intrinsic and inherent silvery brightness of the fibre which is not lost in the manufacturing process, and which cannot be given to goods not made of pure lustre wool'.[18] By the 1860s, the Dishley sheep were seen as too fat by most agricultural writers, and both they and the Longhorns went out of agricultural fashion in both the United Kingdom and the United States.

Bakewell's stock was also in demand from overseas breeders, but there was a legal problem. In 1660, Parliament had banned the export of all sheep and wool, all products made from wool and all the instruments and machinery used in the manufacture of woollen goods, because of the importance of this trade to the British economy. There was pressure to remove the bans in 1806 but Parliament refused to lift any of them. When the export ban was lifted in 1825 it did not mention sheep. Would-be exporters could obtain a licence but it was for specific exports not a general exemption.

Livestock improvement in Ireland

In 1771, Arthur Young wrote that Bakewell had recently sent many cattle to Ireland.[19]

Although Ireland was part of the United Kingdom of Great Britain and Ireland after 1801, when the kingdoms merged, the ban on importing animals still applied there. Sheep farming was not so prevalent in Ireland as in England; cattle were more widespread. Connaught was the primary province for sheep;

they were large, coarse and ungainly. Some English animals were smuggled into Ireland to avoid the ban, where they fetched good prices, but it was financially risky. In 1777, Mr Thomas Johnston petitioned Parliament in Dublin for relief as a result of his unprofitable smuggling. He had brought 'stallions and mares, boars and sows, rams, ewes, and bulls from England; but, owing to the hazard and expense of the undertaking, it had proved unremunerative'. When a ship with 30 rams and ewes was seized at Carlingford and he could not provide documentation for them, they starved to death. There was much sympathy among both Irish breeders and the Irish Parliament for his efforts; the Irish Parliament granted him £200 in relief and 48 'Gentlemen Breeders of the Kingdom of Ireland' signed a memorial certifying that Mr Thomas,

> by his great skill in the choice and unwearied assiduity in the importation of live stock into this kingdom for several years past, and at a vast expense, hath improved the stock in general of Ireland, in a most extraordinary manner, to such a degree, that we are well convinced of the great benefit we have received, and sincerely wish he may meet with Parliamentary encouragement.[20]

Cattle

For a long period, Longhorns very much resembling those in England have been found in Ireland. Ancient records are silent on the matter of whether England or Ireland was the native country for the breed, but they can be traced back to a very remote period. George Culley had his own view on the matter:

> I take the Irish cattle to be a mixed breed between the Longhorns and the Welsh or Scotch, but most inclined to the Longhorns, tho' of a less weight than those in England.
>
> Of late years several of the Irish breeders have bought long-horned bulls and heifers at very high prices from Lancashire, Leicestershire, Warwickshire, etc. particularly the Mr Frenches, and other principle breeders from Roscommon and different parts of the West of Ireland, which have the very great advantage of improving their breed.[21]

Writers in the early 1800s reported that the Longhorn was regarded as an import to Ireland. They were to be found most commonly in the lower and most thickly-populated areas, probably being taken over from Lancashire. The Middle-horn Kerry was the native breed, being found in every mountainous and unfrequented region.

Robert Wallace, writing in 1885, reported that the herd on the Earl of Westmeath's estate at Pallas, County Galway, was said to have been bred pure for over 200 years, fresh blood being imported from England from time to time. Youatt, writing in 1833, said that the Longhorns were most plentiful in Tipperary, Limerick, Meath and a great part of Munster and, particularly, Roscommon, and that in about 1710, some zealous agriculturalists in Meath also started the improvement of the breed (see black and white illustration of an Irish Longhorn). A Mr Waller of Allenstown introduced some of the Old Lancashires at this time. Sixty years later, a relation of his brought over one of the New Leicester breed of cattle. He permitted his neighbours and tenants to have almost unlimited use of this bull and there was scarcely a cottager within three of four miles of Allenstown who did not possess a cow showing some traces of Leicestershire blood. About the same time, Lord Massarene introduced the Longhorn cattle into Antrim.

As regards Bakewell's animals, in 1775, Mr Lesley of Lesley-hill imported one of Bakewell's bulls and the cattle of the neighbouring county were quickly improved. In Langford, the cattle were much improved by the efforts of the Earl of Ross who imported several bulls of 'the best British blood'. On 21st May 1802 he sold 10 six-year-old bullocks at the fair at Ballymaloe for 400 guineas and 10 four-year-old heifers for 300 guineas. For size, shape and

The variety of Irish Longhorn from an engraving by William Youatt, 1838. There is no doubt that they very much resembled the English Longhorn, by which they had been materially improved.

fatness, it was reported, they could not be excelled, and all they had consumed was grass and hay.

Almost every county of Ireland had its successful improver of the Longhorn breed until, it is written, Bakewell himself said that in the richer and more cultivated districts they could even rival the best that the Midland counties of England could produce – excepting those of Bakewell himself. At the fair at Ballinasloe, he found individuals of the breed which he regarded as inferior only to those which he himself had perfected. Bakewell became a considerable exporter of livestock, with animals of his breeding being recorded as far afield as America and South Africa.

Sheep

Irish landowners wanted to improve wool production, and this seems to have been the spur for them to import New Leicester rams to cross with native ewes. Bakewell had already begun trading with Ireland (see the stallion advertisement above) and in 1789 he commented to Culley that 'Mr Baker a Young Man of mine is just returned from a two Months Tour in Ireland was at Ballinasloe finds the Irish very much prejudiced against English Sheep …'.[22] Ballinasloe was the largest sheep and wool fair in the area and Mr Baker probably went there to assess the markets for sheep and wool. The extent of the trade can be seen in the 'Tables of Wool and Sheep sold at the Ballinasloe Fairs' in 1785, compiled by Baker-Holroyd.[23] Culley had visited the fair at about the same time and compared the local sheep unfavourably:

> I am sorry to say I never saw such ill-formed, ugly sheep as these; the worst breeds we have in England are by far superior. One would almost imagine that the sheep-breeders in Ireland have taken as much pains to breed plain, awkward sheep as many of the people in England have to breed handsome ones. I know nothing to recommend them except their size, which might please some old-fashioned breeders, who can get no kind of stock large enough.[24]

But he did see improvement starting, as a result of the import of English rams in some areas, saying 'it is both extraordinary and pleasing to see how much they exceed the native breed'.[25] At this date any improvement would have come from smuggled stock.

When the export ban was finally lifted in 1825, the Connaught sheep breeders, led by those in Roscommon, rapidly took advantage of the opportunity to import large numbers of Bakewell's Leicester rams. Over the next 25 years, their crosses created the improved Roscommon sheep, with a better conformation and better quality wool. They followed Bakewell's breeding methods too, and as a result, 'Selection has been, in fact, the chief means latterly employed in producing the improved Roscommon sheep',[26] (see black and white illustration of a Roscommon Sheep). These smaller and more compact sheep became known as Galways. The overall effect was summarised in 1881 by Thomas Baldwin, Superintendent of the Agricultural Department of National Education in Ireland. He said that the best breed for improving Irish sheep was the Leicester, and that practical experience proved it.

> The fine sheep which are met with in many parts of Ireland under the name of *Ballinasloe* sheep, have been produced by crossing native sheep with Leicester rams. The most skilful and enterprising flock masters in the country regularly use Leicester rams, and by doing so have enriched themselves and benefitted the community.[27]

One such flock master was John Baker-Holroyd, who had considerable agricultural expertise and probably had more influence than anyone else on the improvement of Irish sheep. Baker-Holroyd was born in Ireland but grew up in England, and the family had extensive estates in the lush grazing areas of Roscommon and County Meath. After army service he entered Parliament; he was created Baron Sheffield of Roscommon in the county of Roscommon in 1783 (in the Irish Peerage). He became a member of the Board of Agriculture in 1793, then its president from 1803 to 1806. He visited the family estates in 1784, later writing pamphlets, letters and books on the importance of the trade in wool between Ireland and Britain. He also argued for sheep imports

Roscommon sheep from Sheep and Pigs of Great Britain, *p81.*

from Spain to Ireland to be allowed. In 1796, he bought the Sheffield Park Estate in Sussex, where he ran a model farm with a flock of Southdown sheep; he improved these by crossing them with Bakewell's New Leicesters.[28] He was a significant business associate of Bakewell and a colleague of people like Young and Marshall. He was one of the signatories in support of Bakewell's bankruptcy petition. In 1802 he was created Baron Sheffield of Sheffield in the county of Yorkshire (in the English Peerage), then in 1816 made the Earl of Sheffield and Viscount Pevensey (in the Irish Peerage).

Another significant family importing Bakewell's sheep, cattle and horses was that of Christopher Dexter and his son, who developed the cattle breed which carries their name.

The Roscommon was the most common sheep in the Irish Midlands by the middle of the 19th century and the Roscommon Long-wool Flockbook was established in 1895. But this large sheep's popularity went into decline in favour of the smaller and more compact Galway sheep. In 1923, the Roscommon Breed Society disbanded, and the Galway Sheep Breeders Society gradually took its place, probably with sheep of the Roscommon type to begin with; the Galways acknowledge their Roscommon lineage. Both types are recognised as Irish rare breeds.[29]

Between 1776 and 1779, when there was considerable anti-English unrest, it is claimed that 'the British breed was not well liked by the Irish and some of the sheep were actually killed for political reasons'. An announcement in the *Limerick Chronicle* on 31st October 1785 illustrates something similar which took place on 5th October. Four fine rams belonging to Christopher Dexter were 'killed on the Lands of Newtown, and all the fat and the hind quarters of two of them taken away'. Thirty-six men signed a list of subscribers to raise money to pay for a reward for information and to pay for a prosecution. Given the fraught history of English rule in Ireland, it is not surprising that they were not liked.

The Galway Sheep Breeders Society claims that both the Galway and the Roscommon were 'perhaps closer to the Dishley breed than any of the English breeds which were influenced by Bakewell's stock', and that they bore 'an almost uncanny resemblance to surviving prints and paintings of Bakewell's sheep'. This has recently been confirmed by a group of researchers who have analysed their genomes. Their conclusions are that the 'Galway breed clusters with the Scottish Texel breed and is located close to the Border Leicester breed. This result supports the documented role for the foundational New Leicester breed in the formation of the Galway and Texel breeds'.[30]

Bakewell Livestock in America and the West Indies

The early European settlers of North America had to bring their domestic livestock with them and, as the various colonies grew, their need for more domesticated animals meant a growing import trade, particularly from England. However, with the mounting troubles between an increasingly self-determining North America and a hide-bound British Government, that trade came under pressure and, soon after the American War of Independence broke out in April 1775, the trade was banned by Britain. That war ended in September 1783, but the ban persisted for a while. It is not known how strictly the export ban to the New England colonies, and later to them as independent states, was enforced as trade between England and America grew, but there is evidence of smuggling.

Bakewell and George Washington

George Washington loved farming and would rather have spent his time on this activity than having to be a soldier and statesman; even during those two parts of his life when he was away, he consistently read and wrote about farming, writing to Arthur Young and subscribing to his *Annals of Agriculture*, where he may have first come across the name of Robert Bakewell. This must have encouraged Washington to try various agricultural experiments, planting several crops and cultivating them in different ways.

Washington had a 6,500 acre estate at Mount Vernon, Virginia, where his home farm acknowledges Bakewell's influence on Washington's agricultural practices. He had returned to a neglected Mount Vernon after the end of the American War of Independence in 1783 and remodelled both the house and its estate in 1785-6. He changed the focus from tobacco to wheat and other grains, and to commercial fishing and eventually built up a considerable flock of sheep, numbering from 600 to 1,000, though initially complaining that he could not obtain good rams of the 'English breed', meaning Leicester Longwools, because of the export ban already mentioned. Washington (and Jefferson) was looking to establish a successful stock of sheep in North America because they would provide the foundation for an independent textile industry, thus benefiting the country's economy. His vision was implemented as, during the 19th century, there were textile mills throughout the United States that employed many people.

Leicester Longwools found their way to Mount Vernon, although George Washington was at pains to point out that he had not been involved in any clandestine activity, as he was concerned to maintain his image. Washington's flock was indeed improved by the imported Leicester Longwools and a visiting manufacturer from Leeds told Washington that the cloth he produced was as good as anything produced in England. Washington subsequently sent a fleece to Arthur Young after Young asked for a sample of the wool being produced. Washington expressed excitement about the 'prospect of introducing into this Country a very excellent race of animals also, by means of the liberality of the King of Spain, … I have likewise a Jack and two Jennets from Malta, of a very good size, which the Marquis de la Fayette sent to me. The Spanish Jack seems calculated to breed for heavy, slow draught; and the others for the Saddle or lighter carriages. From these, altogether, I hope to secure a race of extraordinary goodness, which will stock the Country'.[31]

Washington was thus a great proponent of mules, using them for field work, and, whereas at the beginning of his life, plantation work was undertaken by horses and oxen, by the end it was done by mules; these he bred both for himself and for the benefit of the country as a whole.

Bakewell traded directly with the United States of America. In a 1787 letter, he told George Culley that Mr Raike of Bath had passed on an order from 'that Great and Good Man General Washington' for 'some implements of Husbandry which I shall send away next Week', adding 'I shall take the liberty of writing to the General and of recommending some other things than what he has ordered which I think likely to answer his Purpose'.[32] Clearly, Washington had asked Bakewell to get him the farm implements needed for the agricultural changes he intended to make.

Writing to Young in March 1788 Bakewell told him 'I have sold a Bull and Two Heifers which are to put on board this week to go to Mary land N. America and if these please I hope to have further orders'.[33] Washington practised selective breeding on his sheep, aiming for better quality wool, and would have known about Bakewell's methods. There were Dishley sheep at Wood Lawn, part of the Mount Vernon Estate about a mile from the main house, still recorded in the next century. The lustrous wool of the Leicester Longwools would certainly have achieved Washington's aim. Bakewell's sheep subsequently gave rise to what was deemed to be an American breed, the Arlington Longwool; they were developed by George Washington Park Custis, an indirect descendent, from General Washington's Bakewell breed and a Persian Ram previously imported by the General.[34]

Washington could also have applied Bakewell's methods to his plans to breed mules for wider use, using the Spanish Jack he had been given in 1784 by King Charles III of Spain. Certainly, Washington and Bakewell had much the same cast of mind, with a scientific approach to farming; both kept careful and detailed records of work done and its outcomes. The extent of Bakewell's influence is surely underlined by his recognition in successive years by both King George III and George Washington.

After their initial introduction, Leicester Longwools began a steady spread across the United States and, as agricultural fairs became popular during the 19th century, Longwools appeared in all the show records. By the 1870s the East Coast dominated sheep and wool production, but as Americans moved ever westwards, they adopted the Spanish sheep, the Merino, which was better adapted to the increasingly dry climate. Leicester Longwools continued to thrive in the East until the late 1800s, when several factors saw their gradual decline. The first of these was the general move west of sheep rearing as a whole, where the climate did not suit the Leicester Longwool. The second was that the Leicester's crosses with other breeds, such as the Border Leicester, began to take more of the market share. Thirdly, other breeds, such as Hampshires, matured more rapidly, making them better suited to the meat market, which had become important. The heavier wool of the Longwools began to fall out of fashion as more and more people worked inside, a factor which was accelerated with the advent of the motor car. Finally, the Leicester Longwool breeders seem to have lost interest in promoting their breed which had virtually disappeared by the 1920s.

Bakewell's sheep were first imported to Kentucky in the 18th century. The Shaker community at Sabbathday Lake, Maine, kept Border Leicesters and Leicester Longwools from their foundation in 1790 until the community's final closure in the late 20th century. These exports would still have needed a licence.

In September 1806, Honeyborn, Bakewell's nephew, sent six Dishley rams for letting to Ireland, accompanied by their shepherd. All were drowned when the ship carrying them, the *King George*, was wrecked. This shows that the breed was still sought after Bakewell's death but underlines the bad luck of the nephew.

A sample search between 1821 and 1853 of American national newspapers found that the following publications reported or commented on Bakewell's livestock or reputation:[35] the *Daily National Intelligencer* (Washington DC); the *Daily National Journal* (Washington DC); the *National Advocate*; the *National Intelligencer & Washington Advertiser*; the *United States Telegraph* (Washington City); the *Gazette of the United States* (Philadelphia) and the *Emancipator & Free American*. The following American state newspapers carried similar reports and show how widely Bakewell's reputation had spread: the *Aurora & Franklin Gazette* (Philadelphia); the *Maryland Gazette* (Annapolis); the *New England Weekly Review* (Hartford, Connecticut); the *New York Spectator* (New York); the *New Hampshire Statesman & State Journal* (Concord, NH); the *Independent Statesman* (Concord, NH); the *Bangor Daily Whig & Courier* (Maine); the *Virginia Free Press* (Charlestown, VA); the *Charleston Tri-Weekly Courier* (Charleston, VA); the *Pennsylvania Gazette*; The *Atlas/The Boston Daily Atlas* (Boston, Mass.); the *Boston Courier* (Boston, Mass.); the *Morning Oregonian* (Oregon); the *Daily News* (Denver, Colorado); the *Indiana Journal* (Indianapolis); the *Vermont Chronicle* (Bellows Falls, Vt); the *Vermont Watchman & State Journal* (Montpelier, Vt); the *Waukie Sentinel* (Milwaukee). The geographical focus is mainly but not exclusively the New England states, with Oregon and Colorado in the extreme west.

A brief search for sales and agricultural show results between 1821 and 1853 shows that the following places in the United States had Dishley sheep. These references are mainly from the shows and do not give the location of individual farms within the state.

Wood Lawn, Virginia
Freetown, Jefferson County
Charlestown, nr Boston
Harlem, New York
Flemington, Pennsylvania
Baltimore
Charlestown, Virginia

Mentions of Dishley itself were not all associated with sheep. When Smith took over at Dishley Grange from Honeyborn, he also bred Shorthorns. The *National Advocate* for 16th March 1823 carried this report.

> The ship *Magnet*, Captain Mount, has brought out from England a superior yearling bull of the short horned Durham breed. This animal was bred by Mr Smith of Dishley Leics and is descended from the celebrated bull Comet (who sold in the year 1810 for 1,000 guineas) and belongs to Mr J. Brientnall of Woodbridge, New Jersey.

On 28th April 1823, the *Daily National Intelligencer* carried a report from Baltimore of 40 Dishley sheep, mainly one year olds, that had been slaughtered. They produced 3,834lbs of mutton and 588lbs of rough fat. There was 4 inches of fat across the ribs of some of them. These Dishley sheep were 'chiefly remarkable for the fineness of their bone, and their great disposition to fatten at a very early age'. They sold for $15 each.

In 1829, The *American Farmer* published a letter written from Alexandria, Virginia, under the pseudonym of 'Albion', clearly an English emigrant farmer. He had recently gone via Baltimore to 'the *intended* city of Delaware' to buy a ram and 10 ewes from Mr John Barney who bred 'beautiful Bakewell Sheep, and Devon Cattle'. Barney met him 'in the true, hearty style of an English Farmer' despite having been in America for 30 years. They dined on 'a quarter of real Bakewell lamb' which he praised extravagantly for its superior flavour and texture by comparison with other mutton: 'such meat may do very well for those who never tasted anything better; but give me a joint of 'Barney's Bakewell,' I say; here is 'cut and come again,' and those who are so lucky as to get a cut, and taste it, will come again'. Barney had about 200 Bakewell rams, breeding ewes and fat wethers such that the writer 'did not expect to ever see any sheep, in the United States, so very superior, as is this flock'. They fully matched his previous description of 'the best kind of Bakewell sheep in England'. He singled out in particular 'one Ram of Mr Champion's blood' which had been sent from England.[36]

The arrival of New Leicester sheep in Oregon can be dated precisely to 1891 from this report.

> NEW LEICESTER SHEEP. The British Ship Cockermouth Arrives With a New Species of Thoroughbred Stock.
>
> Yesterday an Oregonian reporter, accompanied by Captain George Pope, the Importer, visited the Cockermouth, lying at B street wharf, to have a look at the woolly emigrants, the latest additions to Oregon's fine stock. The sheep, four in number, are in superb condition, and, although only a little over a year old, are monsters in size, and even a novice can see at a glance that it has taken years of painstaking breeding to produce an animal of this class. Captain Pope stated that his agent in England was instructed to spare no expense in getting the best and to make ample provision for their safe transit. The sheep stood the voyage around Cape Horn like veterans, and have become great favourites with everyone on board. They were raised by B. Painter, of Bursley, near Rutland, England, and are from a prize family of what is now known as the Dishley or New Leicester. [A discussion on the relative merits of the New Leicester and the Southdown sheep followed.] Captain MacAdam, the master of the vessel, deserves great credit for bringing the ship as well as the sheep into port in such fine condition.[37]

Until the 1850s in the United States, the emphasis had been on pure-bred Dishley sheep and the pedigrees were often given. They were still praised for their fat meat. After that, while they were generally slipping out of fashion, they clearly still had their supporters in Oregon.

On a visit to Dishley, in 1785, Arthur Young related that Bakewell sold a three-year-old bull, a two-year-old bull and two yearlings to a gentleman in Jamaica who had bought a bull in 1777 which was still in perfect health and vigour, in spite of the change of climate. He had found the breed to suit his requirements so well that he had come again, eight years later, to buy more, having been convinced that the breed was superior to all others of his experience.[38]

Dishley imports to Australia, Tasmania and New Zealand

Australia, Tasmania and New Zealand were also importers of Dishley stock. Because of the export ban, and perhaps also because of the sheer distance between Australia and England, the first sheep sent to Australia were picked up on the way by the First Fleet at the Cape of Good Hope. Later, Bengal sheep, said to be a very prolific breed used for meat rather than wool, were sent from India.[39] A letter in the India office files dated 14th March 1792 contains the first list for seeds and plants to be sent to New South Wales, with a postscript

Negrette Merino Sheep.

recommending sending 'Buffaloe Sheep and Goats' at the same time.[40] Twenty sheep were shipped from Calcutta in March 1792 on the *Atlantic*, but when they reached the colony on 20th June, only two rams and 10 ewes had survived.[41] In the next year another shipment of sheep from Calcutta, probably a speculative venture, was sent aboard the *Shah Hormuzear* and reached New South Wales on 24th February 1793. About 220 sheep had been shipped but over half died.[42]

Although the Merino became the dominant breed there (see black and white illustration), other sheep helped to shape Australia's success, with Bakewell's bloodlines. According to Heazlewood, 'Bakewell's influence had arrived on these shores before any Merinos had set foot here. This was in the form of the Irish strong woolled sheep which were infused with New Leicester. These were here in 1796, and perhaps earlier. Eight years later, in 1804, Bakewell's influence again arrived in the Teeswater'.[43] The export of these Irish sheep with Dishley blood may have been a way of getting round the English export ban. The Revd Samuel Marsden had mentioned getting Irish rams from the captain of an Irish transport,[44] probably before he sent his wool samples back home. The initial 13 Merinos arrived in 1797 from the Cape. The first significant imports of English Merinos came in 1804 when Captain Macarthur obtained an export licence for rams from King George III's flock of Merinos;[45] this was in the face of objections from Sir Joseph Banks, who did not think that sheep would thrive in the colony. But they did, and in 1808 the first large batch of Australian wool was sent back to England by the Revd Marsden.[46] Immediately after the export ban was lifted in 1825, Messrs. Bryant and Gavott (and others) brought English stock over on the *Mountaineer*, as reported in the *Hobart Town Gazette* for 29th October 1825.

> The Mountaineer struck upon a rock at Symon's Bay, at the Cape, and sustained so much injury as to be obliged to unload and undergo repairs, which detained her nearly two months. She brings a valuable importation of stock to the Colony, among which are two cows of the Suffolk breed for His Honor the Lieutenant Governor, and grey cart stallion and a mare, a bull, two heifers, and a cow of the Devonshire breed, belonging to Mr Walton, a thorough bred two year old horse, and a filly one year old; a thorough bred Devon bull, two years old, and a heifer of the same age; nineteen merinos, consisting of 17 ewes, and 2 rams, 7 having died on the passage; three fine Leicester rams, and 2 young ewes, the property of Messrs. Bryant and Gavott. All this stock is in most capital condition, and does the greatest credit to the Commander of the vessel, and importers.[47]

By 1837, a committee enquiring into the progress of the New South Wales flocks found there was a surplus of Merino rams and recommended importing more Bengal ewes.[48] But the New Leicester established its place as the most successful crossing breed, and its position was well summarised by Heazlewood: 'In Australia, however, it became the leading British breed and maintained that position for the rest of the 19th century because it complemented and crossed well with the Merino and delivered satisfactory results to a country whose economy depended on sheep'.[49] The same could be said for South America. A surviving group of certificates of breeding and a pen card show that the Dishley pedigree was highly prized in Australia and Tasmania into the 20th century.

Tasmania Pen Card from 1867

TWO PURE-BRED
Leicester Lambs.

These Rams are descended from Ewes bought about 20 years ago from Mr Forester, Raeselands (one of the greatest prize-takers in England), and crossed by the best Tups which could be procured from Mr Bell, Brompton: Mr Bell, Seale Hill: Mr Robinson, Leakby Palace: Mr Beatie, Neeby House: and the Duke of Devonshire; no expense having been spared in combining good quality with plenty of Wool and Mutton.

Enfield, Tasmania,
1st October 1867.

1905 Certificate for a 1 year old ram, brought from England to Melbourne

The Leicester Sheep-Breeders Association
CERTIFICATE OF THE BREEDING OF A LEICESTER RAM
EXPORTED, OR SOLD FOR EXPORTATION
FROM THE UNITED KINGDOM

Name and Number	*No. 18*
Lambed	*1904*
Breeder	*George Harrison*
Sire	*Lord Sheffield 668*
Grand Sire	*Gainford Reserve 583*
G.G. Sire	*Gainford King 440*

BREEDER'S CERTIFICATE

I hereby declare that I bred the Ram above referred to, named *No. 18* and that the forgoing pedigree and particulars are correctly stated.

Signature *George Harrison*
Address *Gainford Hall*
Darlington

Dated *June 24th* 1905

VENDOR'S CERTIFICATE

I hereby declare that the Animal above referred to, has been sold by me to
Mr *G. Simonds* of *Tasmania*
for exportation to *Melbourne*

Signature *George Harrison*
Address *Gainford Hall*
Darlington

Dated *June 24th* 1905

CERTIFICATE OF THE ASSOCIATION

I hereby certify that I have examined the particulars above stated with the entries in the Flock Book of the Association, and that I have found them to be correct.

DRIFFIELD, Yorks.
June 26th 1905

W.A. Brown
Secretary.

1906 certificate for a 3 year old ewe, brought to Tasmania

The Leicester Sheep-Breeders Association

Ewe
CERTIFICATE OF THE BREEDING OF A LEICESTER ~~RAM~~
EXPORTED, OR SOLD FOR EXPORTATION FROM THE UNITED
KINGDOM

Lady Seamer

Name and Number	*401 (Ear Tag) 58)*	Lambed 1903
Breeder	*E.K. Jordan*	
Sire	*Gilmoor Exchange 764*	
Grand Sire	*Seamer No. 2. 474*	
G.G. Sire	*Rupee 331*	

BREEDER'S CERTIFICATE

Ewe
I hereby declare that I bred the ~~Ram~~ above referred to, named *401 (Ear Tag 58)* and that the forgoing pedigree and particulars are correctly stated.

Signature *E.K. Jordan*
Address *Eastburn*
Driffield

Dated *Nov. 14th* 1906

VENDOR'S CERTIFICATE

I hereby declare that the Animal above referred to, has been sold by me to
Mr *G. Simonds* of *Tasmania*
for exportation to *Melbourne*

Signature *E.K. Jordan*
Address *Eastburn*
Driffield

Dated *Nov. 14th* 1906

CERTIFICATE OF THE ASSOCIATION

I hereby certify that I have examined the particulars above stated with the entries in the Flock Book of the Association, and that I have found them to be correct.

DRIFFIELD, Yorks.
November, 1906

W.A. Brown
Secretary.

Partridge Cochin Cock by Ludlow.

Two New Leicester Rams by Thomas Fairbairn Wilson, 1809 courtesy of Beamish Museum.

Valentine Barford and his pure-bred Leicester Sheep, courtesy of Northampton Museum and Art Gallery.

Robert Bakewell's Black Cart Horse Stallion by John Boultbee.

Two Longhorns and two Leicester Sheep, Thomas Weaver, 1802.

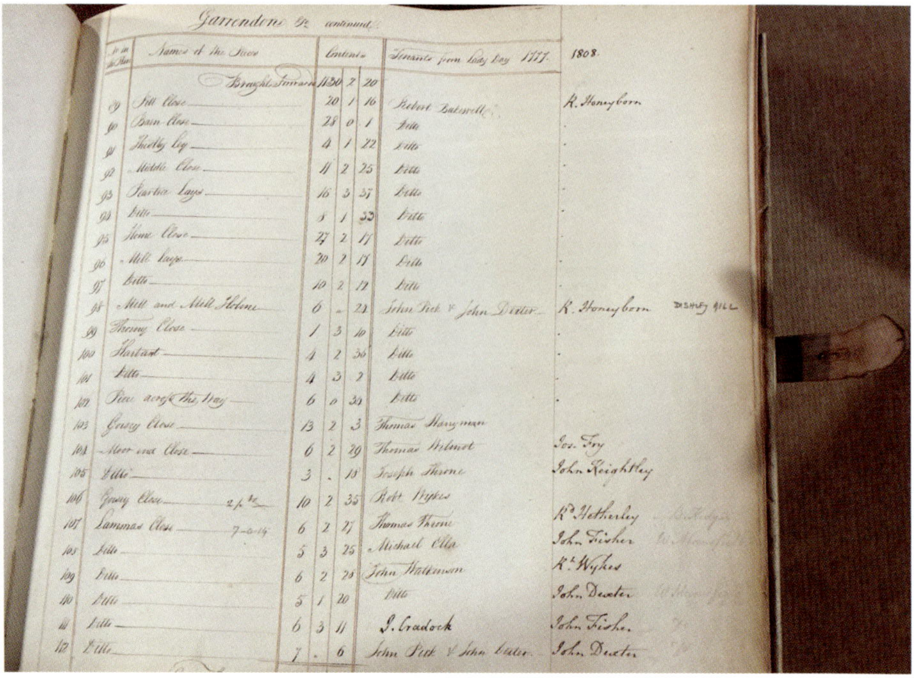

Garrendon &c continued

No. in the Plan	Names of the Pieces	Contents			Tenants from Lady Day 1777	1808	
	Brought forward	1130	2	20			
89	Fill Close	20	1	16	Robert Bakewell	R. Honeyborn	
90	Barn Close	28	0	1	Ditto		
91	Thistly Ley	4	1	22	Ditto		
92	Middle Close	11	2	25	Ditto		
93	Fearties Lays	16	3	37	Ditto		
94	Ditto	8	1	33	Ditto		
95	Home Close	27	2	17	Ditto		
96	Mill Lays	20	2	17	Ditto		
97	Ditto	10	2	17	Ditto		
98	Mill and Mill Holme	6	-	24	John Peck & John Dexter	R. Honeyborn	DISHLEY MILL
99	Thorny Close	1	3	10	Ditto		
100	Hartant	4	2	36	Ditto		
101	Ditto	4	3	2	Ditto		
102	Piece across the Way	6	0	30	Ditto		
103	Gorsey Close	13	2	3	Thomas Haryman		
104	Moor end Close	6	2	29	Thomas Wilmot	Jos. Fry	
105	Ditto	3	.	18	Joseph Throne	John Keightley	
106	Gorsey Close 2 p[illegible]	10	2	35	Robt Wykes		
107	Lammas Close 7-0-14	6	2	27	Thomas Throne	Rd Hetherley	
108	Ditto	5	3	25	Michael Ella	John Fisher	
109	Ditto	6	2	25	John Walkinson	Rd Wykes	
110	Ditto	5	1	20	Ditto	John Dexter	
111	Ditto	6	3	11	J. Cradock	John Fisher	
112	Ditto	7	.	6	John Peck & John Dexter	John Dexter	

Extract from Ledger of Garendon Estate indicating some of the fields let to Robert Bakewell, Lady Day 1777, then Robert Honeyborn in 1808.

Map of Garendon Estate let to Robert Bakewell, Lady Day 1777.

A pair of Longhorns, 1803, by John Boultbee.

A Longhorn ox, bred and fattened at Dishley Grange by Robert Honeyborn, painted by Garrard.

The Longhorn heifer Bandy, *bred by T. Paget, painted by Garrard.*

Longhorn cattle at Calke Abbey in 1880.

Blackbrook Darth Vader, *2025.*

Blackbrook Forget-Me-Not, *2025.*

Blackbrook Zither, *2025.*

1945 certificate for a ram from the Colemans' Speeton flock

TWO FORMS must be filled in and duly signed by the Breeder and Vendor in each case, and must be lodged at the Office, Hunmanby, Yorks. AT LEAST SEVEN DAYS BEFORE THEY ARE REQUIRED.
No. in the Society's Register *G7/a* Name of the person applying for the Certificate and paying the fee thereon *W.A. Coleman & son*
Address Speeton, Filey, East Yorkshire.

Leicester Sheep Breeders' Association.
Offices: ROSELEA, HUNMANBY, YORKS.
CERTIFICATE OF BREEDING AND REGISTRATION
For Sheep Exported from the United Kingdom.

1 No sheep shall be eligible for entry in the Flock Book unless application for entry shall be made while in the United Kingdom, or unless satisfactory reason be given for such application not having been so made.
2 The Council reserve the right of declining any entry if so recommended by the Sales Committee.
3 The statement of the breeding of all Sheep entered must be verified by the signature of the Breeder.
4 No sheep shall be eligible for entry unless from a Registered Flock, and unless the names and the numbers of the Sire and Grand Sire be given, and each Sire and Grand Sire must already be entered in the Flock Book.
5 All entries must be made on the Society's printed forms, and must be certified by the Breeder, the Owner, or their accredited Agents or Representatives.

IMPORTANT RULE – All Pure-bred Leicester Sheep (males and females) exported, must be entered in next Vol. and the Export certificate MUST contain name of BREEDER, VENDOR, CONSIGNOR, CONSIGNEE of every Leicester Sheep exported. Fees for Exportation Certificate to Members – Rams 2/6 each, Ewes 1/- each To Non-Members – Rams 5/- each, Ewes2/- each

[Following information is in columns on the original Certificate]
SPEETON SAMPLE 4289

Registered No. in English Leicester Flock Book (For Secretaries Use only). *97*

EAR OR DISTINCTIVE MARK	E 97 IN Right Ear	
YEAR OF LAMBING	1945	
SEX	Ram	
SIRE	Eastburn Ectype	
NO. OF SIRE	3833	
GRAND SIRE	Haywold	Sunderland
NO. OF G. SIRE	3721	
G.G.S.	Hunmanby Jupiter	
	3370	

The pen card and certificates come from a Tasmanian family who have close connections to the Leicester Longwool sheep. Research has shown that many of these certificates which claim to be in the direct line of descent from Bakewell's own sheep are correct.[50]

Another Australian writer who maintained a stud of English Leicesters from 1916 to 1945 observed of New Zealand in 1925 that,

> in [this] Dominion at the present time, there are more studs of Bakewell's sheep than in any other part of the world. It is curious that at the present time the habitat of the Leicester sheep in England has been transferred from the county of Leicester to the East Riding of Yorkshire, in fact not one registered stud flock is now to be found in the original home county of the breed.[51]

This is a sad comment for Leicestershire, which not only had no major memorial to Bakewell but then also lost its position as the heart of the breed in Britain.

Bakewell's letters and other documentary sources give some idea of the extent of his travels to see other farms, his visitors to Dishley and his business dealings over the livestock. There were regular reports from France in English newspapers for the agricultural shows and exhibitions in Paris which featured Dishley stock. Some Dishley sheep were sent to Jutland in 1850.

He said himself that he had exported his cattle, sheep and horses to Jamaica and Germany. Prussia in particular under Frederick II saw England as having the best farming systems in Europe, and he had sent four students to England in 1764 on a tour of the country to learn English methods. Among other counties, they visited Leicestershire but may not have visited Dishley.[52] He then employed an English farmer, Christopher Brown, to run his state farms. One of Prussia's neighbouring states was Mecklenburg-West Pomerania. Prussia acquired much of West Pomerania in 1720, and occupied Mecklenburg in 1759-64, which nevertheless managed to retain some autonomy. Mecklenburg had predominantly an agricultural economy, growing grain, especially rye, for export. It was a significant breeding area for horses and cattle, also for export, and wool was important too. These factors probably account for the interest in Bakewell's work. Mecklenburg was run as two duchies, and in the Mecklenburg-Strelitz lordship of Rostock and Stargard there is still an isolated farm called Dishley; it is just over 8km from both Rostock and Dummersdorf, a living tribute to Robert Bakewell.

There had clearly been the intention to send his livestock to Russia where the Empress Catherine the Great planned to start an Imperial State Farm. She had sent over seven or eight young men to 'spend some years studying English agriculture',[53] and two were working at Dishley when the La Rochefoucauld brothers visited. But the farm was never established and the Empress abandoned these unfortunate young men. At the end of their time in England they were interviewed by Young, who later noted that on their return, the men were 'turned loose, some to starve, some driven into the army, and others retained by Russian noblemen'.[54] Writing to Culley in 1787, Bakewell recorded that he had received a letter 'from Count de Bruhl Envoy Extraordinary from the Elector of Saxony to the Court of Great Britain, requesting to know the best Methods of Breeding, Rearing and Feeding the best kind of Stock'.[55] He wrote back in great detail. Later that year he was visited by the Earl of Hopetoun and his agent and Mr and Mrs Graham from Aberdeen. His next visitors were the Marquiss [sic] of Graham and 'an Irish gentleman' who came to see his sheep. He had a letter about one of his stallions from Mr Robert Mitchelson from Colwell near Hexham. The King of Denmark's agent was sent to see the sheep; he was an Icelander, 'and was much pleased at seeing his own Country Sheep at this place'.[56]

We may never know the full extent of Bakewell's influence, and where all his stock went, but sample studies such as these can at least give us an insight.

Endnotes

1 Gordon, Revd A., *Robert Bakewell* in the Dictionary of National Biography.

2 Housman, W., Live Stock Handbooks, Vol IV, Cattle, Breeds and Management, (1919), p71.

3 Quoted in Throsby, J., Leicestershire Views, (1790), p415. 'Princess' is a mis-hearing of 'Princep'.

4 We are very grateful to Mr Chris Coleman and his family for allowing us to draw on his history of the flock.

5 Pitt, W., The General View of the Agriculture of the County of Staffordshire, (1813), p193.

6 Coleman, J., Sheep and Pigs of Great Britain, (1877), p35-6.

7 Pawson, H.C., Robert Bakewell Pioneer Livestock Breeder, (1957), p181.

8 Pawson, H.C., Robert Bakewell Pioneer Livestock Breeder, (1957), p134.

9 Pawson, H.C., Robert Bakewell Pioneer Livestock Breeder, (1957), p141-2.

10 Dalton, R., 'John Farey's Derbyshire: Sheep Farming in the Early Nineteenth Century', *Derbyshire Miscellany*, Vol 15, Part 4 (Autumn, 1999) has diagrams showing the distribution of sheep breeds, and the New Leicester tupping linkages, in Derbyshire.

11 Error for Lubenham; the Pagets still had their Ibstock base.

12 Coleman, J., Sheep and Pigs of Great Britain, (1877), p31-2.

13 *Morning Chronicle*, Saturday 11 May 1816.

14 *Derby Mercury*, 28 January 1819.

15 *Jackson's Oxford Journal*, 15 February 1823.

16 *Stamford Mercury*, 28 March 1771; *Manchester Mercury*, 23 April 1771; *Derby Mercury*, 22 March 1771 & 10 April 1778; *Northampton Mercury*, 6 February 1775; *Ipswich Journal*, 5 May 1781.

17 Pawson, H.C., Robert Bakewell Pioneer Livestock Breeder, (1957), p105.

18 Heazlewood, B., *Longwool Origin*, n.d.

19 Young, A., *The Farmer's Tour through the East of England*, Vol 1, (1771).

20 Pringle, R.O., 'The Roscommon Sheep', in Coleman, J., (ed.) The Sheep and Pigs of Great Britain, (1977), p82.

21 Culley, G., *Observations on Livestock*, p91.

22 Pawson, H.C., Robert Bakewell Pioneer Livestock Breeder, (1957), p148.

23 Baker-Holroyd, J., (Baron Sheffield of Roscommon), Observations on the manufactures, trade and present state of Ireland, (1785), p21-2. Digital copy, see openlibrary.org/books/Observations on ...

24 Pringle, R.O., 'The Roscommon Sheep', in Coleman, J., (ed.) The Sheep and Pigs of Great Britain, (1977), p81-2.

25 Pringle, R.O., 'The Roscommon Sheep', in Coleman, J., (ed.) The Sheep and Pigs of Great Britain, (1977), p82.

26 Pringle, R.O., 'The Roscommon Sheep', in Coleman, J., (ed.) The Sheep and Pigs of Great Britain, (1977), p83.

27 Baldwin, T., in Curran, P.L., The Native Lowland Sheep of Galway and Roscommon: A History, (2001), p37.

28 Baker-Holroyd, J., (Baron Sheffield of Roscommon), Observations on the manufactures, trade and present state of Ireland, (1785), p163. Digital copy, see openlibrary.org/books/Observations on ...

29 academic-accelerator.com/encyclopedia/galway-sheep.

30 McHugo, G.P., et al, 'A Population Genomics Analysis of the Native Irish Galway Sheep Breed', in *Frontiers in Genetics*, 8 October 2019, section Livestock Genomics, Vol 10 – 2019. https://doi.org/10.3389/fgene.2019.00927

31 Fitzpatrick, J.C., The writings of George Washington from the original manuscript sources, 1745-1799. US Government Print Office, Washington, Vol 32; letter to Arthur Young, June 1792, p70; letter to Sir John Sinclair, p388-9.

32 Pawson, H.C., Robert Bakewell Pioneer Livestock Breeder, (1957), p107.

33 Pawson, H.C., Robert Bakewell Pioneer Livestock Breeder, (1957), p172.

34 Nicholson, J., *The Farmer's Assistant, being a Digest of all that relates to Agriculture, and the conducting of rural affairs* (Printer: William Dickson, Pennsylvania, 1820), p324.

35 The following overseas newspaper quotations are all from the Gale Historical Newspapers website unless separately attributed.

36 'Albion', 'Barney's Bakewell Sheep', in *The American Farmer*, 2 October 1829, p226.

37 *The Morning Oregonian* for 10 March 1891.

38 Young, A., Annals of Agriculture, Vol VI, p487.

39 The National Archives (TNA) pamphlet ref. 994.02 TUR: Turner, H.N., 'Early Sheep in Australia, Results of Enquiries in Britain', (1978), p8.

40 India Office Library: Public Proceedings 1262, Original Consultations Range 4, Vol 11, No 11, 16 March 1792.

41 Collins, D., An Account of the English Colony in New South Wales, (1798), p181.

42 Collins, D., An Account of the English Colony in New South Wales, (1798), p228.

43 Heazlewood, B., Bakewell's Legacy in Australia, p1.

44 TNA 994.02 TUR, p9.

45 Mitchell Library, Sydney, Historical Records of New South Wales, Vol V, p463, quoted in Turner p8.

46 TNA 994.02 TUR, p6.

47 We are grateful to B. Heazlewood for this reference.

48 TNA 994.02 TUR, p7.

49 Heazlewood, B., Bakewell's Legacy in Australia, p2.

50 We are grateful to B. Heazlewood for this information and permission to quote the documents.

51 Caddy, A., Robert Bakewell and His Times, (1925), p16.

52 Müller, H.H., 'Christopher Brown – An English Farmer in Brandenburg-Prussia in the Eighteenth Century', in *The Agricultural History Review* https://www.bahs.org.uk/AGHR/ARTICLES/17n2a4.pdf

53 Scarfe, N., Innocent Espionage: The La Rochefoucauld Brothers' Tour of England in 1785 (1995), p33.

54 Young, A., The Autobiography of Arthur Young, (XXXX), p125-5.

55 Pawson, H.C., Robert Bakewell Pioneer Livestock Breeder, (1957), p107.

56 Pawson, H.C., Robert Bakewell Pioneer Livestock Breeder, (1957), p113-4.

CHAPTER NINE

The Unitarian Network

Janet Spavold

It has been known for a long time that the Bakewells of Dishley Grange, near Loughborough, were members of the Unitarian congregation. The Unitarians are a Nonconformist group dating from the mid-17th century in England, often originating when the Church of England ejected incumbents who would not subscribe to the Church's current doctrine. A lot of early chapels go back to 1662, and they called themselves Presbyterian groups then. The first Unitarian church in London was established in 1774. Unitarians believe in God as a single entity, and do not accept the Trinity or the divinity of Christ. They would not accept the Book of Common Prayer or the way the Church of England practiced its theology. Nonconformists were legally persecuted and suffered civil disabilities until the Act of Toleration was passed in 1689, though this excluded Unitarians as they denied the Trinity. Dissenting groups were allowed to build their own chapels after the Act, and Unitarians gradually gained acceptability, as Presbyterian groups, during the 18th century. Technically it was illegal to be a Unitarian until the 20th century.

It is difficult to say what exactly Bakewell (4) would have believed in his lifetime. For Unitarians, the difficulty was not the position of Jesus, but that of the Holy Spirit. To practically-minded men of the Industrial Revolution, the idea of the Trinity seemed inherently improbable. They had a preference for freedom of conscience rather than a fixed doctrine, and they claimed the freedom to interpret the scriptures for themselves. They saw (and still see) Jesus as a teacher. They believed in divinity, but not the Nicene Creed; they were not Trinitarians. They objected to prescribed dress, such as surplices, and prescribed forms of worship. Like other Nonconformists, they could not attend the universities because they could not subscribe to the Church of England's 39 articles, which they saw as too prescriptive. They could not be Justices of the Peace or hold any civic or state office. Many employments were closed to them, but they could be lawyers and apothecaries, as several members of Bakewell's family were.

They tended to be rational freethinkers in Bakewell's period, men with a wider variety of social and intellectual interests than other nonconformist sects. Much of their thinking then was strongly influenced by Dr Joseph Priestley, and they ran enlightened schools. They were in general wealthier than other Nonconformist groups, and of a better (and rising) social class. Unitarians, and Nonconformists were generally independent and free thinkers. This affected all areas of their lives, including their entrepreneurial and business ventures. Examples are men like Jedidiah Strutt, Erasmus Darwin or Priestley, and other members of the influential Lunar Society in Birmingham. Leicester did not have a scientific society like the Lunar Society, but individuals did arrange lectures on science and industry, such as the one attended by Bakewell (4) on 23rd February 1785 in Loughborough on experimental physics.[1]

The Unitarian groupings were influential out of all proportion to their numbers, and socially they were a self-reinforcing group through marriages, business ties and intellectual pursuits. Some of this may be because they were excluded from any civic duties. Politically they were Whigs in the 18th-century, then Liberal reformers in the 19th century. However, with political and financial success, some Unitarians later moved to the Church of England and to Conservative politics, to fit in socially.

In the later 18th century, Nonconformist groups set up their own schools, usually called dissenting academies, such as that run by Bakewell's (4) pupil John Ashworth's family in Daventry. They taught a modern curriculum, based less on Latin and theology, and more on science and mathematics. The academies were managed by trustees, who were not theological men as in Church of England schools. It may be that they were appointed because of their experience of managing money and property in their business lives. It was the same for the Unitarian chapels, which were also managed by trustees.

The known names of Bakewell's (4) Nonconformist network echo through this book, and there are probably more of them. Particularly important were the Pagets of Ibstock, especially Thomas (1732-1814), who was probably his closest friend and confidant, and knew more about his breeding methods than anyone else other than Marshall. In 1811 George Culley wrote to Arthur Young about Paget saying,

> I would have written to Mr Paget but have not his address. And he may be dead otherwise if living I am convinced he of all other men, is best able to give you the principal particulars of the late Mr Bakewell's transactions. Because he was not so much younger, had been intimately connected and acquainted with him from his first commencement in business, and I think they were nearly related. However, they were always particular friends and of the same religion. Protestant Dissenters. But very liberal in their way of thinking. Besides, I never knew many if any better man than Mr Paget.[2]

They were in fact second cousins, both descending from the 1681 marriage between their great-grandfather Robert (1) and Mary Paget. The two families acted for each other as executors to wills and as trustees. Thomas was one of the creditors in Robert's (4) bankruptcy. John Paget of Thorpe Satchville, Leicestershire, another Paget Nonconformist, is noted as the man 'who carried the arts of English husbandry into Hungary and helped the Unitarians of Transylvania at a critical period in their history'.[3] Other leading breeders were also relatives. John Stone of Quorn, whose mother Sarah came from Henry Bakewell's line, was Bakewell's (4) great-uncle. John's sons, John Parnam Stone, Thomas Stone and Samuel Stone continued as breeders. In addition, John was the nephew of Nathaniel Stone of Goadby, also a leading breeder.

One of John's executors was Robert Burgess, a breeder at Hugglescote, whose mother Katherine was another of Henry Bakewell's family. In 1777, Robert Burgess married Catherine, elder sister of Nathaniel Stubbins of Holmepierrepont, Nottingham, who kept one of the best flocks of New Leicesters. Nathaniel Stubbins never married, so, when he died it was his sister Catherine's children, Joseph and Robert, who took over the flock. Stubbins belonged to the High Pavement Presbyterian Meeting in Nottingham. John Stone, the Burgess family and the Paget family worshipped at the Bardon Park Presbyterian Meeting near Coalville. John Hood of Bardon Hall was a Presbyterian and the 'leading patron of the Bardon Park Meeting'. Thomas Stone joined the Bakewells at Warners Lane Meeting in Loughborough, along with the Honeyborns. Samuel Stone, John Parnam Stone and Bakewell's friend Thomas Pares and his son John were members of the Presbyterian Great Meeting in Leicester.[4] The Breedons, father and son, were both Nonconformists, and so was Wilkes. This close grouping of family, agricultural and religious interests would have provided both social and business support, unquantifiable historically but clearly significant.

Everything suggests that the Bakewells were part of the Nonconformist network in Leicester and the county which gave the area its nickname of 'Radical Leicester' in the early 19th century. Bakewell may not have been active politically but his grazier and social contacts were all Nonconformists, freethinkers and many were, like him, Unitarians. To the Anglican establishment, they may as well have been atheists. They were in trade and banking, but especially in hosiery, and they were successful self-made men. All the graziers were intimately involved with framework knitting since they supplied the wool needed for the stocking trade, and the knitters had been suffering increasing distress since the end of the Napoleonic wars in 1815 when the army contracts ended. Local and national government excluded all Nonconformists and Roman Catholics from any public office under the Test and Corporation Acts, which were still in force. Parliamentary seats were owned by large landowners or wealthy individuals and there was no universal franchise until the Great Reform Act of 1832. The Borough of Leicester Corporation was a notoriously closed, corrupt Tory group until the Municipal Reform Act of 1835.

The radical Whig opposition, mainly Catholics and Nonconformists of various hues, wanted parliamentary reform, local government reform, education open to all, religious freedom, press freedom and the repeal of the oppressive Corn Laws which kept food prices high. The Nonconformist influence showed strongly in the emphasis on moral force rather than violence to achieve reform. Leicester men were at the forefront of the reform movement, particularly William Biggs (a major hosier who has a statue in Leicester), Thomas Paget junior and Thomas Pares junior. Thomas Paget junior was the son of Bakewell's (4) associate and friend; both families were Unitarians. Thomas junior was a banker, a partner in the Leicestershire Bank along with John Pares and his brother Thomas, and James Heygate. Thomas Pares junior, also a banker and a commissioner in Robert's (4) bankruptcy, was John's son. The bank later became part of the NatWest group. In politics they were all strongly Whig. At least one Dishley Society member, John Breedon of Rotherby, was a shareholder in the bank, as well as using the Leicestershire Bank for personal and business accounts. Two other Leicester Unitarian bankers, William Bentley and William Hodges, appeared as assignees and creditors in Robert's (4) bankruptcy proceedings.

Thomas Pares junior was elected as MP for the Borough of Leicester in the 1818 general election as a reformist Whig, the first time a non-Tory had been elected for the borough. He was re-elected in 1820 but made little progress. Thomas Paget junior had more success. He was a radical reformer who actively supported the framework knitters when they sought to petition Parliament for poor relief. By 1819 he was seen as the leader of the reformist opposition in the county. But there was no prospect of Nonconformist representation in Leicester while the old corporation was unreformed, so he campaigned strongly for both parliamentary and municipal reform. The Leicester Great Meeting of the Unitarians became the centre of reform agitation after 1830, campaigning for peace and reform. Paget and his friends appealed to their supporters to 'refrain from all language and all conduct calculated to create alarm, or to disturb tranquillity' during the election, in line with their belief in moral rather than physical force. They wanted to 'preserve the public Peace, and to prevent the sacred cause of Reform from being wrecked by anarchy, treachery or discord' during the election.

These quotations are from the council of the Leicester and Leicestershire Political Union to the Public in General, published in the *Leicester Journal* for 19th November 1831 prior to the 1831 election, and show the radicals' moral force attitude at a time when only a minority could vote, constituencies were owned by people like the Duke of Rutland, and voting took place in public. Voters were regularly subject to bribery and violence. By then, Paget, Biggs and their friends had become successful middle-class reformers, appealing for support for reform from the local working classes. In the 1831 general election, Paget was elected as one of the county's MPs, in the teeth of Tory opposition organised by the Duke of Rutland who regarded it as his fiefdom. The borough also elected two reform members of the corporation for the first time. The Great Reform Bill that year had strong radical support in Leicester. It was defeated at first but passed in 1832, abolishing the rotten boroughs and providing a measure of male adult suffrage. Paget then turned his attention to getting the Municipal Reform Act passed. This succeeded in 1835, abolishing the old corporation, and Paget was elected the first mayor for the reformed corporation. His first act was to refuse to take a salary. After 1836, Leicester, which had been ruled by the Conservative establishment for generations, had a string of Unitarian mayors; this led to the Unitarian Great Meeting Chapel they attended being nicknamed the 'Mayors' Nest'.[5] It illustrates the influence of a small but intellectually powerful group. This was the political, social and business milieu of the Bakewell family.

Bakewell (4) would fit into the 18th-century Unitarian mould; he thought for himself. His methods of cattle breeding for instance would have been frowned on by the Church of England and the Establishment as incest, prohibited by both secular and canon law. Bakewell's establishment of a blueprint for cattle breeding is described in Chapter 11; similar methods were being employed with racehorse breeding. It underlined the freethinking element in Bakewell's work, and the appeal that would have had for him. Despite this, Bakewell (4) had a good working relationship with Arthur Young, the agricultural writer who was a member of the Church of England.

The Bakewell family's involvement with Unitarianism was not merely limited to attendance at Warners Lane; they took an active part in running it and supporting the neighbouring small Unitarian group. In his history of the Loughborough Unitarians, the Revd Walter H. Burgess noted that the Victoria Street Unitarian Chapel was the oldest Nonconformist chapel in Loughborough, and that its first minister was Samuel Statham, from 1729 until his retirement in 1760.[6] Statham's father was a Loughborough apothecary and a freeman of Leicester.[7] Samuel was apprenticed to Josiah Coleman, another Leicester apothecary, in 1702 and one of his fellow apprentices was John Bakewell, the brother of Robert (3). Samuel was himself admitted a freeman of Leicester in 1724.[8] He must have known Robert (3) and his wife Rebecca very well (see Chapter 2 and Appendices 1.1-1.7). In 1733 Samuel Statham was one of six Leicestershire Presbyterian ministers left 40s each in the will of Archdale Palmer of Wanlip, Leicestershire.[9] Thomas Bruckshaw was the next minister at Loughborough from 1760 to 1769; he had been educated at the Ashworth's Nonconformist Academy in Daventry. He then went, as assistant minister, to the Nottingham High Pavement Chapel and he died in Nottingham. Thomas Bruckshaw was buried at Dishley Church in 1772; this must have been with the support of Robert (4).[10]

Samuel Statham himself gave the land for the first Loughborough chapel, but he retained a right of way from Warners Lane through the chapel property to his other land.[11]

> On ground at the back of Statham's house a modest little Chapel or Meeting House was built about the year 1743. It is described in the deed of 1744 'as lately erected on the land of Samuel Statham with the moneys freely contributed by divers persons thereunto disposed'. The building is still standing, but when the Congregation left it in 1864 for their present church it was sold and devoted to other uses. There was a provision in the deed that stable room had to be kept close by for the standing of the horses of those who came to Divine service on horseback. These were yeoman farmers from the out-lying districts who rode in on Sundays. The Bakewells of Dishley were loyal supporters. Mr Bakewell was famous as a breeder of Leicestershire sheep.[12]

The chapel is no longer standing, but its position is recorded by a blue plaque on the wall of the nearest building in Warners Lane.

The dates here indicate that three generations of Dishley Bakewells were members of the congregation: Robert (3), who died in 1769, his son Robert (4) who died in 1795, and the latter's nephew Robert Honeyborn, who took over the farm at Dishley. They took their religion seriously, and their responsibilities to it; Robert (3) was one of the original trustees for the chapel.

The first deed of the old chapel is dated 29th March 1744, and conveyed the premises from Samuel Statham, clerk, of Loughborough, to John Richards of Normanton-upon-Soar, esquire, and 'Robert Bakewell, of Dishley, gentleman, and their heirs'.[13] This is Robert (3). It is interesting that the Unitarian members had enough faith in the honesty of these families to make them hereditary trustees. Bakewell, a tenant farmer, is described as a gentleman, which indicates that he was comfortably well off and with considerable social standing (though John Richards, esquire, had higher social status). Nine other trustees were named.

When four trustees had died, the survivors were to elect four more within six weeks. There were 11 trustees. The Trust was renewed in 1774, when all except three of the original trustees were dead. The new deed was dated 11th October 1774. The new trustees included 'Robert Bakewell of Dishley, gentleman, a son of Robert Bakewell mentioned in the deed of 1744'.[14] This is Robert (4) the famous breeder, and though a trustee, he was not listed as a warden of the chapel.[15] Joseph Paget of Loughborough was also appointed a trustee; he was a hosier, then Loughborough's most significant industry. By the next renewal in 1816, only Joseph Paget and one other trustee survived. Paget withdrew because of his age, and his son, Joseph junior, was appointed along with William Paget esquire, and John Paget esquire, both of Loughborough, and Charles Paget, the eldest son of Joseph junior. Charles Paget was also a hosier.[16] These families, the Pagets and the Bakewells, provided considerable loyalty and continuity in running the Loughborough Unitarian congregation for two generations of one family and three of the other. By 1863 only one of these trustees survived, Mr Charles Paget, who was by then a Member of Parliament.[17]

From the chapel's account books, Mr Burgess, who apparently led the group, notes that,

> Robert Bakewell, senior, seems to have been treasurer of our Church at one time, judging from this note at the close of the accounts for 1766:-
>
> A deficiency due to Mr Bakewell £2. 4s. 0d.[18]

Bakewell (3) seems to have given or left money to be invested on behalf of the chapel. The account books show that for over 40 years a sum was received annually, described as 'interest of £15, being dividends made on Mr Bakewell's effects'.[19] The interest was paid by Mr John Stone, of Quorn, Leicestershire. On 15th July 1827, the principal was paid into the church account by Stone's son,

> Of Mr Stone money recd [sic] by his Father being two dividends from the property of the late Mr Bakewell £15 0 0 [20]

It was probably a small bequest to the church.

When Robert Honeyborn succeeded to the farm at Dishley he continued to be a generous subscriber to the Loughborough Unitarian Chapel until 1814. The warden's book has an entry,

> 1813 Pd. For Matting Mr Honeyborns Pew 8[s] 3[d] [21]

This is the last mention of him in the Unitarian records; the floor of his pew had new rush matting put down. The close association of the Bakewell family with the Loughborough Unitarian Chapel had lasted from 1729 to 1813. The remnant congregation of the Loughborough Chapel still exists, and they are looked after by the Leicester Great Meeting.

Mountsorrel, a village close to Loughborough, had a granite quarrying industry as well as its agriculture. It had its own Unitarian congregation, founded from a charity 'arising from a plot of land left by Hugh Phipps and vested in 1753 in the names of Samuel Statham, John Stone, Joshua Grundy and Robert Bakewell'.[22] This was Bakewell (3), his relative John Stone, and Statham of course was the Loughborough Chapel's minister. But the Mountsorrel Unitarians were fewer and they struggled to survive, eventually

Two views of the old church at Dishley Grange (above SW and left NE). Nichols, J., The history and Antiquities of the county of Leicester, *Vol III, Part II, (c1811), p763.*

having to be rescued by the Loughborough congregation. In 1761, after his retirement, Samuel Statham was instrumental in getting fresh trustees appointed for the Mountsorrel Unitarian Chapel. They included Joseph Paget senior, the Loughborough hosier, and 'Robert Bakewell the younger of Dishley, grazier'.[23] The Mountsorrel Chapel, like Loughborough's, was described as 'Presbyterian', which at that date meant they were liberal Nonconformists.

The next Mountsorrel trust deed was 1795. The trustees then included Joseph Paget junior and William Paget, both of Loughborough, and 'Robert Honeyborn of Dishley, grazier, a generous subscriber to the Loughborough Presbyterian Meeting'.[24] Robert Bakewell (4) had died on 1st October 1795, so Honeyborn had only just taken over the tenancy at Dishley Grange.

Mountsorrel clearly still relied heavily on Loughborough, as the Pagets were also trustees for Loughborough.

The old church at Dishley (see colour gallery) stands within the curtilage of Dishley Grange, though without its roof and the west wall. It is a small early English building, included on the list of historic buildings by English Heritage. The Bakewell family tombstones lie in the chancel floor. Although they were Unitarians, burials took place in the Church of England as there was then no legal provision for burial for Nonconformist groups.

Endnotes

1 Scarfe, N., Innocent Espionage The La Rochefoucauld Brothers' Tour of England in 1785, (1995), p33.
2 Pawson, H.C., Robert Bakewell Pioneer Livestock Breeder, (1957), p185.
3 Gordon, Revd A., 'John Paget' in the Dictionary of National Biography.
4 Wykes, D.L., 'Robert Bakewell (1725-1795) of Dishley: farmer and livestock improver', in *Agricultural History Review*, 52, I, p41-2 Hood was not, as Wykes thought, the owner of *old blind horse*.
5 Temple Patterson, A., Radical Leicester A History of Leicester 1780-1850, (1954), p214.
6 Samuel Statham (Surman index i.d.25496) The Surman Index http://surman.english.qmul.ac.uk
7 Hartopp, H., Register of the Freemen of Leicester 1196-1770, Vol 1, (1927), p192.
8 Ibid., p236.
9 TNA PROB 11/657/190.
10 Thoman Bruckshaw (Surman index i.d.4032) The Surman Index http://surman.english.qmul.ac.uk
11 ROLLR L288, Burgess, Revd W.H., A History of the Loughborough Unitarian Congregation, (1908), p8 & 11.
12 Ibid., p8.
13 Ibid., p11.
14 Ibid., p12.
15 Ibid., p15.
16 Ibid., p13.
17 Ibid., p8.
18 Ibid., p33.
19 Ibid., p34.
20 Ibid., p34.
21 Ibid., p34.
22 Ibid., p25.
23 Ibid., p24.
24 Ibid., p24.

PART TWO

The Story of the Longhorn

CHAPTER TEN

Longhorns Before Bakewell

Pat Stanley & Janet Spavold

> Where did he come from, this singularly picturesque beast, with the carriage of a lion and the temper of a dove – the one breed, perhaps, in which, the feeder, the butcher, and the artist may equally delight?[1]
>
> (*J. Neville Fitt*)

From the very earliest records of British agriculture there has been a race of cattle set apart from most other breeds by the great length of their horns. In the original types, the horns normally projected nearly horizontally on either side of the head. But, as the cattle were improved, the horns assumed other directions, mostly in the shape known as bonnet or wheel horns.

Longhorns became predominant in medieval times over a large area of the Midland counties and northern England. It is claimed that, like its successor, the Shorthorn, the Longhorn first came from Yorkshire, but from the more western part where the district of Craven was said to be the cradle of the race. But the breed was also very widely spread at an early date in our history throughout what were known as 'the grazing districts' of the Midlands. It is perhaps more feasible to assume that Craven was an area where, for some reason, rather fine Longhorns were bred and, because of their good qualities, they were introduced to other districts to improve local stock, rather than being the one area from which the whole of the Longhorn race first sprang.

J. Neville Fitt reported in 1876 that breeders had written letters to him about their experiences with the Longhorns. One Staffordshire farmer spoke of the cattle being in his family for 280 years, thus putting them there 180 years prior to Bakewell's era.

We learn that there were regional variations in the breed. The Lancashire, Westmoreland, Cheshire, Derbyshire, Leicestershire, Nottinghamshire, Staffordshire, Oxfordshire and Wiltshire were all essentially Longhorn but each assumed a peculiarity of form which characterised it as belonging to a certain district and making it of value there.

Wherever they may have originated, the history of the Longhorn can only be traced with any accuracy from the northern part of Leicestershire and the neighbouring part of Derbyshire. J. Neville Fitt wrote that, 'He seems to have been first brought to perfection on the borders of the wild district known as Charnwood Forest, which, at the time these cattle first came into notoriety, was probably as wild and uncultivated as any spot to be found in England. Right well do the picturesque cattle match with its sylvan beauty'.[2]

Could it be that the early pioneers of agricultural improvement were not only influenced by the breed's abilities to produce excellent beef and cheese, but were also susceptible to the beauty of these animals? It seems an extraordinary coincidence that perhaps the only breed of cattle which would stand out well against such a rugged landscape should have found its home in such an area.

The Early Improvers

The earliest name to be encountered in connection with improvements in the Longhorn breed is that of Sir Thomas Gresley, of Drakelow Hall, Burton upon Trent, 'who took such delight in keeping a dairy of cows similar in colour and shape',[3] in the early 1700s. At this time almost everyone who wished to improve their stock had resorted to using the bloodlines of Sir Thomas, and in general any Longhorn of note coming thereafter can be traced back to his herd.

A blacksmith and farrier by the name of Welby, living at Linton, near Sir Thomas, also goes on record as being a great improver of the breed. His stock were of the Drakelow line, and with great determination and willpower he built up a very valuable herd of cows. Unfortunately, a disease, which at the time was totally incurable, wiped out most of them, nearly ruining Welby and putting a stop to any further hopes he had of cattle breeding.[4]

John Webster of Canley, near Coventry, Warwickshire, next comes on the scene. The family had its origins as landowners in the Trent Valley, and so were also neighbours of Sir Thomas Gresley. The Websters purchased six red Longhorn cows from Sir Thomas in 1710, and when John Webster moved to Canley after the death of his father in 1727, the Longhorns went too.

Webster bred the first historically famous bull of the breed, regarded as the 'purest fountain of the Canley blood', i.e., the improved Longhorns. As a yearling he was considered so unpromising that he was sold on to a Mr Bloxedge. He turned out to be a very good stock-getter and Webster bought him back and used him for several seasons. *Bloxedge*, as he was called, was then sold on to a Mr Harrison of Deakenedge in Warwickshire, and later to Mr Flavell of Hogshill, Birmingham, where he finished his life.[5]

Two eminent breeders of the day gave the following opinions of John Webster's cattle: 'that Mr Webster had the best stock, especially of beace [beasts], that ever were, or (he believed) ever will be bred in the kingdom' and 'That in beauty or utility of *form* they have received improvement little, if any, since Mr Webster's day', except in flesh, 'the more valuable quality, they have been improved'. This was great praise and was recorded as evidence of the superb quality of Webster's breed. A neighbour and good friend of Webster's, Mr Palfrey, in recalling the details of Webster's herd to agricultural writer William Marshall, said that 'With regard to colour Webster favoured his six original red cows, but was much in favour of the brindled finch back and the pied'. The original six red cows were of a whole colour according to this witness.[6]

John Webster became high sheriff of Warwickshire in 1754, and when he died in 1768 he was buried at Stoneleigh and a wall monument in his memory placed in Stoneleigh Church. However, a stern lesson of warning is to be learnt by all cattle breeders from his example. Tempted by high prices, he parted with his best stock, so ruining his herd, whereas Bakewell, Prinsep and Fowler, who followed on, kept their best stock to themselves and became justly famous.

Robert Bakewell laid the foundations of his Dishley herd with a couple of Canley heifers and, like Webster, a Westmorland bull. Through these two original females, there is a direct line to Sir Thomas Gresley's blood and, from Bakewell, that blood was imparted to nearly all the herds in the country.[7]

Another early herd to be mentioned is that of George Chapman of Upton near Nuneaton, Warwickshire, who began to select Longhorns in 1756. These men were the greatest of their generation. They were the outstanding representatives of a small band of livestock improvers who achieved results of inestimable benefit to English farming.

There was certainly a great interchange of blood between Webster, Bakewell and Chapman, and many a ride these old-world worthies must have

Woodbine, *the property of S. Burbery of Wroxhall, Warwickshire, winner of first prize and silver medal at Bingley Hall in 1855.*

had across Charnwood, sometimes called Charley Forest. This was once a noted chase and it is written that the abbot of the nearby Garendon Abbey once complained of John Comyn's actions; Comyn had driven a large number of hogs from the forest to the park in his manor to assess the number that were there unlawfully, claiming the right to kill them. According to the abbot, 100 hogs had been killed and he wanted recompense. A jury subsequently gave a verdict in favour of Comyn, the lord of the manor.[8]

This story is taken from The History and Antiquities of Charnwood Forest, which makes it plain that it must have been no easy matter for strangers to cross this almost pathless wilderness in the days prior to our Longhorn friends.

The charms of Charnwood Forest were extolled in about 1622 by Michael Drayton, as part of his extended poem about England called *Poly-Olbion* (words modernised):

> O, Charnwood, be thou called the choicest of thy kind
> The like in any place, what flood hath hapt to find?
> No tract in all this isle, the proudest let her be,
> Can show a sylvan nymph for beauty like to thee!
> The satyrs and the fawns, by Dian set to keep
> Rough hills and forest holts, were sadly seen to weep,
> When thy high palmed harts, the sport of bows and hounds,
> By gripple borderers' hands were banished thy grounds.

This verse gives a small glimpse into an area of 18,000 acres through which our noted Longhorn breeders often travelled, 100 years later. For many famous sons of Longhorn lore lived around the borders of the forest: Bakewell, Buckley and Stone to the north and Knowles, Astley and Paget to the south.[9]

Letting Out Sires

The improvement in the Midlands was greatly accelerated by hiring out sires for the season. Prior to letting time, the owners held exhibitions of their stock for dairymen and graziers to attend and choose their sires for the coming season. The practice was used for bulls, stallions and rams. There were also public shows in the Midlands for the same purpose. Ashby de la Zouch had its Stallion Show, Leicester its Ram Show and Market Bosworth had its show for bulls. But the sires at these shows were there for selling as well as letting.

The practice of letting meant that, for example, a superior bull, instead of being confined to his owner's herd and those of a few neighbours, became a treasure in which the whole district could share. One year he would be standing in one part of a county and the next year in another part – rather like a circuit judge doing the rounds. In this way one good bull could have an enormous impact on the surrounding countryside and any sons got by him, which shared his good qualities, were also let to circulate his blood still further.

This is the earliest example of a breed structure in which a few famous, long-established herds provided highly sought-after bulls for the use of the breed as a whole, or for upgrading local nondescript stock. As a result, by 1810, the overwhelming majority of cattle in the Midland counties were of improved Longhorn type.

Robert Bakewell was the first man to pioneer progeny testing through the letting of his sires. Through a bull, ram or stallion being used on a wide range of females, on different soil types and in different climatic conditions, a more accurate picture of his worth as a sire could be formed. In this way the benefits of a particular animal could be assimilated far sooner than if he were to remain with one small group of females and his worth probably not fully appreciated until after his death.

Endnotes

1 Fitt, Neville J., *The Agricultural Gazette*, 27 March, 1876, p396.

2 Fitt, Neville J., Longhorn Cattle: their History and Peculiarities. *Journal of the RASE*, Series 2, Vol 12, (1876), p461.

3 Lythall, J.B., The Long-horn Herd Book Vol 1, p9.

4 Marshall, W., The Rural Economy of the Midland Counties, Vol I, p318.

5 Marshall, W., *ibid.* p319-321 and notes.

6 Marshall, W., *ibid.* p319 and notes.

7 Coleman, J., (ed.) The Cattle of Great Britain: Their History, Management, etc., (London: 'The Field Office', 1875), p83.

8 Potter, T.R., The History and Antiquities of Charnwood Forest, Hamilton Adams, (1842), p152-3.

9 Fitt, Neville J., *ibid.*, p467.

CHAPTER ELEVEN

Bakewell and the Art of Breeding

Pat Stanley & Janet Spavold

> The Grand old Longhorn, Bakewell's breed, must ever hold first place in the chronological history of the scientific and systematic improvement of British cattle, and long may the type be preserved in its true grandeur and picturesque beauty.
>
> (*David Lowe*, 1842)[1]

In 1790, William Marshall, commenting on Bakewell's pre-eminence as a breeder wrote:

> It must not, however, be understood, by those who are not locally acquainted with the district, that MR. BAKEWELL, though he has been long, and most deservedly, considered as the principal promoter of the ART OF BREEDING and has for some length of time taken the lead, is the only man of distinguished merit in this department of rural affairs, in the district under survey. It abounds, and has, for many years, abounded with intelligent and spirited breeders. I could mention some fifteen to twenty men of repute, and most of them men of considerable property, who are in the same department, and several of them eminent for their breeds of stock. Nevertheless, it must be said and acknowledged, that Mr Bakewell is at the head of the department; – and, whenever he may drop, it is much to be feared, and highly probable, that another leader, of equal spirit, and equal abilities, will not be found to succeed him.[2] (This and subsequent references from Marshall are from his Rural Economy of the Midland Counties.)

Marshall went on to say that, although he had many opportunities to observe Bakewell's practices and had regular communication with him, he was going to maintain the reporter's impartiality and it was not his intention 'to deal out Mr B's private opinions or even attempt a recital of his particular practice'.

Commendable as this ideal may have been then, it is today a great tragedy for us that he felt compelled to speak only in general terms. It is a lost opportunity which would have given us great insight into Bakewell's thoughts and methods.

Before the improvers arrived on the scene, the accepted practice in breeding farm animals was to select females of the native stock of a county and to cross them with bulls of a different breed, as it was believed that continuing to use the same line of parentage weakened the stock. Ernle says that, prior to Bakewell, 'Stock-breeding as applied to both cattle and sheep was the haphazard union of nobody's son with everybody's daughter'.

As George Culley later observed,

> The great obstacle to the improvement of domestic animals seems to have arisen from a common and prevailing idea amongst breeders ... that no bull should be used in the same stock more than three years, and no tup more than two: because, (say they) if used longer, the breed will be too near a-kin, and the produce will be tender, diminutive, and liable to disorders; some have imbibed the prejudice so far as to think it irreligious; and if they were by chance in possession of the best breed in the island, would by no means put a male and female together that had the same sire, or were out of the same dam.
>
> But fortunately for the public, there have been men, in different lines of breeding, whose enlarged minds were not bound by vulgar prejudice, or long established modes, and who have proved, by many years' experience, that such notions are without foundation.
>
> Mr Bakewell has not had a cross (from any other breed than this own) for upwards of 20 years: His best stock has been bred by the nearest affinities; yet they have not decreased in size, neither are they less hardy, or more liable to disorders; but on the contrary, have kept in a progressive state of improvement. ...
>
> It is certainly from the best males and females that best breeds can be obtained or preserved; to breed in this manner is certainly right, so long as better males can be met with, not only amongst our neighbours, but also amongst the most improved breeds in any part of the island, or from any part of the world, provided the expense does not exceed the proposed advantage. And when you can no longer, at home or abroad, find better mates than your own, then, by all means, breed from them; whether horses, neat cattle, sheep, etc. for the same rule holds good through every species of domestic animals: But, upon no account, attempt to breed or cross from worse than your own; for that would be acting in contradiction to common sense, experience, and that well established rule ... 'That best only can beget best;' or, which is a particular case of a more general rule, viz. That 'Like begets like'.

> On this simple axiom the whole mystery of improving stock seems to depend, and, like many other valuable truths, has been neglected, most probably for its simplicity; and other modes pursued as whim or fancy directed, without either reason or experiment to support, or even give the least colour of plausibility to the practice.[3]

When the improvers came along they used practices directly contrary. They bred not only from the same line but from the same parentage. The term used at first for the practice was 'breeding in-and-in', and later 'line breeding'. It became the ordinary practice of superior breeders to mate father and daughter, son and mother, brother and sister. It was by these means that Bakewell produced his celebrated stock, for he and his fellow improvers held the belief that there could be only one best breed, and that if you crossed it, you would be doing so with an inferior breed, thus adulterating and not improving.

Marshall, again writing in 1790, pointed out that great strides had been made in the livestock breeding of the time by 'breeding in-and-in' and by 'crossing superior branches of the same breed'. The rapidity with which these improvements were made was quite remarkable when compared with the years that had gone before. George Culley said that 'the kind of cattle most esteemed before Mr Bakewell's time were the large, long-bodied, big-boned, coarse, flat-sided kind and often lyery (having little fat in the flesh) or black fleshed!'

Bakewell's improvements in all manner of farm livestock were little short of magical and they were brought about using line breeding, as has just been mentioned. This went totally against the old and extensively practiced idea that no bull should be used on the same stock for more than three years, and no ram for more than two years, otherwise it would lead to deficient offspring. Bakewell had to overcome almost overwhelming resistance to change of any sort. The age-old attitude of 'what was good enough for my father is good enough for me' was predominant. Also, this was a time when, after all, religious beliefs were sacrosanct and Bakewell's breeding policy went against everything the church supposedly stood for.

Pawson quoted George Culley: 'Some have imbibed the prejudice against Bakewell's breeding practice so far as to think it is irreligious',[4] as providing a valid reason for Bakewell's fabled secretiveness regarding the breeding of his stock, but in truth, he did not hide his methods. It is also clear that his methods were passed on to his pupils such as Culley and the Colling brothers for, in using the principles laid down by him, they were even able to surpass him.

However, it was said that other than his faithful shepherd, a man called John Breedon, whom he employed in 1760, no one knew the exact secret of how he obtained such outstanding specimens of the various types of livestock which graced the pastures at Dishley. Breedon played a most active part in the formation of the New Leicester sheep, but the manner in which the two men chose which animals to mate was their special skill.

The goal for which Robert Bakewell strove, both in the production of livestock and in the general management of his farm, was economy. His principal target in cattle breeding was to refine and reduce the bone and offal of a beast and to increase the proportions of consumable meat. He is quoted variously as saying, 'All is useless that is not flesh' and similar sentiments regarding 'beef' and his ideal was to obtain three-fifths to roast against two-fifths to boil. In his travels, to extend his knowledge of farming practices, he found much to admire in the methods used by the farmers of Norfolk and, from abroad, those of Holland and Flanders who were characterised by their efficiency and thrift. It was upon these principles of management that Bakewell is believed to have founded his own system at Dishley.

The many visitors to Dishley gave accounts of scrupulous neatness, order, ingenious time-saving devices and the economic ways of efficiently carrying out the ordinary work of the farm. In other words, in every way, rigid economy was practiced and in view of the path that Bakewell was destined to walk, this was no bad thing, for the many and varied experiments which he was to conduct were to prove both costly and time-consuming. The more successful the farm, the better able it was to support the testing of Bakewell's theories.

Bakewell's decision to adopt the already well-established cattle breed of the Midlands, the Longhorn, was not taken lightly. He tried many breeds before deciding that the Longhorn was capable of doing better in Leicestershire than any other breed.

So his work of improvement began, work which had already been started for him by the likes of Sir Thomas Gresley of Drakelow and the Websters of Canley. For, although he had not been there when the first seeds were sown, he certainly had the imagination to use knowledge from those who had gone before and carry it to its conclusion. The original, pre-1700s Longhorn had been a slow-growing, slow-moving, lean-fleshed creature which was light in the hindquarters and a fair but not outstanding milker. The breed which Bakewell took on had already risen to a high degree of advancement in both its beefing and dairying qualities.

An idea of the improvements to livestock brought about by Bakewell can be seen by the following figures given by Ernle, quoting Sir John Sinclair, of the average weights of sheep and cattle sold at Smithfield Market in 1710 and 1795.

> In 1710 the average weight for beeves was 370lb., for calves 50lb., for sheep 28lb., for lambs 18lb. In 1795 (the year of Bakewell's decease) the average had risen for beeves to 800lb., calves to 148lb., sheep to 80lb., and lambs to 50lb.[5]

A Longhorn bred by Mr Minion of Stedford, near Tamworth, the property of T.S. Burgess of Holmepierrepoint (Holme Pierrepont), Nottinghamshire. This animal was 10 years old when exhibited at Smithfield in December 1839 and won the first prize of 20 sovereigns as the best fatted cow of five years and over (from Farmer's Magazine, *April 1840).*

Although no record survives to tell us exactly how Bakewell achieved his ultimate goal, we must assume that he started by breeding from animals as close to his ideal as he could find. By breeding them in-and-in, he fixed in them those characteristics that he esteemed the most. It is known that Bakewell kept four principal points steadfastly in mind:

1. Breed
2. Utility and beauty of form (to include fineness of bone, lightness of offal and the greater weight in the best parts).
3. Quality of flesh (a point which had not before been considered by breeders).
4. Disposition to fatten.

Robert Bakewell's Longhorns

It has always been acknowledged as a fact that Robert Bakewell started his herd of Longhorns by the purchase of one red and one yellow heifer from Mr Webster of Canley, Warwickshire,[6] and a bull from Westmorland. The often-quoted line 'From these pure Longhorns he bred the whole of his herd' has been passed on from one writer to another down the years. However, when one considers this statement, it seems certain that there must have been cattle at Dishley before the Webster Longhorns were introduced and they must surely have been of the long-established breed of the area, the Longhorn.

An interesting disclosure by Arthur Young related the following, published in 1811: 'I learn from Mr Culley that some errors have been afloat relative to the origin of Bakewell's cattle, the chief merit of which has been assigned to the stock of Mr Webster of Canley; but this stock were in several ways inferior to the yellow breed, before at Dishley. They were greatly deficient in disposition or propensity to make readily fat, which the yellow cattle were so eminently possessed of'.

As we have already observed, Longhorns, although the most predominant of the breeds throughout the Midland counties, took on localised characteristics, colour being one of them. We know that one of the heifers purchased by Bakewell, from Mr Webster, was of yellow colouration. It is purely conjecture, but could she have been picked out by Bakewell not only for her breeding ability, but also because she was of a colour favoured by him, and already to be seen at home? He is on record as having said that 'Pale colours indicate finer meat than darker ones'.

This quotation from Arthur Young is the first mention we have found that there were superior cattle at Dishley *before* the purchase from Webster, although the area was already well known as a stronghold of the breed. If it is true that Bakewell was in possession of superior stock before the Webster purchases, he is much maligned by his detractors who allege that he gained the glory for Webster's deeds.

According to Pawson, Bakewell's famous bull *Twopenny* got its name 'because someone remarked that he was not worth twopence'.[7] There are also two conflicting stories of how Bakewell's famous bull, *Twopenny*, was bred. The first and most widespread, which Pawson follows, is that his dam was one of the two heifers bought from Webster, the heifer later known as *Old Comely*. *Comely* was born in 1765 and lived to the great age of 26 years. According to Pitt, writing in 1809, 'She died when life had become burdensome to her'. She is known historically as *Old Comely*, and years after her death parts of her were still to be seen, preserved in the hall at Dishley, which Bakewell kept as a museum. It is recorded that her sirloin was 4 inches thick.[8]

Bakewell kept pickled joints and skeletons of his best stock in order to compare one generation with another – ancestors with their descendants. The fineness of bone, size and shape of frame, thickness of muscle-layer and depth of outside fat were all factors he compared. All this information was recorded to serve as a guide. It must have greatly helped Bakewell's assessment of the living animals, enabling more accurate estimates of their value for breeding purposes.

A visitor to Dishley named Throsby described *Old Comely* in 1790 as standing

> like a venerable ruin on props of magnificent architecture, bulging fine limbs, enfeebled with age. I will not attempt to describe what she has been, only in general terms observe that she is now 25 years old, and has been esteemed by judges as one of the finest animals of that species ever bred. She now lives in an asylum, a meadow full of keep (in other words, a rich meadow full of feed), set apart to soothe her passage to the earth, for in the slaughterhouse she is not to make her exit. She eats but little and yet retains upon her back, which is now a yard over, broad cushions of fat. She seldom moves; she stood like a statue while I went round her, upon her legs bowed at the joint like those of a decrepit old man in the last stage of his existence.[9]

The other story about the illustrious bull *Twopenny* was that he was out of a cow bought by Bakewell from Sir William Gordon of Garrington. Gordon was married to the widow Phillips, of the Garendon Estate, of which Dishley was a part. (Garrington is clearly a corruption of Garendon.)

Arthur Young, who saw *Twopenny* in 1770, described him as being a very big bull, 'most truly made' on the barrel principle, circular, but broad across the back. He also stated that Bakewell refused an offer of 200 guineas for the bull. A fee of 5 guineas per cow was charged for the service of *Twopenny*, standing at Dishley, but his sons could be hired for the season for between 5 and 30 guineas.[10] In later years Bakewell was to relinquish his hold on his pride and joy for a while by letting both Mr Chapman of Upton, Nuneaton, Warwickshire, and Mr Fowler of Little Rollright, Oxfordshire, hire him for a season each.

Bakewell bred his Longhorns in-and-in as a means of fixing the type he desired for beef production. Being fully aware of the risks attached to such a course, he was prepared to face the consequences and cull hard where necessary. Lowe recorded that 'Amid many disappointments he never despaired of his ultimate purpose, but bore up against ridicule, neglect and predictions of failure till the end'.[11] Nevertheless, Bakewell's practice was not solely one of inbreeding but one of breeding *mostly* from within his own group of whichever type of animal he was dealing with. Occasionally he bred from closely related animals, but also sometimes from unrelated animals.

The resulting cattle of his long endeavours were exceptional for their fineness of bone and excellent fleshing qualities; John Lawrence described them as:

> Sound, tight, cylindrical carcase; wide in the hips, but with very little prominence of the huckle bones; straight back, well filled behind

> shoulders; neck long and fine without any superfluous skin or dewlap, horns long, taper downwards, and of a deep yellowish colour; head fine and smooth. The barrel form, gradually tapering towards the ends, was the model, as in sheep. The cattle being peculiarly light in the belly and offal.[12]

Marshall described Robert Bakewell as being in possession of many valuable individuals:

> His bull, D. generally known by the name of the 'mad bull', is a fine animal; and is a striking proof of the vulgar error, that breeding in-and-in, *weakens* the breed. He was got by a son of Twopenny, out of a daughter and sister of the same celebrated bull; she being the produce of his own dam. Nevertheless, D. is the sire of Shakespeare, by another daughter of the same bull, and is probably the most *robust* individual of the longhorned breed; while D. himself, at the age of twelve or thirteen years, is more active, and higher mettled, than bulls in general are at three or four years old.
>
> This has long been esteemed Mr Bakewell's best bull; and has been kept, principally, for his own use. He has never let, except part of a season to Mr Fowey [Fowler]; but has individual cows brought to him, at five guineas a cow.
>
> Mr Bakewell's cows are of the finest mould, and the highest quality: and his HEIFERS beautiful as taste could well conceive them: clean and active as does. Mr B's exhibition of cattle would gratify the most indifferent spectator, and could not fail of being highly satisfactory to every lover of the rural science.[13]

Youatt described the Dishley Longhorn as being

> unrivalled for the soundness of its form, the smallness of the bone and its aptitude to acquire external fat; while they were small consumers of food in proportion to their size; but at the same time their qualities as milkers were very considerably lessened. The grazier could not too highly value the Dishley, or New Leicester Longhorn, but the dairyman and the little farmers clung to the old breed as most useful for their purpose.[14]

In his second visit to Dishley, in 1785, Arthur Young noticed considerable 'improvement' had taken place with the Longhorn. This consisted of the development of enormous masses of fat over the hip bones and around the tail head. Formerly, Bakewell had been content to breed animals heavy in the hindquarters, now he was breeding for abnormal, almost grotesque, protuberances of fat. This at the time was seen as a most desirable point for, until this advent, the great difficulty had been to breed animals that would fatten easily. The working man's diet being very poor, fat meat then formed an essential part of it. Fat also provided a large part of the energy intake for rich and poor alike in the days when physical activity, including walking, was a much larger part of life than now. Fat by-products like tallow also added to the value of an animal, because of their use to provide oil for light when no other was available.

Young recorded that the hip bones of one cow were 'buried in a mound of fat fourteen inches in diameter, but she continued to calve yearly'. High condition was the rule of the day in the breeding herd at Dishley and Young went on to describe the cattle as being 'fat as bears' and 'if the degree of fatness in which he keeps all these cattle be considered, and that he buys neither straw nor hay, it must appear that he keeps a larger stock on a given number of acres than most men in England'.

In more detail about this visit, Young, presumably referring to himself as 'AY', went on:

> AY 'the 16th' to Dishley, where I found the enterprising Mr Bakewell, amidst improvements that will reflect a lasting honour on his name.
>
> The fifteen years that have passed since I had seen Mr Bakewell's cattle have not elapsed without his gaining considerable experience, and carrying every part of his stock much nearer to perfection.
>
> The time alone is sufficient to explain the reason of some difference that will be found between the present minutes, and those taken here before.
>
> **General Principle**
>
> To explain the principles, which have guided him in breeding a beast or sheep (for the same rules are applicable to both) for the butcher, will at the same time explain his own stock, which are in the highest perfection, when examined with an eye to those principles.
>
> The leading idea, then, which has governed all his exertions, is to procure the breed which in a given food will give the most profitable meat ... that in which the proportion of the useful meat to the quantity of offal is the greatest; ... also in which the proportion of the beast to the inferior joints is likewise the greatest.
>
> The propriety of the rule is obvious, and at one stroke cuts off many common notions that will not stand the test of that critical examination, which may on this principle be instituted. Thus the short leg, when the result of a great heaviness in the belly and the shoulders, indicates no more than the weight of the beast being in the worst joints. Some are at present fond of a great dewlap with Virgil.

But, as it is mere offal, yet undoubtedly demands that nourishment which might better [go] to a better place, it is to be rejected as an absurdity and classed with the folly of a Norfolk sheep master, who admires a ram's horn, three feet long and nine inches in circumference. (Mr B. when in Suffolk, measured the horns of one of Mr Macro's rams of those dimensions). For the same reason, a thick hide, a great head, or in a word any part of the animal being heavier than ordinary, except those joints which are the most valuable, are to be considered as breeding offal not meat; and on the contrary, those best joints cannot be too heavy, under which idea Mr Bakewell has bred some beasts to be so exceedingly fat on the rump as to appear ... monstrous to the eye. Of this sort was the ox shewed at Mr Tattersall's, and he has many bulls and cows of the same kind. The experiment is remarkable, and shews to what perfection skill and attention will carry breeding.

POINTS OF A BEAST

On this plan, the points to attend to in a beast, are those where the valuable joints lie, the rump, the hip, the back, the ribs, and after these the flank ... that is to say, the backward upper quarters, but the belly, shoulder, neck, legs, and head should be light, for if a beast has a disposition to fatten, or be heavy, in these it will be found a deduction from the more valuable points. It has been said, but improperly, that a barrel on four short sticks would represent the true form; but that shape swells at the top and the bottom, whereas the back of a beast should be square, straight, and flat, or, if any rising it should be from a disposition to fatten and swell about the rump and hip bones. And the belly should likewise be straight, for if it swells, it shews weight in a bad point. Again, the shortness of the leg is what Bakewell calls a non essential; under which title he classes all those points, which fashion, custom, or prejudice have at different times, and in different places, called attention to, but improperly. Head, neck, horn, bone, leg, skin (thick hides however are generally to be esteemed worse thriving beasts than thin), colour (Pale colours, as white, yellow, etc Mr Bakewell thinks are indications of finer meat than the darker ones), etc have all been considered as important; but, in fact, none of them are so, for let every one of these circumstances be condemned, in a beast, he should, notwithstanding that, prefer him, if his carcass was well made, and shewed a disposition to fatten in the valuable points.

This doctrine is new, and of very great importance to graziers; as far as reason will permit one to judge the principles on which it is founded are just, and whoever has the pleasure of viewing Mr Bakewell's cattle, will see them powerfully exemplified in actual practice, with a success that cannot permit many to remain infidels.

> **FEELING**
>
> In order to judge whether a beast has the right disposition to fatten, Mr Bakewell examines by feeling. His friend Mr Culley of Fenton, near Wooler, in Northumberland, who has had infinite number of beasts, both fat and lean, go through his hands, agrees entirely with him in this circumstance, and when in Norfolk and Suffolk with him could scarcely believe his ears, when he was told, that lean bullocks and sheep were always bought there by the eye only. So absolutely necessary is the hand in choosing either, that they both agreed that if they must trust to the eye in the light, or to the hand in the dark, they would not hesitate a moment in preferring the latter. The form of the bone in sheep is quite hidden; it is the hand that can tell whether the back is flat, and broad and free from ridge in the backbone, or examine correctly, if the other points are as they should be.
>
> In a bullock the situation of the bone is seen, but there are other circumstances which are essential, that the eye quite escapes. The disposition to fatten is discovered only by feeling; and in this habit in the grazier goes very far. Mr Bakewell and Mr Culley when in Suffolk, endeavoured to make me understand this object by feeling many beasts that were quite lean, and presently convinced me that there was a manifest difference in them. The hip bone should be covered with something under the skin that feels soft and oily; the same along below the backbone, and on the ribs; with a good flank. In beasts that will not thrive well there is nothing of this softness to be perceived, but the hide is tight and hard. A distinction is however to be made between a certain frothy looseness, which is different from that soft mellow feel, if I may venture the expression, which indicates the true disposition to fatten. I can give by words, but an imperfect idea of it, but any person that examines many beasts in a fair will by marking some difference presently be convinced, that it really exists in the degree mentioned; and very great and extensive experience has convinced the graziers in various parts of the kingdom, that these points decide the future thriving of the beasts, and with so few exceptions, that the rule may be considered as general.[15]

Writing a few years later, the importance of assessing a beast by touch was reinforced by George Culley's own comments.

> In this age of improvement, I apprehend we examine more narrowly; we distinguish with more perspicuity, and consequently we judge more accurately. We are not content now with judging by one of our senses, (as I believe used to be the case) by looking only; but we now join the sense of feeling to seeing: the farmer or breeder of this day is not ashamed

> to learn from the butcher, to feel with his fingers, – that touchstone of knowledge in regard to judging of animals, already prepared, or to be prepared, for the shambles. We, undoubtedly, first judge by the sight; which, being pleased, we bring the sense of feeling to its assistance; and if this also approves, we then conclude that the animal suits our purpose, or is answerable to the idea we have formed of it.[16]

William Pitt's description of two bulls he viewed at Dishley in 1809, under the ownership of Bakewell's nephew, Mr Honeyborn, who retained the farm after the death of his uncle was: 'As fine in the bone, clean-made and free from offal, and from every appearance of coarseness as many heifers'. This description is re-enforced by images portrayed by the painters of the period, who painted true to life and had not been touched by the later fashion of the mid-19th century to exaggerate an animal's form for the gratification of the proud owner.

The description of the bull, *D*, giving details of the abnormally high tail head as a point, would be frowned upon today. The Longhorn has come a long way to be accepted in today's world, with our greatly different standards.

The following is a description of the Improved Leicester Longhorn of Bakewellian fame, as written by William Marshall:

> The *forend* long, but light, to a degree of elegance. The neck *thin*, the chap clean; the head fine, but long and tapering. The *eye* large, bright and prominent. The *horns* vary with the sex, etc. Those of the bulls are comparatively short – fifteen inches to two feet: those of the few oxen that have been reared of this breed are extremely large: two and a half to three and a half feet long: – those of the cows, nearly as long, but much finer; tapering to delicately fine points. Most of them hang downward, by the side of the cheeks, and, if well turned, as many of the cows are, shoot forwards to the points. The *shoulders* remarkably thin and fine, as to bone, but thickly covered with flesh; not the smallest protuberance discernible. (The Dishley breed, I think excels in this point: some of the heifers have shoulders as fine as race horses.) The *girt* small, comparatively with the shorthorned and middlehorned breeds.
>
> Nevertheless there are some individuals, more particularly, perhaps of Mr Fowler's breed, that are tolerably well let down in the girt. The *chine* remarkably full, when fat; but hollow, when low in condition. The *loin* broad, and the *hips* remarkably wide, and protuberant. The *quarters* long and level; the *nache* [rump] of a middle width; with the *tail* set on variously, even in the individuals of the highest repute. The *roundbones* small; but the *thighs* in general fleshy; the *legs* small and clean, but comparatively long. The *feet* in general neat, and of the middle size.

> The *carcase* as nearly a cylinder as the natural form of this animal will allow. The *ribs* standing out full from the spine; the *belly* of course appears small. The *flesh* of the superior class I am describing seldom fails to be of the first quality. The *hide* of a middle thickness.
>
> The *colour* various: the brindle, the finch back, and the pye, are common: the *lighter* they are, the better they seem to be in esteem.[17]

An anonymous journalist, writing for a newspaper after visiting Dishley soon after Paget's great sale in 1793, recorded: 'The famous white bull is a noble animal but I found many preferred that sold at the Paget's sale'. Does this suggest that the term 'white', when used by a layman, meant a pale roan? Alas, we shall never know. The Paget bull mentioned is obviously the son of *Shakespeare*, which lived up to his expectations when he realised the price of 400 guineas.

Bakewell conducted many experiments throughout his life with all forms of livestock to ascertain the return they gave him on his capital expenditure. As well as testing one Longhorn against another, he also pitted Longhorns against other breeds. As the results always ended in favour of the Longhorn, they confirmed for him that the path he had chosen was the right one. Although none of the records of these experiments have survived in his own hand, we are indebted to Messrs Young, Marshall and Pitt who tell us that no one ever made more comparisons between the different breeds of cattle than Bakewell, and, accordingly, 'No one that was able to tell us so much, has told us so little'.

Melcombe at two years old. Bred by J.H. Burbery, he was the property of James Davis of Melcombe, Horsey, Dorchester, and the winner of the first prizes at the Battersea and Worcester meetings of the RASE in 1862 and 1863.

Bakewell did not make public the results of his own experiments, his reasoning being that as they were carried out by his own men on his own farm, they would be invalid to anyone other than himself, for fear that he would be accused of loading the results in his own favour. Young, writing in 1771, remarked that 'Mr Bakewell has several comparisons between other breeds of cattle and his own which I purposely omit taking notice of because such experiments are impossible to be accurate from the great differences in certain breeds in feeding, fattening, etc. Besides, supposing such accuracy, still other people, and particularly those of the countries compared, would never give credit to such comparisons unless the very best breeders in the very best countries themselves chose certain beasts to represent their breed. Nor does Mr Bakewell's breed want any such experiments to recommend them'.

In his letters to George Culley, Bakewell expressed the wish that Arthur Young should not be so open in praising him to the press, fearing that it would only help to alienate him from the rest of the farming community, and so damage his cause for the progression of agriculture. This reveals the modest and sensitive side to his nature.

The following is just one of the many and varied trials carried out at Dishley. Three newly-calved cows were individually housed – a Holderness, a Scotch and a Longhorn. The Holderness ate the most food and gave the greatest quantity of milk. The Scotch ate less food and gave less milk but produced the most butter, while the Longhorn ate the least food, gave less milk and butter but laid on the most flesh, thus proving that the Dishley cattle were best adapted for the needs of the grazier in the production of beef.

Always one to be thinking of practicality, a most forward-thinking innovation of Bakewell's was to use his three-year-old heifers for draught purposes, replacing the traditional oxen which were then released from their labours for the purpose of slaughter. Bakewell thought oxen the most unprofitable animals to keep and only retained exceptional specimens for draught or slaughter. Bull calves were generally fattened and killed for veal at around two months of age, producing 160lbs of veal. He felt that more profit could be gained from fattening and killing heifers either before or after their first calf, compared to the costly practice of keeping oxen for slaughter. As soon as heifers came to breeding age they were broken to the yoke, calving for the first time well into their fourth year, this being the accepted age at the time, because it was thought best for the health and strength of both cow and calf. Draught cattle were kept stalled throughout the year, just like horses, and in the summer fed on mown grass while in the winter they had a diet of straw or hay with turnips or cabbages. They were used for all the light carting work, the heavier tasks usually being left to horses. Heifers were easily trained and managed, and from 12 to 20 of them were constantly ready for use. Three of them would work a plough or cart, and they were found to be most tractable and docile. They could work for up to nine hours a day, usually from 6.00 a.m. to noon and, after a two-hour rest, from 2.00 to 5.00 p.m.

Bakewell devised a cattle crush specifically to aid in the shoeing of oxen and heifers, the largest specimens being easily dealt with in this manner. He is said to have refused the sum of 120 guineas for a working team of six heifers.

In all Marshall's travels, Bakewell was the only man of his acquaintance to keep cows solely for the purpose of rearing their own offspring. Elsewhere, with the cows' milk being required for the manufacture of cheese, calves were butchered either at birth or at four days old and sold for the price of 3s or 4s. A ready market for the meat was to be found in manufacturing towns or collieries. Replacement stock were reared by two methods. In the first, the calves were left on a cow for anything from six to 12 months of age, a cow of little value being bought for the purpose. Such a cow would generally rear either one bull calf or two heifer calves. The second and more usual way was to allow the calf to suck from a cow for one to two weeks before being transferred to bucket rearing. The bucket feed was gradually changed from whole milk to a mixture of skimmed milk and ground oats. In periods of milk scarcity, calves in some areas were kept quiet by being dosed with three paste balls the size of walnuts, made from wheat flour and gin. These were administered 15 minutes before a feed and kept the animals in a comatose state most of the time. Calves were reared on the bucket for up to four months.

When we summarise the contribution made by Bakewell to the Longhorn breed we find they were capable of growing fat on less feed than the original stock. They fattened quickly and in more valuable parts of the body, and the flesh itself was improved in its eating qualities. They retained their ability to withstand harsh conditions and poor pasture, justifying Bakewell's faith in the Longhorn to keep itself in good condition on less food than other cattle of equal weight. The only drawback to his practice was the loss of ability to produce enough milk for the dairy. This one point made the choice of the Longhorn untenable for the small farmers of the day. They required a dual-purpose animal and therefore clung to the old, unimproved type.

The various improvements that Bakewell had made to the Longhorn were confirmed as still relevant and evident more than 100 years later, in an 1878 summary.[18] But, when compared to the animals we have today, and the aims we are striving to achieve, the considerations of those days are entirely different. The need for fat meat in Bakewell's time has been wholly eclipsed nowadays by the call for leaner and yet leaner carcases. Today's fashion is for an animal of large proportions, a good frame and bone, with bulls needing to show width, depth and quality, including being level in the back and sound of foot. Good hindquarter development is also keenly sought.

Arthur Young once said of Bakewell: 'I am proud to be able to pay a tribute of respect to the memory of this great man who by his genius has created and scattered over the green hills and verdant meadows of his native land an animal whose beauty and usefulness will ever be a living monument to record his name'.

Bakewell may be said to be the first agricultural patriot, 'for the man who endeavours by incessant exertions to improve the animal creation is as fully deserving of that distinctive appellation, as is the statesman whose exertions are employed in human rights' (*Farmer's Magazine*, August 1842).

Endnotes

1 Lowe, D., On the Domesticated Animals of the British Isles, (1842).
2 Quoted from Pawson, H.C., Robert Bakewell Pioneer Livestock Breeder, (1957), p9-10.
3 Culley G., Observations on Livestock, (1794), pviii-xi.
4 Pawson, H.C., Robert Bakewell Pioneer Livestock Breeder, p30.
5 Pawson, *ibid.*, p66.
6 Information given on a catalogue for the sale of Mr Wyatt's cattle in 1840 to Mr Nugent.
7 Pawson, H.C., *ibid.*, p54.
8 Pawson, H.C., *ibid.*, p53-5.
9 Pawson, H.C., *ibid.*, p54.
10 Young, A., The Farmer's Tour through the East of England, Vol 1, (1771), p110-3.
11 Lowe, D. *ibid.*
12 Pawson, H.C., *ibid.*, p53.
13 Marshall, W., The Rural Economy of the Midlands, Vol 1, p321-2; repeated in Pawson, p55-6.
14 Pawson, H.C., *ibid.*, p56.
15 Longhorns, p542, Chapter VI, *RASE Journal*, (1878) – 'Breeds of Cattle, Sheep, Pigs and Horses'.
16 Culley G., *ibid.*, pxiii.
17 Marshall, W., The Rural Economy of the Midlands, Vol 1, p327-31 & notes.
18 *RASE Journal*, (1878), Chapter VI, 'Breeds of Cattle, Sheep, Pigs and Horses', p542.

CHAPTER TWELVE

Breeders After Bakewell

Pat Stanley, Janet Spavold & Brenton Heazlewood

> His celebrity as a breeder of long-horned cattle will ever shed a halo over his name.[1]

The lack of national recognition of the achievements of Robert Bakewell has been a subject of comment for many years. The following appeared in the *Farmer's Magazine* of August 1842:

> Memoirs of utilitarians are very seldom written. The quiet tenor of their lives, unlike the glare of that which sheds a halo around the heads of the warrior and the statesman, affords but little field, for the display of the talents of the biographer: hence it is, that the real benefactors of mankind are but too frequently passed over in silence, and we walk over the green sward that flourishes where the remains of those whose energies were devoted to the improvement of the race of animals, that contribute largely to add to the creature comforts of man, moulder into primitive dust, without even thinking that we tread on hallowed soil; for the earth beneath which a good man rests, is hallowed by the remembrance of his virtues and his usefulness.
>
> Mr Bakewell, who was the pride of the agriculturalists of the last century, and justly celebrated for his improvements in the breeding of long-horned cattle and sheep, has been strangely neglected by those who ought to have been his warmest and most grateful advocates. But such is the way of the world: neglect, almost universally, is the reward of merit. ...
>
> Cultivators naturally look to the amount of profits that are likely to arise from the practice of any separate system; that of Mr Bakewell did not place him among the wealthy, but had he been less liberal in his ideas, less given to hospitality, less visited by the patrician portion of the community, who entailed enormous expenses upon him, he must have lived and died a rich man, whether it was his desire to do so or not. ...
>
> A practical agriculturalist is of far more value to a nation than even a practical manufacturer, for the benefits the one confers upon the

> human race are lasting – the other but transient; yet both deserve to be honoured by their country …
>
> It ought to have been remarked that the liberality and high mindedness of Mr Bakewell never evinced itself more fully than in his incessant application to the improvement of the Dishley Estate. He was but a tenant, but he yearly increased the value of the land he rented for the benefit of the landlord.[2]

For 200 years we have subconsciously followed Robert Bakewell's methods, and the breeds we have today have been cultivated with the use of the knowledge we have gained from him. Arthur Young, writing after Bakewell's death, said, '... his system was established with such completeness that men forgot not only the existence of any different conditions, but even the very name of the most active pioneer of change'.[3]

Some at least of the first Dishley Society members brought up their sons to follow in their footsteps, maintaining Bakewell's methods and flocks. John Breedon's son, also John (1767-1850, of Rotherby, Leicestershire), was one of them. He was the last surviving member of the Ram Club, formed in 1789 to follow Bakewell's methods into the next generation of breeders. He had bought five Dishley ewes at the 1793 sale of Paget's Dishley stock. He was also part of a group of Leicestershire sheep breeders who 'have agreed to make a public shew of their Rams, at Leicester' in June and at the September Fair in 1795. 'The Rams of each Breeder to be viewed at home as heretofore, any Time previous to or after the Shews'.[4] There were 16 breeders in this group: 3 from Cossington, 2 from Rothley and 1 each from Quorndon, Woodhouse, Kirby, Hoton, Shoby, Rotherby, Gaddesby, Lockington, Hoby, Barkby and Tilton. They had bought some of Bakewell's rams after his death in order to maintain the quality of their New Leicester sheep.

As second-generation sheep breeders, they followed Bakewell's methods in their Ram Club, encouraging potential customers to visit their farms to inspect the rams for tupping that year, but taking the rams to Leicester for the show and the fair. The 16 Ram Club members also followed the original society's secrecy over their rules, which were disclosed by Dixon in the fourth volume (New Series) of the Royal Agricultural Society's *Journal*. They were to

> keep the transactions secret upon their honour. Mr Paget was the president of the club, which held its earlier meetings at the Bull's Head and the Anchor, at Loughborough, alternately, and fined each member a guinea for non-attendance. There were twelve members, and the rules were made and kept with Draconian strictness. No member might sell ewes and lambs to breed from, unless he sold his whole flock or dealt with members only. Only forty ewes could be taken in to keep, and

> those must be the property of one person; not more than two dozen rams could be shown to any person or company at one time; and even members could only show their rams to each other between the 1st and 8th of June, when the general show commenced. On July 8 they were bound to rigidly seal their pens for the space of two months. Certain flockmasters were not to pay less than 100 guineas in their first contract, and after that 30 guineas for wether-getters. Not more than 30 rams might be let by one member in one year; and it was further enacted that there were to be no dealings with flockmasters who showed rams in the market, and that the much-dreaded members of the Lincolnshire Society should not have a ram unless four joined and paid 200 guineas for him.[5]

Although Dixon tells an excellent story, it has proven difficult to verify his information, so caution should be exercised in relation to the above quote. If the Ram Club's aim was to continue the quality of the New Leicester sheep, it succeeded. When John Breedon's flock was sold in 1844 it was advertised as a

> truly valuable Flock of RAMS and EWES, which have been purely bred from the flock of the late Mr Bakewell, of Dishley, for upwards of 60 years.[6]

But not everyone became a disciple of Robert Bakewell and some years later, criticisms came to the fore.

Objections to the Dishley Sheep

Mr Wright of Chesterfield wrote in 1830 that at one point Bakewell's sheep became too small, fine, light-woolled and effeminate, the result of constantly using light-boned, small rams. The great strides forward in mutton production came at no little cost, for milk production fell and, with the short bone of the legs, lambs found it difficult to suckle. When Bakewell realised that his breeding policy was having a detrimental effect on his flock, he made a judicious cross to a larger ram and thus amended the problem. The length of leg was increased again, making it easier for the nursing lamb and also making the walk to market less stressful for culls. Thus, he left at his death two very different types of Leicester sheep.

Dickson's principal objections in 1805 were that the New Leicester fattened too much, which affected the flavour of the meat; that the quantity of their wool was deficient and they did not thrive when folded. Some people thought they were too small, and their skins were too thin as well. A more serious objection was that they were not reliably prolific, and that they seldom had twin lambs.

In 1809, Pitt reported a remark made by Bakewell to a gentleman who was complaining 'That his mutton was so fat that he could not eat it'. 'Sir' was the reply, 'I do not breed sheep for gentlemen but for the public'. Not many years after his death this answer would not have served him, for the public as well as the gentlemen gained a taste for leaner mutton.

John Lawrence, writing in 1809, was clearly no follower of the Leicester or Dishley sheep. He reveals that the first thing that sprang to mind on inspecting a prize winner of this breed was 'the facility with which a certain form is converted into a living grease tub'.[7] He also related the following:

> I once dined where a leg of prize mutton made a dish. It weighed 17lb. At my desire, the fat which dripped in cooking was measured, and it amounted to between two and three quarts; besides this, the dish was a mere bog of loose, oily fat, huge deep flakes of it remained to garnish that which we called, by courtesy, lean, being itself also thoroughly interleaved and impregnated. It struck me forcibly, that an addition of a reasonable quantity of bone, and an exchange of seven or eight pounds of fat for lean meat, would have contributed much to the actual value and good character of the joint; but little of which was eaten at the first table, and I have reason to think, not much more at the other tables.[8]

Lawrence researched the subject and found that by 1805 labourers usually rejected over-fat mutton if there was a choice, for from necessity they had to eat up all they bought, whereas the rich could admire, taste and throw away that which they did not require. He laid the charge that the Dishley sheep provided rich men's mutton. 'The late fashionable excess in fattening even pork and bacon, has generally disgusted the labourers; in consequence, much wanton waste has ensued, and the dealers in such articles, now find difficulty in their disposal'.[9]

As observed above, not everything written about Leicester meat has been favourable as this crank story indicates.

> DANGER OF EATING LEICESTER SHEEP
>
> An astonishing discovery, made by Mr Justice Ward of New Zealand, is recorded in the *Wellington Evening Post*. The learned gentleman has ascertained that crime is attributable, in a great measure, to the consumption of Leicester sheep, and he sets forth his theory in the following words:
>
> 'As a machine for the production of wool and tallow, I believe the Leicester sheep to be unrivalled; but the creature is utterly unfit for human food. The human stomach was not intended to be turned into a

tallow vat, except in the polar regions. I attribute a large proportion of the crime in the colony to the abominable animal, the Leicester sheep. In the best form it is simply an animal mass of tallow. Its consumption leads inevitably to dyspepsia, and dyspepsia to drink. Drink takes a man direct to the assizes, and the assizes lead him to lower depths still'.

The Post suggests that 'if eating Leicester sheep leads a man to drink, and to commit assaults, burglary, forgery and murder – because that in truth is all implied in the assertion of Mr Justice Ward – then an Act of Parliament should be passed forthwith, providing that Leicester sheep should henceforth be disposed of solely by being boiled down'.[10]

Robert Fowler

Of the many breeders who followed hot on the heels of Bakewell, Robert Fowler, of Little Rollright, on the borders of Oxfordshire and Warwickshire, became the most prominent. After purchasing two cows from Mr Webster and hiring Bakewell's bull *Twopenny* he set the seal on what was to be one of the most successful and famous herds in English agricultural history, second only to Bakewell's, for when his Rollright herd of Longhorns was eventually put under the hammer it was to become the first pedigree cattle sale ever recorded. (See Appendix 2)

Robert Parry of Shipston-on-Stour, writing on 14th May 1791, informed us that from the mating of the Webster cows with *Twopenny* came the famed matriarchs of his herd, *Old Longhorned Beauty* and *Old Nell.* He hired several bulls from Dishley, finally ending with *D* in 1778, a very inbred grandson of *Twopenny*. Fowler sealed the fortune of his herd when he bred *Shakespeare*, who was by *D,* the grandson of *Twopenny*, out of a daughter of *Twopenny*. From this point on he never looked outside his own herd for sires. *Shakespeare* was to occupy a position among the Longhorns equivalent to *Favourite* and *Comet* in later years among the Shorthorn tribe.

William Marshall, writing in 1790, gives us the following description of *Shakespeare* and the Rollright herd.

> This bull is a striking specimen of what naturalists term ACCIDENTAL VARIETIES. [Although he is greatly inbred] he scarcely inherits a single point of the longhorned breed, his horns excepted. When I first saw him in 1784, then six years old, and somewhat below his usual condition, though by no means low in flesh, he was of this description.
>
> His head, chap and neck, remarkably fine and clean. His chest extraordinarily deep; his brisket down to his knees. His chine thin; and rising above the shoulder points; leaving a hollow on each side, behind

> them. His loin, of course, narrow at the chine; but remarkably wide at the hips; which protuberate in a singular manner. His quarters long, in reality; but, in appearance, short; occasioned by a singular formation of rump. At first sight, it appears as if the tail, which, stands *forward*, had been severed from the vertebrae, by the chop of a cleaver, one of the vertebrae extracted, and the tail forced up to make good the joint: an appearance, which on examining, is occasioned by some remarkable wreaths of fat, formed round the setting on of the tail: a circumstance, which, in a *picture*, would be deemed a deformity; but, as a *point*, is in the highest estimation. The round bones snug, but the thighs rather full, and remarkably let down. The legs short and their bone fine. The carcase, throughout, (the chine excepted), large – roomy – deep and well spread.
>
> His horns apart, he had every point of a Holderness or a Teeswater bull. Could his horns have been changed, he would have passed, in Yorkshire, as an ordinary bull of either of these breeds. His two ends would have been thought tolerably good; but his middle very deficient. And I am of opinion, that had he been put to cows of these breeds, his stock would have been of moderate quality. But being put to cows, deficient where he was full (the lower part of the thigh excepted), and full where he was deficient, he has raised the longhorned breed to a degree of perfection which, without so extraordinary a prodigy, they never might have reached.
>
> No wonder that a form so uncommon should strike the improvers of this breed of stock; or that a carcase they have been so long striving in vain to produce, should be rated at a high price. His owner, however, happened to be among the first of his admirers and could never be induced to part from him, even for a season; except to Mr Princ[s]ep; who had him two seasons, at the extraordinary price of eighty guineas a season. A price at which no other bull has yet been let.
>
> This extraordinary animal is now (1789) eleven years old, and firm in his constitution; but so lame, in his hindquarters, as to render him at present, and during the last season, entirely useless.[11]

In 1793 a bull called *Shakespeare* was sold at the dispersal sale of Mr T. Paget of Ibstock, Leicestershire, for the sum of 400 guineas. Although in recent years it was assumed that this was the *Shakespeare* of Mr Fowler's, the above statement of his lameness gave little credence to the tale, and on finding the records of the Paget sale it was seen that the bull sold in 1793 was bred by Fowler, by his bull *Shakespeare* out of *Young Nell*. He was bought by a syndicate of local gentlemen and the price which was paid indicates the great regard which was held for his sire. The gentlemen after their purchase charged 25 guineas per cow for his services, but Paget retained the right to send two cows to him yearly.

Fowler was left in a very strong position after the passing of *Shakespeare*. He had many valuable females by him as well as a bull by the name of *Garrick*, whose dam had been got by Robert Bakewell's *Hampshire Bull*.

Marshall witnessed 25 cows at Fowler's and found many of them to be of an extraordinary stamp, especially in the fineness of the fore-end and width and fatness of the hindquarters. 'A daughter and sister of *Shakespear*, being got on his own dam, is amongst the first of his herd: another evidence of the good effect of breeding from the same family'.[12]

Initially Fowler conducted his business on the old principles of selling, and not the new way of letting. He not only sold his surplus bulls, but also heifers that were not required for his own dairy purposes. During the years 1785-89 Mr Coke of Norfolk had all the heifer calves he could spare at 10 guineas each, and in October 1789 he declined an offer of 500 guineas for 10 bull calves by *Garrick*.[13]

During the latter part of his life, it was said of Fowler,

> If he did not buy, on the other hand ... he would not sell (at least, his cows) while there was a chance of breeding; and the Arab sticks to his mare no more lovingly than did Mr Fowler to his Longhorn queens after the sale of the three Twopenny heifers ...[14]

He had sold these heifers to Mr Guy of Toddington for 80 guineas in 1771 as the foundation stock for Guy's herd. Fowler, regretting his sale to Mr Guy, almost immediately offered to buy back one of the heifers, *The Painted Lady*, at the amount he had received for the three, but to no avail. At one point, Fowler refused an offer of 1,000 guineas for three cows and a bull, almost certainly as a direct result of the regret he always felt for that transaction with Mr Guy.

Mr Guy, although unfortunate in the choice of some bulls, which he afterwards made use of, and therefore not so successful as he might otherwise have been, the prices of his stock at his dispersal sale in April 1790 are enough to convince us of their superiority to most others, with some of his cows by a bull of Mr Fowler's selling for up to 40 guineas.

It was eventually the auctioneer's hammer that broke the bond between Fowler and the mothers of his herd (see Appendix 2) for 'a most true and exact record of this historic sale of pedigree stock'. The prices obtained reached as high as £220 10s for *Sultan*, a 2 year old by *Shakespeare*.

At the request of some friends of Fowler, and for others wishing to possess a copy of the extraordinary sale, a reprint was devised by the auctioneer in May 1791. It did not go to print until a full 66 years later, in 1857. The probable reason for this was that, just after the sale, requests were made for the reprint to carry copies of engravings taken from paintings Fowler had commissioned of his cattle. This proving difficult, the task was shelved at the time. It is also worth noting that the stock not only produced the extraordinary prices shown at the sale, but that offers of a considerable advance were made afterwards to many of the purchasers.

Thomas Paget

In 1793, Mr Thomas Paget of Ibstock, Leicestershire, on deciding to retire to a smaller farm at Market Harborough, put his herd of highly-improved Longhorns up for sale by auction. The highest price achieved was for *Shakespeare* out of his father *Shakespeare* and *Young Nell* at 400 guineas (see Appendix 3 for a listing of the prices achieved.)

Shakespeare, lot 34, was bought by a partnership from Leicestershire and afterwards served cows at 25 guineas each. Pitt saw him in use at Mr Stone's of Quorndon. All the heifers offered at the sale were by lot 34 and, together with the cows, had been bulled by him, except for lot 52 which had been bulled by a son of *Garrick*. Five four-year-old bullocks, perfectly quiet for work plus their working harness, made 11 guineas each. The 64 lots of Longhorns averaged £31 15s 3d, but unless they were by *Shakespeare* they made low figures.

Paget was also an 'ardent and successful' breeder of New Leicester sheep and an associate of Robert Bakewell's. Paget sold his flock at the same time as his Longhorns, on 16th November 1793. The top sheep made 62, 52 and 45 guineas – prices still within the range of comprehension of sheep farmers over 200 years later. Overall, 130 females, including 20 shearlings, averaged £25 16s 11d each and grossed £3,200, an enormous sum at the time.

Mr Thomas Prinsep of Croxall

Pitt, in his Agricultural Report of Staffordshire, of 20th May 1794, said:

> May 20, 1794, viewed the cow-stock of Mr Princ[s]ep of Croxall. His land is mostly, but not wholly, in Derbyshire, and therefore more properly belonging to the Report of that county. I shall just observe, that they are of the long-horned breed, and by long attention have been brought to a very high degree of superiority; large, thick, heavy, and well made, with a pretty good show for milking, and such a disposition to fatten, that Mr Princ[s]ep observes, the young stock are obliged to be almost starved by short pasturage, otherwise they run fat and never stand the bull: the cows give upon the average about eight quarts of milk each, which the owner thinks equal, from its superior quality, to a much greater quantity from inferior breeds. Mr Princ[s]ep's bull, named Bright, which always has, and will invariably be kept for his own stock, is a majestic noble animal; large, thick, heavy in the valuable points, with the least imaginable proportion of

Engravings of the bull Bright *and a cow, the property of T. Prinsep in 1794.*

> offal; with a skin handling soft and sleek. This majestic animal is so gentle and docile, that three or four persons at once may handle him without the least sign of ferocity or even notice, on his part. Bright-Eye, the son of Bright, now three years old, is a beautiful and most complete animal.[15]

Marshall, writing in 1784, informed us that *Bright* was by *Shakespeare* and therefore of the celebrated *Twopenny* line.

Not everyone agreed with Pitt's assessment of Prinsep's herd; 12 years later Mr Curtis of Lynn Regis, Norfolk, visiting Mr Prinsep said,

> Mr Prinsep has, perhaps, the first dairy of cows in the county, where that pre-eminence is defined to mean symmetry, size, and aptness to fat. These milking cows are, every one of them, fit for the butcher. The bailiff told me his master once killed a cow, seven years old, which weighed 104 stone. ... Mr Prinsep may be proud of what he calls by the misnomer of a dairy of milking cows, but they were more fit to be turned into Christmas beef than make butter and cheese.[16]

In 1784 Marshall said that a German prince was probably less respected in the environs of his residence than Mr Prinsep in the neighbourhood of Croxall. On a later visit, Marshall wrote that, 'Although the cows of his breeding were of very fine quality in the past, his present herd show evident marks of improvement. Every cow and heifer of the Shakespear blood is distinguishable at sight; by the extreme fineness of the forend, the width of the hips, and the formation of the rump'.[17]

Prinsep had hired *Shakespeare* from Mr Fowler for the amazing sum of 80 guineas per season for two seasons. Prinsep's foundation cow, also by the name of *Bright*, had been purchased from a Mr Chadwick of Castle Bromwich, and was by Webster's bull *Bloxedge*. The Prinsep herd was sold on 25th-26th September 1811 at Croxall, Derbyshire, the sale conducted by Mr Boot. The 65 head of cattle averaged 61 guineas each, the best prices being 220 guineas for the cow *White Lupin* and 160 guineas for a bull calf under six months. (See Appendix 4 for a list of prices achieved.)

Prinsep's grandson, Mr T. Levett Prinsep, reverted to Longhorns in 1873 when he purchased foundation animals for his herd at Mr Chapman's sale. He went on to preside over the meeting held at Birmingham Fatstock Show in 1876 to establish a Longhorn Herd Society and Herd Book.

Richard Astley of Odstone Hall

Mr Richard Astley of Odstone Hall, Market Harborough, Leicestershire, was another noted breeder in the early years of the breed's zenith. He was a large exporter of Longhorns to Ireland, where several of his bulls were held in very high regard.

His fame was mainly based on being the owner of a celebrated cow which was exhibited all over the country, being transported by horse-drawn van.[18] She must have been an exceptional animal and it is unfortunate that her pedigree, with those of the rest of Astley's cattle, has not survived.

Other than the fact that he was a large buyer at the Rollright sale in 1791, and therefore dipped deeply in the *Shakespeare* blood, no other records of how his herd was bred have been found. Thomas and William Knowles, his neighbours, were partners with him in the Rollright purchases.

As indicated above, there is little evidence that Astley was a member of the Dishley Society from its formative years; he supposedly later fell foul of its rules and was ejected. Notwithstanding this doubt, he was one of the founder members of the Smithfield Club, being at the inaugural meeting in London, in December 1798.

In 1797, *Washington*, a bull, and *Lady Washington*, the bull's dam, both from Rollright, were in his possession but pensioned off. He kept a dairy of 30 Longhorn milking cows, which were the largest and best, except for those of Mr Prinsep, that William Pitt, the agricultural author, had ever seen.

Richard Astley's farm consisted of 500 acres and belonged to his brother, Dugdale Astley, of Everly, Wiltshire. Pitt declared the farm to be 'very productive, the crops being generally very full and good. Mr Astley has removed the ant-hills from his pasture, and levelled the surface of the meadow land, by removing uncouth ill-formed banks, taking care to reserve the surface soil in its proper place'.[19] One wonders if this work was carried out after Astley had seen such procedures at Dishley.

Following the passing of the legendary breeders, Webster, Bakewell, Fowler, Prinsep and their fellows, the Longhorn rapidly declined. Their empire began to crumble and decay, not totally from the onslaught of the up and coming Shorthorn, but rather from within their own walls. By 1815 their battle was all but lost and the 'Druid' (H.H. Dixon) came very close to the truth when he wrote in 1815, 'When the Durham ox (Shorthorn) began his six years of caravan (exhibition) life the fate of the Longhorn was sealed'.[20]

By 1834, it was noted that so far from wishing to emulate Bakewell, 'although the general advantage which has resulted from crosses with the new stock must be acknowledged, yet some breeders are now as anxious to avoid all mixture of the Dishley blood, as they formerly were to obtain it'.[21]

Between 1859 and 1861, Mrs Beeton started publishing her ideas on Household Management as monthly supplements in *The English Woman's Domestic Magazine*; these were then brought together as a single volume in 1861 with subsequent reprints and enlargements. Having written at length about the virtues of the Shorthorn, she dismissed the Longhorn in a few sentences, thus:

> LONG-HORNS. – This is the prevailing breed on our midland counties and in Ireland; but they are greatly inferior to the short-horns, and are fast being supplanted by them. Even where they have been cultivated with the nicest care and brought to great perfection, they are inferior to the others, and must ultimately be driven from the farm.[22]

During the years when the Longhorn fell from favour, it could quite easily have slipped out of existence, but it still held captive a few faithful hearts who, having seen the mistakes of their predecessors, endeavoured to do better for their chosen breed. Biding their time, they waited for the breed's re-emergence to popularity in the mid-19th century. The ancient fame and glory of the Longhorn was once more to cast its spell over many of those who had deserted it. The 'Leicester Curly Coats' were again remembered for their hardy constitution and ability to thrive on poor pasture. The light of former days was beginning to shine once more, and from sales such as Mr Chapman's of Upton in 1873, many new herds were to spring to life (see Chapter 13). The second age of the Longhorn had come.

Endnotes

1 *The Farmer's Magazine*, August 1842, Second Series, No. 2, Vol VI, p82.
2 Ibid., p81-5.
3 Pawson, H.C., Robert Bakewell Pioneer Livestock Breeder, (1957), p45.
4 BNA, *The Stamford Mercury*, 15 May, 1795.
5 Coleman, J., (ed.) The Sheep and Pigs of Great Britain, Part II, (1877), p31.
6 BNA, *The Stamford Mercury*, August 1844.
7 Lawrence, J., A General Treatise on Cattle, The Ox, The Sheep, and the Swine, (1809), p332.
8 Lawrence, J., A General Treatise on Cattle, The Ox, The Sheep, and the Swine, (1809), p328.
9 Lawrence, J., A General Treatise on Cattle, The Ox, The Sheep, and the Swine, (1809), p329.
10 *The Warwick Examiner and Times*, Queensland, 25 March 1876, in Heazelwood, B., *The English Leicester Sunday Roast*, n.d.
11 Marshall, W., The Rural Economy of the Midland Counties, Vol 1, p322-5.
12 Marshall, W., *ibid.*, p326.
13 Marshall, W., *ibid.*, p326 and notes.
14 Young, A., Annals of Agriculture, Vol XV, p391, & Fitt, Neville J., *The Agricultural Gazette*, 27 March 1876, p396.
15 Pitt, W., General View of the Agriculture of the County of Stafford, (1813), p176.
16 Lythall, J.B., The Long-horn Herd Book, p13 – Reprint, & Fitt, Neville J., *ibid.*
17 Marshall, W. *ibid.*, p326.

18 Lythall, J.B., The Long-horn Herd Book, p12 – Reprint, & Fitt, Neville J., *ibid.*, p397.

19 Pitt, W., General View of the Agriculture of the County of Leicestershire, including a survey of the County of Rutland, (1809), p39.

20 Fitt, Neville J., *Journal of the RASE*, Series 2, Vol 12, 'Longhorn Cattle: Their History and Peculiarities', p460 quoting the *Mark Lane Express*, 7 January, 1967.

21 Library of Useful Knowledge, British Husbandry, Volume the First, (1834), p28.

22 Mrs Beeton's Book of Household Management, (Chancellor Press, 1986), p273.

CHAPTER THIRTEEN

The Chapman Family of Nuneaton

Pat Stanley

> There is something very quaint and grand in a field full of Longhorns, and Mr Chapman's grand old bull, the Earl of DERBY, with 32 milking cows around him in the great meadow, was a sight ever to be remembered by all admirers of live stock.
>
> (*The Druid – H.H. Dixon*)[1]

One of the oldest herds of Longhorns on record was the Upton herd, maintained by the Chapman family near Nuneaton for 117 years. A George Chapman went to live at Upton in 1745 and started buying Longhorn cows in 1756. He hired bulls, first from Robert Bakewell and later from Robert Fowler of Little Rollright in Oxfordshire. The last bull he ever hired was Bakewell's noted *Twopenny*, afterwards using his own home-bred sires. He acquired the reputation of being a breeder of sound judgement and Bakewell once said of him, 'that he had one of the best herds of Long-horns that he knew anywhere'.[2]

The Upton herd had, since its foundation in 1756, been carefully bred from the best strains of blood that could be procured. It rightly took its place as one of the oldest and best herds in existence. Many people went to the sale as first-time purchasers looking to lay foundations of future herds.

> It is well known that similarity of characteristics in cattle always denote purity of race, and careful selections of a recognised type during a long course of years is the only way to secure this greatly desired effect. Like the Drakelow Baronet, ... the Chapman family succeeded in breeding 'a dairy of Longhorn cows, alike in colour and shape', for the 51 offered at Upton showed a remarkable likeness to each other, all possessing the orthodox white back.[3]

The herd was bred for its good milking qualities and for the use of the milk in cheese-making.

A few years before the dispersal of the Chapman herd, a contest was held with a friend, a Mr S. Craven Pilgrim, a Shorthorn breeder, whose animals hailed from the famous Bates blood, and were cultivated for their milking properties. At the height of the summer grass in June, both men selected six of their best cows. The Shorthorns produced 152lb of milk against the Longhorns' 135lb, but the weight of curd for the Longhorns was 19½lb as against 14½lb for the Shorthorns. The trial was repeated again in September of the same year, the whole of Mr Pilgrim's cows, numbering 36, against the 32 of Mr Chapman's Longhorns. The Shorthorns produced 605lb of milk, making 66½lb of curd. The Longhorns produced 553lb of milk, making 69lb of curd.[4]

On the death of George Chapman in 1802, the herd was taken over by his son, Samuel, and later by his grandson, Richard Hemming Chapman. They continued to cultivate their herd's usefulness for dairy purposes, for which it was famed. The name of R.H. Chapman became well known in the show ring, the herd having many successes over a period of 30 years. Calke Abbey records for the late 1870s reveal that Mr Chapman on three occasions in the past few years had won the laurels from the Shorthorns, although the animals had been bred mainly for dairy and cheese-making purposes, and not for exhibition especially.[5]

In 1850, Mr Chapman's bull *Marquis of Exeter* took the first prize of £20 at the Royal Agricultural Society Show, held at Exeter. In 1851, his *Lord Windsor* took the prize of £10 at the Royal Show held at Windsor, and at Birmingham in 1853 he took £20 and a silver medal.[6]

When Chapman exhibited his five-year-old bull *Old Sparkenhoe* at the Plymouth Royal Show in 1865, he collected the first prize of 15 guineas. The men of the west were reported to show how impressed they had been with the

Sparkenhoe, bred by Colonel Inge in 1860 and winner at the RASE in 1865.

Marquis of Exeter 15 years before, by exclaiming that *Sparkenhoe* was the same bull – thus suggesting how true to type both bulls must have been, as well as how vividly the illustrious stranger of earlier years was remembered.[7] At the later show, *Old Sparkenhoe* was considered a great curiosity and became the main attraction in the cattle lines for sightseers. Many people in that area had not before seen Longhorns and it was wondered what bovine tribe the bull could belong to. A reporter, giving an account of the stock classes for the day, said that 'Not one person in 500 could guess his breed'. The catalogue for the show states that *Sparkenhoe*'s sire was *Tom*, bred by Col. Inge, a Longhorn breeder of great note in Staffordshire, and his dam was *Fillpail*, also a prize cow at the Royal and bred by Mrs Baker of Rollright, Oxfordshire, where the great Longhorn breeder Fowler previously lived.

R.H. Chapman was one of the founder members of the Longhorn Cattle Society in 1876, even though he had, after great deliberation, retired in 1873, and a sale had been called for 16th December that year. The event was extremely well attended, with one estimate being that 600 were present. Amongst the distinguished company were Lord Howe (the landed proprietor of the district), Mr Heygate (MP), Mr T. Cox, Mr Pickering (agent for Lord Bagot), Mr Bosworth (agent for Sir John Crewe), Mr T. Nichols (agent for Mr T. Prinsep), Mr J. Burbery, and Mr Hanson Sale of Atherstone. The event was recorded in detail as described below.[8]

> The tables for luncheon were set in double lines in an outbuilding amidst a profusion of banners and evergreens. The pillars and roof timbers were decorated with the heads or skulls and polished horns of prize-winning specimens of the Longhorn breed. On the walls were hung many oil paintings of past favourites, amongst the most outstanding being *Shakespeare*, the brindled bull, which made 400 guineas in 1793 at Mr Paget's sale. The picture was the property of Mr T. Paget of Humberstone at the time and had been loaned for the occasion.
>
> Hanging near to *Shakespeare*, and looking very like him, was a portrait of Mr Fowler's *Brindled Beauty*, lent by Mr Warner, this cow having made £273 at the Rollright sale in 1791. Then, by way of contrast in depth of carcase and colour, was an oil painting of a pair of Mr Warner's red-sided white cows. There was a portrait of a cow bred by Mr T. Prinsep, with horns curling up several feet in advance of her head, and two other pictures showing specimens of the Croxall herd which was dispersed in 1811 when prices were very high.
>
> It was noted by onlookers that if these paintings were true likenesses of their subjects, then the Longhorns of Mr Chapman's breeding had made great strides since that time. The comment was made that the breed was much improved in a hundred years and that *Shakespeare* and

This steer was bred and fed by R.H. Chapman, who considered him one of the best of the breed ever exhibited. In 1863 he won £25 and two silver medals. He was sold to the butcher at three years eight months for 50 guineas.

Brindled Beauty of former days were no more to be compared with the *Lady Calke* or *Lady Rollright* of the later age than 'chalk is like cheese'.

An exact likeness, again in oils, was on display of Mr Chapman's *Old Sparkenhoe* bull which won first prize at the 1865 Royal Show at Plymouth. Another very striking canvas portrayed the Upton herd's *Marquis of Exeter*, which first appeared at the Royal Show at Exeter in 1850. In 1862 and 1863, Mr Chapmen won first prizes at London and Birmingham with oxen which were both shown to perfection on one canvas, displaying proportions that were firm evidence for the assertion that one of them was the best Longhorn ever exhibited up to that date.

Lord Howe occupied the chair for the proceedings and, reflecting the interest shown in the event, letters of apology, regretting inability to attend, were received from Lord Leigh, Mr C.N. Newdegate (MP), Mr A. Pell, MP, the Marquis of Queensberry, Lord Westmeath, Sir John Crewe and others.

Mr Heygate (MP), proposing the toast to 'The breeders of Longhorns', said that 'All in attendance that day would regret the breaking up of this noble herd which has become so famous in history'.

Lord Howe, proposing the main toast of the morning, 'Success to Mr Chapman', said he did so with mingled feelings of regret and satisfaction – satisfaction in having to uphold the reputation of one who through life had proved himself a straightforward and honest neighbour, and regret at the loss of one who had endeared himself to them by so many manly qualities. He was perfectly certain that there was not a soul within ten miles of Upton who would not vouch for the intense interest and enthusiasm he had thrown into everything he took in hand. He regretted his loss from the neighbourhood very much, and he wished him all the success in his future undertakings that an honest man deserved. The herd had descended from father to son for several generations and, while he regretted the loss the neighbourhood would sustain through Mr Chapman's retirement, he wished him all the success, prosperity and usefulness that could attend him.

Mr Chapman, who responded with considerable feeling, said the present position was forced on him after very serious consideration. He had had many thoughtful days and sleepless nights in coming to the conclusion he had arrived at, and though he had to take that public occasion in expressing his obligations to his landlord, who had even allowed him to choose his successor, he took the liberty of saying that farming was not what it was. He said that some of his friends had asked how he could break up the home and herd of his ancestors? His reply was that labour was now a most serious item in expenditure, making cheese production more difficult and less profitable. He praised the men who worked for him as being a well-conducted and skilful body of men, but past middle age, and it was in vain that he looked for the new generation that was to take their place. He said it was his intention to retain all the young stock and so would not be losing his links with the breed entirely, when he retired to St Asaph in Wales.

Earl of Upton 7th, calved March 1872 and bred by R.H. Chapman.

> Mr J.H. Burbery gave an address which was very sound. 'Do not starve your animals, but do not pamper them; give them good, wholesome food, with plenty of fresh air and water; keep them not too closely housed but let their natural hair protect them; and then you will find, as I have done, that for general purposes no breed will beat the Longhorn'.
>
> The sale then proceeded with the top price of 50 guineas going to the two-year-old prize-winning heifer, *Lady Calke*. The sale was conducted by Messrs Lythall and Clarke and the best prices achieved were 50 guineas for the two year old *Lady Calke* and 46½ guineas for Lady Rollright, a prize heifer at Birmingham. (See Appendix 5)
>
> The yearlings and calves were reserved and subsequently sold as two-year-olds during the Birmingham Show weeks in 1874 and 1875. In 1874, the 12 heifers reached an average of nearly £37 and *Brindled Beauty*, the star lot and said by Mr J. Neville Fitt to be one of the most beautiful heifers he had ever seen, was knocked down to Mr Townley-Parker at £67 4s. In 1875, 15 very even heifers and in better condition than those of the year before, but with no *Brindled Beauty* to outshine the rest, averaged £34 15s with the top price of £49 7s being for *Uplands Last Link*.

Considering the smaller number of Longhorn buyers at this period, these prices compared very favourably with the large number of Shorthorn sales held during that time. The 78 head of cattle which Mr Chapman sold over a two-year period averaged £34 to total £2,700. And he still retained a few heifers to breed from in his retirement.

Endnotes

1 Fitt, Neville J., 'A History of the Longhorn Cattle', *The Agricultural Gazette*, 27 March, 1876, p398.

2 Lythall, J.B., The Long-horn Herd Book, p11 – Reprint.

3 Lythall, J.B., *ibid.*, p26.

4 Fitt, Neville J., *Journal of RASE*, 2nd Series, Vol 12, (1876), p483. Quoting an article by H.H. Dixon in the Mark Lane Express, 7 January 1867.

5 Lythall, J.B., *op. cit.*, p39-47 and newspaper cuttings in a scrapbook at Calke Abbey.

6 Lythall, J.B., *ibid.*, p37.

7 Lythall, J.B., *ibid.*, p40.

8 A shorter version is by Fitt, Neville J., *The Agricultural Gazette*, 27 March, 1876, p397.

CHAPTER FOURTEEN

The Longhorn Revivalists

Pat Stanley

> The Longhorn may now be said to have become something of a fancy beast in England.
>
> (*Sporting Gazette*, 1874)

Any livestock breeder would testify to the pleasure gained from viewing a fellow-breeder's stock, be it on their home ground, or at a show or sale. This phenomenon is not restricted to our own time, but it is one which has taken place since men first kept animals. During the 19th century, when the Longhorn had taken a great dive in popularity, its cause was championed by a few brave men who were willing, and admittedly able, in monetary terms, to fly in the face of popularity and carry on with the breed of their forefathers.

Wonderful records of herd visits have come down to us from this period, courtesy of some excellent articles in publications such as the *Sporting Gazette* and *Bell's Weekly Messenger*. Even then, it was a rarity for a Longhorn herd to feature in their pages. It is therefore a very special privilege to join with these gentlemen on their expeditions, to those hallowed grounds, and to be given our own guided tour of herds which existed so many years ago. But we start with an extract from *The Agricultural Gazette* of 13th March 1876, advertising an important sale of Longhorn cattle belonging to Mr R. Brown of the Old Manor House, Farewell.

> **LONGHORN SALE**
>
> The breeders of this old established sort will probably be found mustered in force at Farewell, near Lichfield, on Wednesday, March 15, when a herd bred with the greatest care for purity of pedigree and great milking qualities, the property of Mr R. Brown, whose health compels him to relinquish business, will be offered for sale. The herd came into possession of the present family, through Mr J.L. Brown, in 1807, and was founded at an earlier date than that, the farm having been in possession of the family for 300 years. The late Mr J.L. Brown was a successful exhibitor for many years, and a winner of no less than 10 cups at the Royal and local shows; and amongst the animals now offered will

> be found the hardy constitution and other valuable qualities for which this breed of cattle is especially prized. Space will not allow us to give a list of the bulls that have been of service at Farewell during the century, but we may say that the most noted of their day are, with exceptions, to be found in the list of sires there used, and amongst their breeders are to be found the names of Meek; Robinson, of Coton; Prinsep, of Croxall; Ward, of Tixall; Astley, of Odstone; Smith, of Smitterfield [Snitterfield]; Salisbury, of Dordon; Shaw, of Fradley Old Hall, near Lichfield; Chapman, of Upton; Burbery, of Kenilworth; Warner, of Weston; and Twycross, of Canley. In 1841, Mr Brown's father sold to the Hon. Mr W.B. Nugent, of Higham Grange, near Hinckley, six Longhorns which averaged £82 8s 2d, amongst them being the SHAKEPEARE bull, £127, and the *Cherry* cow, £100. The herd has for many years been bred entirely with a view to producing good milkers rather than show animals, and this is a fact that purchasers requiring really useful dairy stock should bear in mind.[1]

The herd having been in the possession of Mr R. Brown or his family for 70 years, most of the principal names in the Longhorn world were present or represented at the sale. The prices achieved were not so high as at Mr Chapman's sale (see Appendix 5), perhaps because the herd had not recently been bred for the show ring, or for the characteristic finching or white mark on the back, which is much sought after today. Dairy qualities were of greater importance for him than colour, and a bystander at the sale gave witness to the superior fleshing qualities of the animals, 'While their milk was so good, the cheese made from it was vastly superior to ordinary cheese'. Cows made up to 26 guineas, heifers to 24 guineas and calves to 10 guineas. Purchases at the sale included the bull *Nelson* to Mr W.T. Cox for £35 10s; the bull *General* to Sir John Crewe; and the heifer, *Lady Prinsep* to Mr T.L. Prinsep for £25 4s. Cows averaged £20 each but were only in store condition.

Mr Brown had used a bull called *Washington* by Mr Meek's *Washington* out of Prinsep's *Brindled Beauty* from 1808 to 1812.

Brown won seven £10 silver cups and three £5 cups at the old Staffordshire General Agricultural Society Shows at Lichfield, when competing against Herefords, Shorthorns and Devons, often with 20 entries per class, from 1813 to 1820. Prizes were also taken at the Royal Shows at Derby and Shrewsbury.

A friend of Mr J. Neville Fitt's attended the sale and reported to him:

> The great revival in favour of what is now called the old Staffordshire Longhorn induced me to go and see a herd sold on its native soil. The morning, like all the rest in this month of March, was a truly cold and wet one; however, I resolved to make the best of it and, after admiring the

> lovely cathedral of Lichfield, went on my way to Farewell; on the road I was joined by a working man and we soon got into conversation about the stock to be sold. He said that Mr Brown's were the only beasts of the sort about, and when pressed for the reason of this difference in farmers' tastes, added: 'Why the people here know nothing about them, they all run after the new fashioned sort, but Mr Brown has tried them and knows they are best for him'. Being questioned as to the reason why they were better, he said: 'Why, they give such good and rich milk, that the cheese made from this is as much different as between chalk and cheese, and they are such good feeding stock, and are not gluttons at eating, but will eat anything you give them and, what is more, they will do on poor land, and this Mr Brown has found out many years ago. I have seen them for years and know it is quite true'. I think this is a good character reference of the breed, spoken without any interest to serve, and after a real honest Staffordshire fashion.

The Calke Abbey Estate

In the south Derbyshire village of Ticknall lie the gates to the park which surrounds Calke Abbey, the seat of the Harpur-Crewe family since 1622. The history of Calke can be traced back to the 12th century when it was selected by Augustinian monks as the site for a priory, but it did not come into its present name of Abbey until 1808. It is now owned by the National Trust.

The two members of the family of interest in the history of Longhorns are Sir George Crewe, who came into his inheritance in 1819, and his son, Sir John, who succeeded him in 1844.

Sir George took his responsibilities as 8th baronet, and one of the largest landowners in the country, most seriously and devoted his time, as steward of the family estates, to the welfare of his tenants and land. He brought about great improvements, both to the land and to the conditions in which tenants had to live. That was something which had been ignored by the generations who had gone before him, their main interests being of a sporting nature.

Sir George was living at a time when gentlemen had become keen enthusiasts of agriculture, their imaginations fired by the great improvers who had gone before. Sir George had a genuine admiration for Robert Bakewell and his Longhorn cattle and collected as much information and as many mementoes as he could gather on the great man. He also built up a large and very fine herd of Longhorn cattle. Although his improvements were largely confined to the outdoors, Sir George purchased some remarkable paintings, including some by John Ferneley, the Leicestershire horse artist. Also amongst the collection is a fine painting of a Longhorn bull at Dishley Grange by John Boultbee, Boultbee being the only artist known to have painted Bakewell himself. Sir George carried on with Longhorns for a good many years, until they finally gave way to the breed very much at its zenith, the Shorthorn.

Sir George died at the age of 48 in 1844, honoured and beloved by all who knew him. He was thereafter dubbed the 'Honourable Baronet'. As the great mantle passed from father to son, the passion for agriculture remained undiminished. Sir John Harpur-Crewe shunned public life and duties and spent most of his time on his estate at Calke. In 1860, he reinstated Longhorns there, founding his herd on the best strains that could be obtained. They, together with the Portland sheep which had been introduced into the park by his great-grandfather, Sir Henry, became his great passions in life. Like his father before him, he was a great admirer of Bakewell and, in championing Bakewell's breed of cattle, he gave powerful assistance in preventing the pure-bred race from slipping quietly into extinction.

At this time it was common policy amongst the farming fraternity to cross Longhorn cows with a Shorthorn bull to obtain an excellent dairy animal, beloved of the small farmer. The first crosses were said to be very hardy cattle, fattening quickly. But their calves were considered uncertain and in the majority of instances the third or fourth-generation animals, crossed back to the Longhorn, were considered to be Longhorns but did not have the good qualities of the original stock. They were smaller animals, poor milkers, poor fatteners and were not always certain to hold to the bull. It was claimed that a third of these crosses proved infertile.

Having arrived at this sad situation, many breeders wished they had kept with the original Longhorns. But good Longhorn bulls were almost impossible to obtain at that time. Sir John, like men before him, found the Longhorn too beautiful and useful to be allowed to fade away forever. It is our good fortune that he had the courage to follow his convictions and not the path to the Shorthorn, which was so popular at the time. He also left us a unique legacy of records at Calke Abbey.

Sir John Harpur-Crewe and his Longhorns

On 21st February 1874, the *Sporting Gazette* published the following article on 'Sir John Harpur-Crewe's Longhorns'. A copy of the article has been preserved at Calke Abbey.

> The Longhorns may now be said to have become something of a fancy beast in England. Gradually ousted from his position on the broad Midland pastures by the quicker feeding, and therefore better paying, Shorthorn, he has retired from the homestead of the tenant to the foldyard of the landlord where in a few instances his picturesque appearance has caused him to hold his ground against the Oxfords and

the Duchesses (well-known families of Shorthorn cattle). True, some tenants still stick to the old sort but the lover of Longhorns, when he looks around and enumerates the few holdings on which they are yet to be found in Leicestershire, Derbyshire, Lancashire and Westmorland, and then reverts to the times when nearly all those grazing districts owned their sway, may well say, indeed, 'The old order changeth for the new'. They bear the character of a very old and original race about them and, perhaps, with the exception of the Highlanders and Welsh cattle, approach nearer in appearance to the wild types than any breed that still remains amongst us. Their long horns, some drooping in long forward curves by the side of their cheeks, others standing out in graceful turns or, as they are occasionally seen, standing almost straight out at right angles to the head and hoisted like those of an antelope, combined with their rough, shaggy coats which hang in winter time, at any rate, like manes on their forequarters, and are especially thick on the foreheads, give them certainly a most weird appearance. The colour also, generally brindled and ranging from almost black to dull brown or dun on the sides, becoming mottled as it approaches the back, belly and quarters which, for the most part are white, certainly detracts nothing from the idea that they must come of a wild and ancient race, and in looking at the bulls we can well fancy that it was one of the kind who had taken a mad freak in his head, that the sturdy old Baron saw chased by the butchers and their dogs at Stamford, and forthwith instituted the once-fashionable sport of bull baiting. Might they not well recall the poet's description of bovine duel in old times –

> The stooping warriors, aiming head to head,
> Engage their clashing horns. With dreadful sound
> The forest rattles, and rocks rebound.
> They fence, they push, and pushing louder roar;
> Their dewlaps and sides are bathed in gore.

Their history, however, rests in private hands, as there is no *Herd Book*, and they have received but little encouragement, as compared with other breeds, at shows either of fat or breeding stock. Nevertheless, when brought into direct competition with others, they have held their own very fairly and, in some instances, taken prizes over the heads of Shorthorns and other more favoured kinds.

Having been granted permission to see the herd of Sir John Harpur-Crewe at Calke Abbey, we not long ago made a pilgrimage into Derbyshire for that purpose.

Although early in February, mild and balmy as the May morning of the poets, or the day that first lures the flyfisher to the stream, was the day on which we found ourselves on the road to Calke. What a pretty drive it was over the four miles that separates the quiet little town of Ashby de la Zouch from the Abbey! Unfortunately, beautiful as was the weather, there was a thin haze in the air which just served to shut out distant objects and confine the view within a distance of a mile or two, so that the most inveterate day-dreamer would have been puzzled to fix in his mind the spot most suitable for the renowned lists, or settle the route by which Rowena started on her homeward journey to the Saxon Grange. Worse still, it shut from view the bold range of hills known as Charnwood Forest where, long years ago, we have enjoyed, in our 'youth's first prime', sport with the Quorn which, if it wanted the steeple-chasing attractions of the other side of the [River] Wreake, was not wanting in either interest or pleasure. However, our blood is tamed now, and we look at the surrounding country more with a view to its stock-carrying capabilities than its likelihood for holding a scent, and descant more on the value per acre than on the practicability of the fences. So, as we pass the little village of Smisby, where Sir John's boundary commences, we are content to learn that the land is good and forward, bearing heavy crops of hay. As the park is neared we come on a lot of thriving Scotch beasts, which the mildness of the season still allowed to be foddered in the open field, and then beyond the stone wall we enter the park, and almost at once find ourselves driving through a magnificent herd of deer. Right well they look, too, though the park is somewhat heavily stocked and our guide and charioteer points with pride to a big buck lying down amongst some rushes, with a magnificent 'head', whose turn it will be to die when the season comes round once more, while another just beyond him is scarcely inferior in any way and will most likely both hear and feel the crack of doom from the rifle ere the summer is over. Much as we admire them, however, we must not linger amongst the deer, for Mr Bosworth, the steward, is already on the lookout to conduct us to the farm where the dairyman is waiting with his pets all in order for inspection. We must just linger, however, as we go, to note the peacocks, turkeys, guinea fowls and Dorkings whose name is legion and whose abode is at the snug steading where the cows are provided for. Anything nearer or more commodious than the accommodation here is seldom seen. Each cow is stalled well, bedded down, and honoured with a small board on which her name is painted, hung over her head, while behind them there is a broad passage, wider than is found in many stables; and, as regards freedom from offensive smells and general cleanliness, the most fastidious stud groom need not be ashamed to compare notes with this well kept cow byre, which is light and airy, yet warm.

Sir John Crewe's favourite cow, Tulip, *who took first prizes at both Birmingham and London in 1868 and for whom a butcher offered the price of 60 guineas. Sir John, however, declined the offer and took her home to Calke, where she went on to found the outstanding* Tulip tribe.

First in honour, as perchance in quality, came Tulip III by a bull named Thorpe out of Tulip (see black and white illustration), who carried off the only prize offered in which she could compete both at Birmingham and London in 1868, and for whom a butcher was ready to give the high price of 60 gns. Sir John, however, preferred to bring her home and she is now certainly one of the finest cows in the herd and has since bred him four calves, though a certain amount of patchiness about the rump still remains to tell of her training days. A right good looking cow she is now and one that any man might be proud to own. Next to her stands Brindle III by Dairyman (a bull, I believe of Sir John's breeding) out of Brindle, a nice level cow and a good specimen of her kind. Lovely IV, somewhat older, is a big cow by Young Sampson out of Lovely III, and Beauty III, with her widespread horns and lengthy frame carried one back more into the old fashioned type. Fanny, sixth of the name, is reported as being of a first rate family; and Trimmer II, light coloured though she be, with her fine barrel and bowed horn, is considered a rare specimen of the sort. Tulip IV keeps well up to the prestige of that family and the dark brindled Upton is a grand beast; while Bright Eye drops back to the old Lancashire sort. Brindle II, a good coloured cow, would probably ere now have committed 'the happy despatch' if left in a state of nature but civilisation has stepped in and prevented her horns, by the aid of the knife, from growing right through her head, as they inevitably must have done. Canley II is well worthy of the name she bears, and Joan III does no discredit to her family history. Leaving her, we come on yet a younger Tulip, good as the rest of the name.

In the next shed we find Snowdrop, the sole remaining memento of Sir John's Shorthorn fancy and we can well appreciate the fiât which reserved her when the rest were sold. Next to her is Titney, a good thick beast, and somewhat of a curiosity, as she has one horn growing up and the other down. The next worthy of mention are Daisy and an unnamed companion, both of the real, rough-looking Longhorn character and bought on account of the sort. We then come on a nice two-year-old out of Tulip, a black or nearly black and white (somewhat of an unusual colour). On then to Brindle, and Patty, a heifer for whose good looks we cannot say a great deal, and so ends the home-bred ones.

There were yet, however, half a dozen purchased at Mr Chapman's sale to see, and right good ones they were, full of that mysterious thing called quality, whatever it may be, but scarcely of the size or weight, notwithstanding more flesh, of the home-bred ones. However, Young Dairymaid, Crossredy (who has strangely wide open horns), Dishley, Brindled Beauty, Cumberland Duchess and Croxall cannot fail to improve any herd they enter, and will no doubt do good service to the one which, as was recently shown in the columns of *The Gazette*, they have been implanted. We think we are correct in stating the present strength of the herd to be 41, not counting the young stock yet to calve or the bulls.

Our next visit was to Old Time, the yearling bought in the autumn at Mr Chapman's sale, who has both grown and improved since he came to Calke, and promises to make a valuable bull. He has a rare head and horn, is a good colour and is a nice level young beast, being bred in Westmorland by Mr Matthew Hewertson and got by Tom Boy out of Daisy. He was bought by Sir John principally, we believe, with the view of introducing fresh blood into the herd. There is another young one of the same age by the prize bull Sampson who promises well and has an exceeding well-sprung rib and capital quarters, though his colour, being much lighter, we fancy somewhat detracts from his appearance. A nice young bull calf completed our inspection of this side of the yard and then we went away by the neat poultry houses in one of which, by the way, a sea eagle has held solitary state for upwards of twenty years, whose screams at strangers are a better warning of intrusion than even a dog. He is said to have got loose once but was captured ere again he had played the havoc amongst his feathered neighbours that might reasonably have been anticipated.

Beyond him we came on a fine two-year-old bull by Earl of Upton out of Joan, now promoted to the post of king of the herd, replacing Sampson who, a time ago, went to the butcher at a goodly price, although by no means fatted to the weight he might have been. This is a very handsome, dark-coloured bull, with a capital head and horn and a fine frame and, in looking at him, we can quite think that the breed was at one time a favourite one with graziers, for there is every indication of a good feeding bullock about him and, on the other hand, with his curly front and rough, bison-like mane, he is certainly a grand-looking animal to have about a gentleman's place, although we fancy that it is quite as well for strangers that he has a house and well walled-in yard allotted to his domain for, though the herdsman says he would do nothing except in play, he looks as if he was capable of playing somewhat roughly if he liked.

Some weaners in a close, hard by, next engaged our attention and amongst them a very promising cross between the Shorthorn and the Alderney; while amongst the young calves they have one crossed between the Long and Shorthorn, marked exactly like the Chillingham cattle, white with chestnut ears.

Perhaps the prettiest sight of the day was the dairy itself, with its cool, clean-looking stone shelves, glass milk pans, and such a collection of curious ware ranged round the walls as could scarcely [have] been seen together elsewhere, while it is shaded from the sun by a chestnut tree so large that the limbs have to be supported by props. A walk by the pretty elm-shaded church which stands in the middle of the park brought us to the yard where the deer are fed with their hay and a stroll thence to the farm showed us some more very useful heifers and these, with the exception of a fatting beast or two at some outlying buildings, who are doing well, completed our Longhorn survey of Calke and ended as pleasant a stroll amongst the stock as it was ever our lot to enjoy.

At one time Sir John was a breeder of pedigree Shorthorns but about ten years ago he sold them off and made up his herd of Longhorns of which he had always kept a few. With the exception of not feeding quite so quickly as their more fashionable rivals, they have earned a capital character as useful beasts and we hear that when well made out, the butchers are very fond of them, for they always die as good as they look, and show a preponderance of capital lean meat. For those who wish to combine ornament with usefulness, the Longhorn is certainly a capital beast, and we are in no way surprised that though nearly expelled from the homestead of the tenant they are yet found lingering in more aristocratic precincts.

(*Augeas*)

In an address given by Sir John Harpur-Crewe in 1876, to the tenancy and yeomanry cavalry on the arrival back at the abbey of his son Vauncey and his new bride from their honeymoon, the great pride and depth of feeling he had for his animals is tangible. 'Perhaps you will allow me to add that if, after you have viewed all the beautiful presents which the bride and groom have received, and anything else which may be worth looking at in the house, any of you who would like to see my herd of Longhorn cattle, I shall have great pleasure in showing them to you, and I will meet you in the dairy yard for that purpose when I venture to think you will find me somewhat more in my natural element'.

This having been said, it was met with great cheers from the assembled company. It can be imagined that this will strike a chord with the sentiments of many proud Longhorn owners who, like Sir John on that day in 1876, could not wait to shrug off the fetters of the day and to stand once more at peace, surrounded by these majestic cattle (see the illustration of *Longhorns in Calke Park, c1870* in the colour section).

Sir John kept a scrapbook of press clippings relating to all matters Longhorn, reports of sales, of shows and of articles on Bakewell. This scrapbook was finished by his son, Sir Vauncey, after the death of his father in 1886, with the report of the sale of the Calke Abbey herd and the reduction of the flock of Portlands. Sir John had left instructions in his will that on his death, his beloved herd of Longhorns was to be dispersed, as his son did not share his passion for agriculture and he knew that the estate would revert to its former sporting ways. Sir Vauncey's all-consuming interest was birds, both alive and dead, and he was to go on to form one of the largest collections of taxidermy in the country, amounting to several thousand cases. He was rarely to be seen without a gun and became a virtual recluse. So, Sir John had been wise in his decision, giving other Longhorn enthusiasts an opportunity to make use of his valuable bloodlines and not leaving the herd to deteriorate in the hands of his disinterested son and heir.

The following description is taken from a press clipping in the scrapbook. It comes from *Bell's Weekly Messenger* and was written by Joseph Darby, after the death of Sir John but before the sale. The report on Portland sheep was included.

> Among the ancestral homes of our principal landed gentry Calke Abbey, the abode of the late lamented Sir J. Harpur-Crewe, is possessed of features of great interest, and the deceased baronet not only filled the spacious mansion with art treasures, including a collection of rare bird specimens, calculated to feast the eyes of any naturalist, but took great delight in the arts of agriculture and stock breeding.
>
> In the magnificent undulating park not only were to be seen old and beautiful trees and immense herds of red and fallow deer but also

Portland sheep and Longhorn cattle, both of which breeds he brought to a great height of perfection. In his patronage of the Portlanders the later Sir J. Harpur-Crewe may be said to have stood alone, and to him must be accorded the high honour of having preserved Portland sheep from the fate which overtook the old Berkshire, Norfolk, Mendip and many a once famous breed. Portland sheep are to be found today nowhere in perfection, but at Calke Abbey. In their native island, they have been so neglected as to have become not only few in number but indifferent in character.

On the contrary, the large flocks, numbering altogether nearly 400, which have so striking an appearance in the park of Calke Abbey, and which will be offered at public auction there by Messrs German, German and Cooper (who have sold them for many years at their sales at Ashby, realising from 1s to 1s 3d per lb) on the 19th May next year, display such meritorious characteristics no less than a pleasing appearance that it may naturally be expected, they may attract the admiration and regard of some of our nobility and country gentry, who will have a noble opportunity at the auction to possess themselves of a few of the ewes for propagation.

No prettier animal can be a denizen of a park. They have short, thick, rotund bodies on short legs like the best-bred Exmoors but are far less wild. In this respect they have most likely altered their original character by having been domesticated for so many generations in this good district. At any rate it has enabled them to attain larger size and greater weight of carcase than they are originally reputed to possess in the Isle of Portland. The sheep were brought direct from the Island of Portland by Sir Harry Harpur in the year 1770, who was ancestor of Sir John Harpur Crewe and the tradition is that they are descended from a few hapless animals which were cast ashore there, when the Spanish Armada was demolished in good Queen Bess's glorious days. Though their flesh is so excellent, they are not a dainty sheep; on the contrary, they are very hardy and will do well on scanty pastures.

The deceased baronet also brought his magnificent herd of Longhorns to very great perfection, and as the whole of it will be offered by public auction by Messrs Lythall, Mansell and Walters on Wednesday 19 May 1886, a description of it may be useful to those who think that the world should not be parcelled out entirely of one pattern, and that Bakewell's favourite breed may even now, in many respects, hold its own against the widely popular Shorthorns. The latter may, perhaps, come to earlier maturity, but, if so, that is the only superiority; as for hardihood, constitutional vigour and longevity few British breeds can equal the Longhorn. The distinguishing feature of the herd at Calke is its uniformity in high quality. It would be difficult to find in England so many animals of the breed brought to so high a standard of perfection, with noble, picturesque heads; a large proportion with tapering wheel-horns, encircling the neck itself.

The colour is a dark handsome brindle or red, with a sheet of white all down the back and white spots on the thighs. They are also well ribbed and mostly deep and compact in form, with a quality of skin and flesh a grazier would certainly not despise. The high standard of excellence for which the herd is so remarkable is only what might be expected considering that it was formed in the first place by the best strains possible to be obtained; that it has been maintained with such persistent care, skill and good management is qualified by the many prizes obtained from the showyard.

Some very grandly-shaped pure-bred oxen and steers of the Longhorn breed are also to be sold, which will afford an admirable opportunity for graziers and exhibitors to get animals of the right stamp, fit to feed on for winning prizes at the leading fatstock shows of November and December.

The late Sir John's name was well known at Smithfield and Bingley Hall. In 1861 he showed a beast at Smithfield which took the gold medal as the best cow in the show of any breed. His Longhorns have beaten the Shorthorns at local county shows, Shorthorn men being the judges, and in 1864 his Longhorn was the greatest attraction at the London show.

Sir John Crewe's pair of Longhorn steers, which took the first prize at Birmingham in 1868.

> The head of the animal now adorns the entrance hall of Calke Abbey, along with many others of a remarkable character. Some very fine heads are preserved at Calke, in some cases the horns being nine feet from tip to tip, taking in the measurement of the skull. In 1875 Sir John exhibited at Bingley Hall a Longhorn ox, and in the whole show only two animals were as heavy at the same age, the £100 prize Hereford steer, 19 months older, weighing only 136lb heavier. In ten years Sir John sent no fewer than 11 of the prize oxen in the class at Bingley Hall, twice standing first and second.

So we come to the report of the sale itself. It is interesting to note that of the auctioneers who sold the herd on 19th May 1886, Mr John Lythall was the very first secretary of the Longhorn Society, and the family of Mr Walters was to become one of those that helped the breed to survive in later years, when numbers had fallen to a very vulnerable level.

The early pedigree cattle sales were great occasions. The following is an unidentified press clipping from the family scrapbook.

> The weather on Wednesday, though heavy clouds threatened in the morning, turned out beautifully fine, and there was a large attendance at the sale, large numbers of breeders or their agents travelling from a distance to be present. Most of the farmers, dealers and butchers from the district were also there. The sale was preceded by a luncheon in the riding school, of which some 500 or 600 partook, the reputation of the family for open-handed liberality being fully maintained. Mr H. Chandos-Pole-Gell, High Sheriff of Derbyshire, presided.

The writer then gives a list of all the gentlemen present whom he was able to identify. Of these, perhaps the most recognisable today are: Mr T.L. Prinsep, Mr Carswell (agent for Mr W.J. Legh, Lyme Hall, Cheshire), Mr Oxley (agent for the Duke of Buckingham), Mr Turnbull (agent for Mr Mundy, Markeaton), Mr John Gretton (Drakelow), Mr Matthews (Foremark), Mr George F. Brown, Mr Wragg (Ashby), Mr S.H. Chapman (steward), Mr Sale (Smisby).

> After luncheon, the chairman proposed a toast to 'the Queen', followed by one for 'Success to the Sale'. He said all those present knew the circumstances under which the Longhorns were to be sold. They also knew the great interest Sir John had taken for many years past in the breed of cattle. They would agree with him [the speaker] that it must be very painful to his successor to part with them, but they were left to be sold, and they would be. Individually he had taken a very great interest in the Longhorn breed – the breed of our forefathers long before any

other kind was known in the Midland counties. There was no class of cattle more hardy and more useful in their way, but owing to many circumstances their cultivation was confined to a few, enthusiastic people.

He had been born and bred amongst them, his father having been a Longhorn breeder all his life. He had often seen Sir John Crewe's Longhorns in different parts of the country and he was sure they would meet with the approbation of all present. [Applause.] Unfortunately the times were not very good but he would remind them that if they wanted to secure good Longhorns they must not wait until times were good. They must get them when the opportunity occurred, and such an opportunity as this might not occur again – for many years.

He was pleased to see so many Longhorn breeders present from different parts of England, and he was sure they did not mean to go back without some of the plums in their charge. He could only hope they would fix their minds on the same animals. [Laughter.] There were two good men at a sale, the first and the last bidder, and if they took pains to be the last, they were certain to get what they wanted. [Laughter and Cheers.]

Mr Shaw, who had acted as confidential agent of Sir John Crewe for many years, acknowledged the toast on behalf of the family. He expressed the deep regret felt by Sir Vauncey at being unable to attend the day's proceedings but felt that it was impossible so soon after his father's death. He thanked the large company for their presence, and Mr Pole-Gell for his kind remarks. He asked his audience to understand the awkwardness of his position. He felt gratification in meeting them but, on the other hand, the death of Sir John and the dispersion of the famous herd of Longhorns and the unique flock of Portland sheep made it a day rather for sadness than rejoicing. [Hear! Hear!]

He spoke on the merits of Sir John saying that he was a thoughtful English Gentleman [applause] and was deeply interested in everything connected with his estates. He was on the most friendly terms with his tenantry, both rich and poor [applause], and sympathised with them, not only in their successes but in their troubles. Sir John was a man of very great observation, great judgement and great decision of character, and those were all qualities he brought to bear in breeding and rearing the splendid herd of Longhorns and the fine flock of sheep which they would have the opportunity of purchasing that afternoon. He hoped that every gentleman would be satisfied with his purchases, and look back with pleasure to his attendance on that day. [Applause.] He concluded by proposing a toast to the chairman.

The company then adjourned to the sale rings at the dairy.

The sheep were sold first by Mr George German, who began by giving a brief history of the flock. First, 110 Portland ewes and 125 lambs were offered in 22 lots, the amount realised being £424 7s 6d. The top price was given by Major Burgess of Hendon who secured five ewes and 10 lambs for £25. Two similar lots made £23 15s and £22 10s respectively. The remainder were sold in lots of five ewes and five lambs. The price per ewe and lamb averaged £3 10s but where a ram lamb was included it rose to £4 12 6d, the latter price being given by Mr J. German. Fat sheep went up to £3 7s, wethers went to £3 4s, ewes up to £3, paid by Mr Matthews of Foremark. Two Portland rams were £4 each, one going to Major Burgess.

At a quarter to three a move was made to the cattle ring where Mr Lythall took up the running. He said, valuable as was the flock of sheep which had been offered by Mr German, it was more as a breeder of Longhorns that the late Sir John Crewe's fame as an agriculturist [sic] was established. This was an opportunity which might not occur once in a century, and it should be taken advantage of by those who did not wish to see the oldest, most profitable and most picturesque breed of English cattle become extinct. All the animals had been out during winter and were not in any way prepared for sale.

Mr Lythall then commenced the sale, which took 2½ hours, but was watched with unabated interest. Some of the prices obtained were exceedingly good, some of the bullocks making 8½d a pound, the top price of 44 guineas being given by Mr Shaw of Fradley for a splendid brindle and white ox, and the well-known dealer of Melbourne, Mr Toone, purchased the two best bulls, Cornet and Farmer, the sires of much of the young stock where were sold, for 35 guineas and 28 guineas respectively. Mr Watts of Melcombe took away a sound young animal, Robin Hood, for crossing, it is believed with Shorthorns, and Mr Bates of Malvern secured a gem in Little John. This gentleman was a large purchaser and paid the top price for calves, namely 14 guineas for a capital young heifer by Cornet called Cumberland Duchess which he secured after an exciting contest with Mr Bradley who was also a considerable buyer. Mr Griffiths secured a good young calf for the classic groves of Hawarden and Mr Oxley made several purchases for the Duke of Buckingham. Mr Carswell also secured a couple of heifers to join the celebrated herd at Lyme Park. The last animal on the catalogue was a yearling heifer, a cross between an Alderney and a Longhorn, supposed to be the only one in England. The experiment was purchased by Mr Matthews of Foremark for 12 guineas. The total figure for the cattle was £1,906 1s 6d, giving an average, including calves, of £19 1s 2½d.

The top price for the day was the fat four-year-old ox by the Abbot of Calke which went to Mr Shaw of Fradley, for 44 guineas. (See Appendix 6 for a list of the prices achieved.)

Sir John died in 1886, and an agricultural correspondent of the time wrote:

> The late Sir John Crewe's position as an agriculturalist was quite unique in the Midlands and, in many respects, may be compared with Bakewell in Leicestershire, Bates in Yorkshire and Coke in Norfolk, for the great interest he took in improving domestic animals. He loved, as the great breeders named did, to study their habits and improve their best characteristics. … Truly Calke Abbey and its surroundings may be termed 'classic ground' for domestic animals, and agriculture generally.

The Duke of Buckingham's Longhorns at Stowe Park

Arriving at the quaint old town of Buckingham by train we had a pleasant drive through the beautiful avenue which leads from the town to Stowe Park and which, after a mile and three quarters, brought us to the hospitable roof of Messrs Oxley who are the Duke's agents and in whose hands is placed the keeping of the Longhorn herd. How well the trust has been carried out our subsequent remarks, as well as the long lists of prizes that have been won, will show.

We were shown to the yard where the herdsman soon led out the grand bull Conqueror the third, the Birmingham RASE winner, for us to renew our acquaintance with. A magnificent specimen of the breed and, as we look over him, we can well understand how it is that he has never been beaten. He has a fine head, beautiful neck, short legs, carcase as straight as a gun barrel, good quarters and is a rich brindled colour shading off lighter towards the quarters, and shows all the orthodox markings of the breed. He is by Young Conqueror, a son of Old Conqueror, and out of a cow of the breed of Mr Twycross of Canley. He came out as jauntily as ever and, having been let down a bit from showyard condition is, we hear, as good in service as ever he was in his life and, since we saw him, has again taken first honours at the Royal, at Liverpool.

Leaving him, we came on Wild Rose, a very beautiful in-calf cow. Her head and horn are capital, she has a back on which you could make up a bed, and is altogether a very beautiful animal. Moreover, she is well worth consideration as an example of the aptitude of the Longhorn race for quick feeding, as we were informed by Messrs Oxley that she had been got into show condition for the Royal at Liverpool within a very short time, and had been in low condition previously. When we saw her many a butcher would have been glad to have the chance of driving her to the slaughterhouse. She was second at the recent Royal Show.

Our attention was next turned to quite a juvenile, a bull calf by Conqueror lll out of a heifer that took second prize at Birmingham, and a smarter one for his age than Young Sambo [apologies for this now inappropriate name], as we believe he is to be called, is seldom seen. He is very level and true made all over, and a capital colour. With him also was White Rose's calf, another very neat one.

A couple of heifers next claimed our attention and, of them, Waterloo, a nice useful beast by the old bull, particularly good behind, looks altogether like framing into a prize taker. Next to them was Earl of Temple by Conqueror lll out of Duchess, never yet exhibited. He has not been sent along and forced as some of the rest have and consequently he is not so shapely in appearance, while he is somewhat coarser in the head. There is no doubt that he is a very useful bull but we are inclined to doubt whether he will ever show the grand quality of his sire, even if he is made up for show.

The Marquis, a first-prize taker at Birmingham, took our eye much more. He was two years old on 27th March, and certainly has made capital use of his time while he has been in the world. He has a rare loin, is good in colour, shows a deep forehand, has a powerful set of thighs and quarters on short legs, and altogether struck us as one of those thick, deep fleshed animals that are sure to come out quite as good as they look

Conqueror *3rd, bred by the Duke of Buckingham and Chandos and winner of first prizes at RASE shows in 1876, 1877 and 1878.*

and prove well to the butcher, while the great amount of quality he shows tells at once that he must be a good feeder.

In a small paddock close by was what we took to be the gem of the herd, Countess of Temple, a daughter of the old bull who also took a first at Birmingham. Many are the animals of various breeds we have run our eyes over, Shorthorns, Herefords, Devons, Sussex, etc., but we have never seen to our mind a more beautiful beast than this Longhorn heifer. For depth and squareness of frame no Shorthorn could beat her and, for horn, countenance and quality, she is not to be surpassed. We saw her just at her best and, when shown at Liverpool where she obtained a second prize, she was too near her time of calving to show to anything like the same advantage, as the bloom was considerably gone off. With her was another heifer out of a cow of Mr Godfrey's who took a prize at Wolverhampton, called Buttercup ll, and an undoubtedly good beast she was but, as we told Mr Oxley, she should not have been seen at the same time as Countess of Temple, for that heifer so filled the eye that it was impossible to turn to anything else when she was present.

A bull bought from Mr Godfrey and called Earl of Wigston is higher on the leg, plainer in the quarters and shows less quality than the home-bred ones, but no doubt he will answer a good purpose in introducing a change of blood. A nice lot of calves, of capital colour and good strains of blood, completed out inspection of the home premises before we started for the Park to see the herd of cows.

Here, as we turned to the wide expanse of greensward, and saw the mothers of the herd spread out and grazing quietly over its slopes, the picturesque characteristics of the Longhorn came out in all their beauty and it was impossible to help feeling that there is no breed of cattle which so well becomes a nobleman's park as the Longhorn, for nothing can equal their grand dignity and stateliness.

First we saw the heifer Lady May, own sister to the Countess of Temple, but at the time of our visit she was by no means looking her best. Then there was her dam, a fine finch-coloured cow that had taken second at Wolverhampton. Lady Twycross, first at Birmingham, came forward as if anxious that her showyard honours should not pass unnoticed, but it was in vain as far as commendations on our part were concerned, for the Messrs Oxley prefer the 2nd to her on that occasion, Barmaid, an opinion in which we concur, as did a great many good judges who saw them in the Royal enclosure.

Baroness, a very good dark red cow, then caught our eye, 'neat as paint' as the saying is, notwithstanding she had roughed it all winter, a thing which tells little on this hardy race. Then there was her dam, Lady Caroline, a fine old cow of the same colour, who took a second prize at Oxford. Wildfire, dam of Wild Rose, and The Marquis, was pointed out to us as one of the cracks of the herd and, pointing to Emma, a rare dark brindle, Mr Oxley remarked: 'We would like to get them all of that colour'. Then we strolled away across the hill towards the great house and into what are called the pleasure grounds.

This is an area of 300 acres where cedars and other trees of extraordinary dimensions are to be seen. It is an area quite distinct from the park and appears to be a veritable forest. Within this area one might well fancy from the number of temples which are erected, that one had been transported to Greece and that the whole heathen mythology had dropped their temples in this chosen spot. Not that I would for one moment insinuate that the noble owner of Stowe and his ancestors are, or have been, given to Pagan rites, but that no expense has been spared in decorating these extensive grounds. In fact, some new view with a temple or grotto is continually opening on you, while the screams of the peacocks, which run wild here, would lead us to believe that Juno was the presiding goddess and that Argus (fabled to have 100 eyes) was kept here on guard to watch the movements of Io (loved by Zeus and changed by Hera, wife of Zeus, into a cow) in the shape of a Longhorn heifer in the pastures beyond. Then as we wandered through the whole coppices of choice shrubs, suddenly we came on a secluded little valley low down amongst the wood-covered hills where grazed a lot of choice young Longhorns. Strange sight, perhaps, but as we listened to the old herdsman's recital of the merits of the place, we no longer wondered that the descendants of Lady Caroline, the Buttercups and the Primulas were here, for we found it was the best place for wintering young stock to be had, and far before the yards and sheds for health. Here our task was done, and returning through the pleasure grounds, hares scudded away from under our feet as we made our way to Messrs Oxley's hospitable roof. After draining a stirrup cup we started once more for Buckingham. So we finished our inspection and a very pleasant day altogether, coming away with the impression that, although Messrs Oxley have not very long held their present post, the herd has fallen into hands that will do it justice, as well as with the conviction that we never saw so much quality in Longhorns before.

(From a Calke Abbey scrapbook, provenance unknown)

Endnotes

1 *The Agricultural Gazette*, 13 March 1783, p338.

CHAPTER FIFTEEN

Breed Recognition

Pat Stanley

The Longhorn Cattle Society

In December 1875, a meeting was held in Bingley Hall, during the Birmingham Fatstock Show, with the intention of starting a Longhorn Cattle Society and herd book.

> Mr T.L. Prinsep, of Croxall, presided, and a good number of breeders were present, amongst them being Col. Dyot, MP, Mr W.P. Cox, Mr Oxley (from the Duke of Buckingham's), Mr Townley Parker, Mr R.H. Chapman, Mr Godfrey, Mr T. Setchwell, Mr Taverner, and many others.
>
> The Chairman said he could bear evidence to the worth of the breed, for having tried it, he was perfectly surprised at the produce they gave. Devonshire cream could be as well made in Derbyshire as in the western county. He moved:
>
> 'That in consequence of the increasing interest manifested in Longhorns, and with the view of securing this old-established breed its proper position amongst English cattle, this meeting is of the opinion that it is desirable to form a Longhorn society.'
>
> This was seconded by Mr Setchwell [whose herd had been in existence for three generations, having been started by his grandfather].
>
> Mr Townley Parker then moved:-
>
> 'That believing purity of breed is essential to improvement, this meeting is further of opinion that the pedigree of Longhorns should be duly recorded, which object will be best obtained by establishing a Longhorn herd book.'
>
> Mr Forrest proposed, and Mr Taverner seconded a third resolution, appointing a committee, consisting of Sir J.H. Crewe, Bart., Mr W.T. Cox, Mr T. Levett Prinsep, Mr J.H. Burbery, Mr Townley Parker, Mr R.H. Chapman and Mr Oxley, to consider and report to a general meeting to be held at Birmingham during the Royal Show week in 1876.[1]

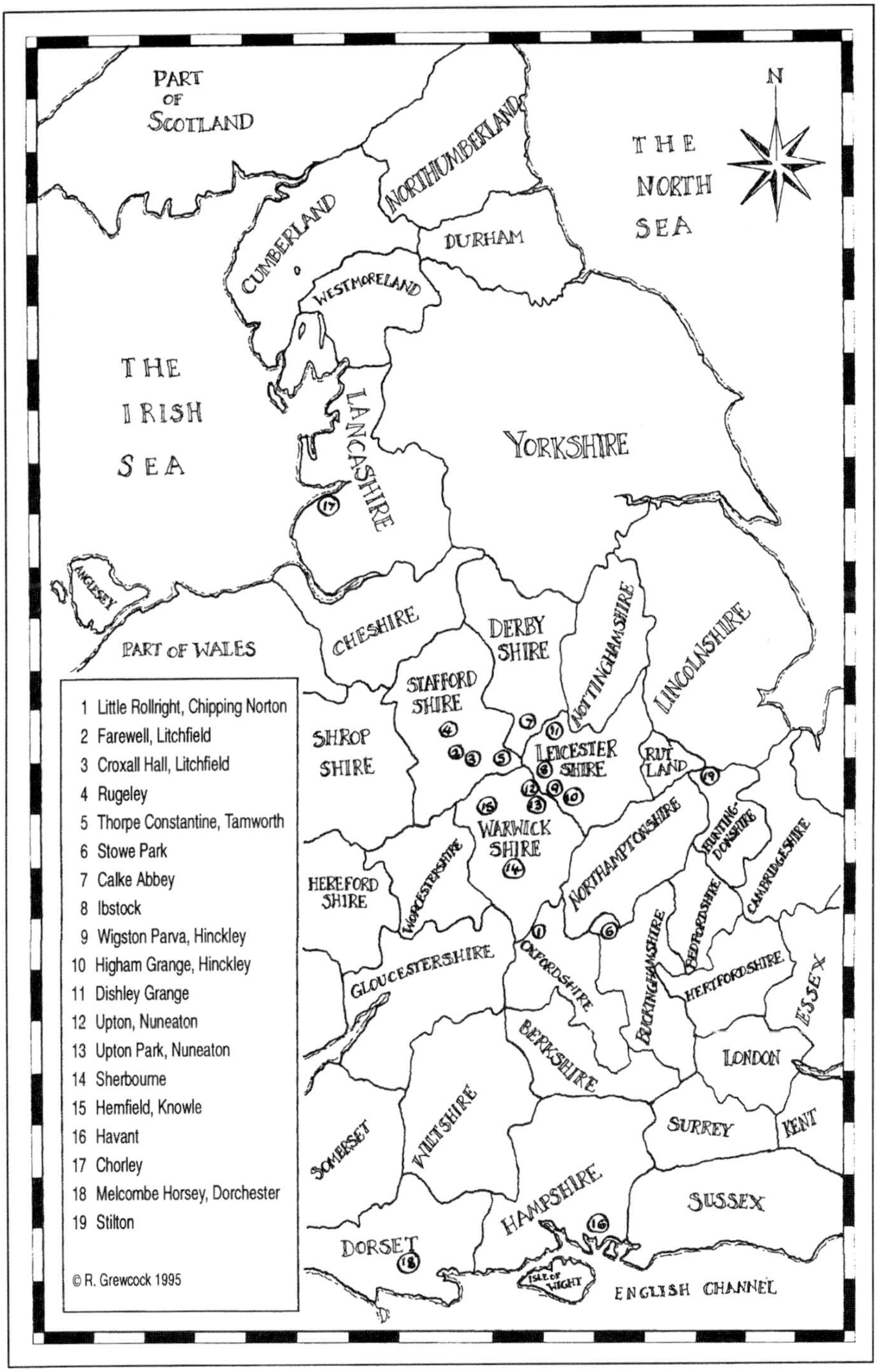

Homes of some of the legendary Longhorn herds. Note the concentration in the Midland counties.

The men responsible for the formation of the society were mostly from families who had supported the breed for generations. For example, Mr S. Forrest and Mr J.H. Burbery were nephew and uncle, Mr Burbery, on his retirement, having put his Wroxhall herd, which was started by his ancestor Mr Jackson in 1750, into the hands of his nephew for safekeeping.

The Birmingham Royal Show in 1876 provided a great boost for the breed, as there was an entry of 60 Longhorns (see Appendix 8). The following article shows the resurgence of interest in the breed at this time.

Mr John Godfrey's Longhorns

Being in Leicestershire shortly after the unusual number of more than three score Longhorns at the Royal Show at Birmingham, I paid my respects to the breeders in this county who were placed first. At Stoke Golding, where the church spire rises in the midst of many pastures and a few arable fields, we pulled up for an hour in driving from the station at the comfortable farmhouse of Mr Berry. A mixed herd of Shorthorns and Longhorns was feeding in the cool evening in pastures before the windows, unconscious of any rivalry, past, present or to come, between their races, but not unconscious of short commons this season in consequence of the drought. I have seen no pampering of cattle in Leicestershire nor was there any here. When the pastures run short of grass the cattle suffer. You may order linseed cake as Mr Berry told me he did for the cattle last year, and presently things are not quite right in the cheese room. You may manipulate as usual with all possible and prudent care, and paste peppered paper over the cracks when the cheese rises and its skin breaks, but you cannot escape the effects of such a season as the present. If you feed your cows artificially your cheese will be the worse for it, and if you do not feed them your cows will be so. I have never seen such bare ribs in Leicestershire as now. On arrival, being unaware of the state of the pastures, I remarked of one herd I had seen before, 'Why, I thought your cows were bigger!' 'So they were,' was the short reply. I found the ribs of this dairy as distinct as the keys of a piano; and the point of my stick passed along them produced sounds far too bony to be pleasant. Heavy rains are sadly wanted. Longhorns pick up rapidly in good grass and they are hardy and can rough it well in winter as the thick hides and coarse limbs of the common sort would indicate. The merits of the breed have been summed up concisely, Mr Berry showed me some animals of excellent quality: and I then passed on to visit next day a herd with no Shorthorns alongside, placing it, as it were, on trial.

Mr John Godfrey of Little Wigston Farm, near Hinckley, owns, rents and occupies 120 acres of pasture and meadow and 20 acres of arable land, all for sale at the present time, worth from 50s to 60s to rent, and worth as grass 25 per cent more than as arable. The soil is good sand loam with water-worn pebbles in it, a drift geologically covering the New Red Sandstone strata on which it rests to a considerable depth. The meadows are on the banks of a little stream running through a disproportionately wide valley into the Soar – one of the few rivers of Leicestershire – below the farm. The holly and yew have grown upon the farm immemorially and may still remain about the hedges and in clumps and rows here and there, supplying with other trees, that ample blind which I must say is very pretty in this part of the country, though it must cost the occupiers of the land a great deal annually to maintain it. The elm is, however, the prevailing tree and in Brick Kiln Close a pasture of 8 or 9 acres we found 20 cows and a bull beneath the shade of a hedgerow of stately English elms growing to a great height on the rich loamy soil. Here was the best blood of Dishley Grange, which can be traced through the herds of Chapman of Upton, Green of Odston(e) and other eminent breeders. This best blood has lately flowed direct from Mr Godfrey's herd to that of the Duke of Buckingham and Chandos, the winner of the 1st prizes for bulls in the two classes at Birmingham, and to that of Mr Satchwell who was commended in two instances for cows, and of Messrs Burbery, Chapman and others.

The natural colours of a Longhorn are black, red and white. The sides and head are brindled by the admixture of the black and red, or they may be red only (pale or deep) but the back is always white. The bull which might be observed dallying with a wanton heifer in this secluded pasture, on the cool north side of a hedge, is as dark on the sides as a lion. Old Brindled Beauty, a grand cow, grazing near him is almost as dark. Fairy is a rich deep red without black, and Wide Horn is equally perfect and a pale red. These are cows of long descent, and of a beauty which has captivated the professional and public eye more than once. Fairy's deep, large frame is perfection. She is descended from the 'Sampson' breed, and won the 1st prize at the Warwickshire show in 1873. Wide Horn is no less famous. She comes towards us swaying in her walk, a barge of beef on legs, not too stout, carrying her prodigious horns low, and stretched across the grass, 66 in. from tip to tip. She is a beautiful cow, pale red upon the sides and without a black hair on her. Brunette, her daughter, a 3 yr old, with her first calf, and all the promise of a milker in her square well formed bag, shows a perfect breeding with a brindled coat and soft skin.

There are others which might be mentioned, and, in fact, there is not a cow that might not be sketched as a good specimen of the breed, but I must finish here with the Last Rose of Summer, sister to the Duke of Buckingham's First Conqueror, bred by Mr Godfrey. She has the slightly Roman nose which is proper in this breed, the horns of a great length, firm in texture, placed low on the skull, nicely sloping in perfect character, and particularly fine from head to point. May her horns be exalted some day to their deserved position in some breeder's hall. There are at least a dozen pair of horns in this herd worth polishing, and there is not a coarse horn in the lot.

We found Tiger, the 2nd prize Young Bull at Birmingham, in a suitable lair, an animal with all the colours and with the spotted thighs which denote purity of breeding. He has the admired Roman nose, and I should have estimated the present weight of beef of this 12-month-old bull at 75 stone, worth £22 10s at 9d per lb or 9s per week from birth. The owner thinks this under the mark, and it may be so, but what a great return it is! If the feeders of common stock would take these figures to bed with them and meditate upon them, and compare them with such returns as they get from common cattle, our fairs and markets would soon be swept of their rubbish. Length of frame is the boast of the Longhorn breed.

There are, at present, 21 cows on the farm, four two-year-old heifers, four yearlings, and eight calves. About 3 tons of cheese is made in an average season. Profits have been reduced by the Foot and Mouth disease. Mr Godfrey agrees with Mr Berry as to the cheese going wrong when the cows receive artificial food.

(*H. Evershed, published 14th September 1876*)

The Longhorn Herd Book

In the early days of showing livestock, it was the animal which counted – not the breed. Hence there were no separate classes for breeds which, in modern terms, would have been difficult to arrange before most breed societies and their herd books had been established. The first volume of *Coates Herd Book for Shorthorns* appeared in 1822, other breed herd books were not started until the 1870s.

As a result of the general meeting called to form the Longhorn Society, the first volume of the herd book was published in 1878. It had a preface by Mr John B. Lythall, the honorary secretary appointed to the new society, who had obviously gone to enormous trouble to research the history of the breed and to record important sales of the previous 45 years. Lythall is the same man who conducted the sale of Sir John Harpur-Crewe's Longhorns in 1886.

Details laid down for the new society were a subscription of 1 guinea per year, a registration fee for bulls of 5s, for cows of 2s 6d and for calves of 2s 6d. Non-members were charged £1 to register bulls and 10s to register cows, and £1 for the herd book.

The first herd book of the society, published in dark navy leather with gold lettering in 1878, announced its officers as:

President: His Grace the Duke of Buckingham and Chandos.

Vice-presidents: Sir John H. Crewe, Bart., Major General Sir F. Fitz-wygram, Bart., W.J. Legh Esq., MP, G.F. Muntz, Esq., T L. Prinsep, Esq.

Treasurer: Mr Joseph H. Burbery.

Committee:	Mr W.P. Burbery	Mr T.J. Oxley
	Mr R.H. Chapman	Mr T. Satchwell
	Mr S. Forrest	Mr W.S. Shaw
	Mr J. Godfrey	Mr V.B. Watts
	Mr R. Hall.	

Honorary Secretary: John B. Lythall.[2]

Herd books have been published as follows:

Vol. 1 – 1878, Vol. 2 – 1899, Vol. 3 – 1901, Vol. 4 – 1903, Vol. 5 – 1905, Vol. 6 – 1907, Vol. 7 – 1909, Vol. 8 – 1911, Vol. 9 – 1915, Vol. 10 – 1917, Vol. 11 – 1919, Vol. 12 – 1921, Vol. 13 – 1923, Vol. 14 – 1925, Vol. 15 – 1927, Vol. 16 – 1930, Vol. 17 – 1931, Vol. 18 – 1933, Vol. 19 – 1935, Vol. 20 – 1937, Vol. 21 – 1939, Vol. 22 – 1951, Vol. 23 – 1955, Vol. 24 – 1962, Vol. 25 – 1966, and from there to the present day.

Although the animals entered in the first herd book were not great in number, they consisted for the most part of some of the greatest blood the breed had ever possessed, coming as they did from the remnants of some of the purest and oldest herds in existence.

During the 1880s the breed took another dip in popularity, with most of the agricultural societies ceasing to have classes for the Longhorn. But by 1898 the tide turned once more in their favour, and the Birmingham Royal Show again opened its gates to the breed, stimulating the second herd book in 1899 and the revival of the society. Several new herds were formed, and the society was able to boast 21 members. This enthusiastic band of men, apart from keeping the spirit of the Longhorn alive in the show ring, pitted their skills in milking trials against other breeds. Amongst those taking part was Mr W. Hanson Sale of Arden Hill, that herd lasting well into the 20th century under Mr Jim Beechey.

Recent Years

The breed continued quietly until the Second World War, when it once again virtually disappeared. The society ceased operation, and the breed itself was only kept alive by enthusiasts such as Messrs. M.E.H. Walters, 'Grendon', W.E. Nokes, 'Hill', R. Wales, 'Stoke', and J.S. Stanhope of the 'Watling' herd, who re-formed the society in 1948.

A high point for the Longhorns in the show ring came at the RASE meeting in 1981 when the breed produced the winning pair to take the coveted 'Burke' trophy against all other breeds, in the form of *Eyebrook Richard*, bred by the late Mr T.B. Johnson, and *Rousham Carnation*, bred by Mr C. Cottrell Dormer. This sparkling pair in the ownership of Mrs B. Weiner thrust the breed into the spotlight, since which time it has gone from strength to strength.

In 1995 the Longhorn Cattle Society had grown to 392 members, 206 herds and a total of 2,500 breeding stock and by July 2023 it had 509 members (429 full member status) and 352 herds.

Longhorn Breed Standards

Standards for the Longhorn breed have, as with most breeds, varied down the years, with the present Longhorn Cattle Society setting out the following standards:

> The HORNS to be fine in texture with traces of pink towards the base and free from any black colouration. The shape to come either straight from the head with a forward sweep or turned with an incurve towards the mouth. Thick or coarse horns, or a tendency to grow backwards, are seen as undesirable.
>
> The EYE to be prominent and of docile expression.
>
> The BODY to be level in the back and well sprung over the ribs. The shoulder to be neat but full behind. The animal to be wide over the sirloin, level over the pins and to have a neatly set on tail. It should have a level underline, not to be coarse in the brisket, be full in the thigh and lengthy and level in the steak piece and not split up between the hind legs. The cow to have a level, silky bag, well up under the tail, with teats well placed, of good size but not coarse.
>
> The HIDE to be thick but soft to the touch and pliable.
>
> The COLOUR is optional, but all to have a white line down the back, and with a slight preference given to red brindles and rich reds and to a white patch on each thigh. Cattle with black on the horns, black hooves or with no red hair in the coat are considered undesirable and are to be discouraged.

Endnotes

1 Fitt, Neville J., *The Agricultural Gazette*, 27 March, 1876, p398.

2 The Long-horn Herd Book, Vol 1 – facsimile.

CHAPTER SIXTEEN

Livestock Specialists in Bakewell's Period

Hilary Matthews

During the late 18th century, many country gentlemen and larger farmers interested in agricultural improvement regularly corresponded and visited one another, creating a sense of solidarity between them. Location was no deterrent, and many travelled great distances to visit like-minded agriculturalists professionally and socially.[1] During Robert Bakewell's working life and the two decades after his death, when his nephew Robert Honeyborn was at the helm, Dishley Grange continued as a mecca for visiting agriculturalists. Apart from friends and associates, many of those who came to Dishley were titled and influential men from home and abroad. They came to look, learn, hire, and buy livestock. But Dishley also hosted visitors of a far lower status. These not only included local ancillary tradesmen, such as blacksmiths and butchers, but men from further afield who performed a variety of specialist livestock husbandry roles that agriculturalists like Bakewell and Honeyborn either did not have time to execute or did not possess the requisite expertise to perform themselves. These men could advise on livestock, buy and sell stock on the farmer or landowner's behalf, perform husbandry tasks including castrating and speying (neutering) farm animals and provide services such as veterinary treatment and wool grading. They were also a conduit of news, passing on information, gossip and messages between farmers, locally and further afield. But although much is known about Dishley's eminent visitors, little is known of these journeymen livestock specialists and the various tasks they carried out for agricultural improvers at the end of the 18th- and into the early years of the 19th century.

Treating animals

Experienced men like Bakewell carried out many routine husbandry tasks themselves. For instance, they would likely have castrated the tup (ram) lambs at a few days of age, tying a strong cord around the top of the testicles, cutting them off, or their shepherd may have bitten them off. They would also have undertaken 'twitching' (castrating older tups) by tying off the codpiece with a strong cord, the end coated with shoemakers' wax and turpentine spirit applied to the ligature.[2] But as new methods evolved, such as removing testicles by excision, specialists, experienced farriers or 'cow doctors' were required to undertake these new castration techniques. Richard Dovey in the Staffordshire area went from farm to farm castrating livestock for clients ranging from large and wealthy landowners such as John Crewe of Crewe Hall to smaller farmers, like George Tollet, who was very hands-on running his 500-acre farm.[3]

Although castrating farm livestock is commonplace in Britain today, the procedure of sterilising (ovariectomising) female livestock is no longer a common practice. But during the late 18th century, female cattle, sheep, and pigs routinely underwent a surgical procedure (speying) to stop them from coming into season. Specialist English graziers, whose trade was fattening cattle on grass and turnips for the London markets, particularly liked to purchase speyed heifers, being prepared to pay a premium for them. They considered these speyed heifers settled down quicker and fattened more efficiently than their breeding counterparts. Still, some Welsh drovers exploited this desire for speyed heifers, and anecdotal tales abound of English graziers' expensive Welsh 'speyed' heifers miraculously calving![4]

Speying cattle could be carried out at different ages: at two to three weeks, two to three years, or at the end of a cow's productive life. The benefits of speying or castrating in the first few weeks after birth were a larger frame, a greater capacity to lay down fat more quickly and a quieter animal that was more docile to handle than an entire one. Still, if a farmer did not want to commit to speying calves at such a young age, he could wait until animals reached two or three years of age before deciding whether to breed from them. The surgical risks in older animals were less than in young calves, and whilst speying at an older age did not increase the animal's overall frame, it did remove the distraction of them bulling (coming into season), every three weeks. This meant not only were speyed heifers quieter workers under harness, but they put weight on faster. Older cows could also be speyed at the end of their productive lives or earlier if they were poor milkers. Again, the ability to stop these old cows from bulling appealed to farmers, wanting to remove any distraction that got in the way of fattening them at the end of their lives.

Male pigs not required for breeding were routinely castrated at around seven weeks, whilst older sows not considered good breeders or that were inclined to fat could be speyed.[5] The well-known farmer and author Richard Parkinson, who visited Dishley during both Bakewell and Honeyborn's time, recounted how he bought a sow from his brother that had recently farrowed. He intended to mate her but could not get her into pig. It later transpired that the sow had been mistakenly speyed at his brother's farm.[6] Ewes could also be speyed, particularly at the end of their productive lives. It was a procedure that farmers like Bakewell found useful for animals considered not good enough to breed from or that had become too old to breed.

As the procedure involved surgical intervention and incurred more risks than castration, a person trained in the technique was required. One such person who speyed animals was William Bailey from Mildenhall. He speyed livestock in the East Anglian area around 1787 with his son. Still, as the procedure was reported in the agricultural press in 1782, they may have been carrying it out from around this time. The operation, which appears to have been the same in all species, involved removing some or all of the ovaries. The technique Bailey employed was to remove 'the lusts' located at the ends of the ovaries. After their removal, Bailey or his son would throw a handful of salt into the surgical incision before stitching up the wound with a double wax thread.[7]

As veterinary surgeons were not widespread, if a heifer became infected after the procedure or any livestock became ill or injured, there were limited options for farmers and landowners. One veterinary surgeon Bakewell may have known was Richard Lawrence, who worked as a vet in the Birmingham area around 1790. Around 1801, Lawrence relocated to London and wrote the first of his treatises on veterinary medicine. Like most of his books, this focused primarily on horses.[8] He promoted his publications at major agricultural shows and sheepshearing meetings and would have known Honeyborn, another regular attendee.[9]

Because men with specialist veterinary knowledge, such as Lawrence, were thin on the ground, Georgian farmers often consulted human surgeons to help with their livestock's ailments. Indeed, the Smithfield Club's first committee not only included Honeyborn and two other influential Dishley men, Nathaniel Stubbins and Joseph Wilkes, both friends of Bakewell, but also Stephen Kent, a Southwark surgeon.[10] Another medical man involved with a group of farmers was Joseph Vickers, a Loughborough surgeon. Vickers was a good friend of Bakewell, frequently at Dishley and likely advised him on veterinary problems.[11] As the Dishley Society awarded Vickers a silver cup in recognition of finding a cure for sheep suffering from the complaint of 'overflowing of the Gall', he doubtless also helped other Dishley members with their livestock's ailments.[12] Indeed, Vickers was so well in with the Dishley contingent that he added his name to the protest letter that Honeyborn and other Dishley members within the Smithfield Club instigated in 1800 after they felt the Club's president, the 5th Duke of Bedford, had reneged on a resolution members had voted on the year before.[13]

But, vets and surgeons cost money, and in the late 18th century, most good farmers and landowners would have possessed a reasonable degree of practical knowledge, enabling them to deal with many livestock complaints themselves. There were several books they could have consulted, such as Joshua Rowlin's The Complete Cow-Doctor, to help them home-medicate their animals.[14] Even some landowners, such as Francis (Frank) Sitwell, who owned 3,000 acres in Northumberland, could treat their livestock. Sitwell wrote to *the Farmers'*

Magazine in 1806 to describe how he cured a young calf, a particular favourite of his, that had contracted Navel ill.[15] His successful treatment included dosing the calf with a concoction of port wine, powdered bark and laudanum, bathing its belly with an infusion of camomile flowers and marshmallows and feeding it on tiny amounts of milk laced with rhubarb powder.[16] Georgians were prodigious letter writers, and home-produced remedies like Sitwell's were often exchanged in letters between farmers of different social levels. When Thomas William Coke, the great Norfolk landowner, sent the Sussex tenant farmer John Ellman his own 'recipe' for eradicating worms in sheep, it was not considered unusual but rather a gesture of friendship between two keen Southdown sheep breeders.[17]

But not all ideas came free. Georgian agricultural improvers loved ingenuity, and many were hungry for ingenious ideas and inventions that might aid their farm enterprises and bank balances. Innovative men designed and marketed various products, hoping to gain the patronage of one or more eminent agriculturalists. Others spread their net more widely and marketed their inventions and ideas through subscriptions. The type of product or invention marketed in this way could be as diverse as a new design to catch rats, a better way to eradicate turnip root fly or a cure for scouring (diarrhoea) in cattle. Interested parties added their name to a subscription list circulated by the 'inventor', and each subscriber pledged a sum of money, usually a guinea, to get the idea into production. A typical list contained 500 or 1,000 names and would fill up more quickly if a renowned agriculturalist endorsed the idea or product. Only when the subscription list was complete would the inventor release his product, idea or remedy. Subscription lists were usually circulated at farming gatherings, including shows, meetings and the weekly livestock markets, and often advertised in agricultural journals and newspapers. The post-show dinners, when the men had been drinking and were in a genial mood, were always a good time to hand around subscription lists; the instigator hoping that alcohol had sufficiently loosened the men's wallets.[18]

It was not only novel ideas and inventions that were marketed this way but also agricultural and veterinary books. This was the method that both Lawrence and Parkinson adopted when promoting their books.[19] Often these subscription lists came to nothing or took years to gain the requisite number of subscribers. One subscription list that took three and half years to fill was *Bellamy's Bath Powder*. Mr Bellamy developed a remedy to stop scouring in cattle, and his subscription list offered this 'recipe' along with other remedies to cure cattle diseases. But even after involving several notable agriculturalists to trial and endorse it, Bellamy struggled to fill his list.[20] It was only after he advertised in *Annals of Agriculture* that Bellamy obtained the 500 subscribers he required.[21] Honeyborn may have subscribed; he attended many meetings where Bellamy promoted his remedy and circulated his list.

The Wool Trade

Although the significant advancements Bakewell and others had made in livestock breeding meant that meat had taken the upper hand over fleece, wool production remained important. Farmers sold their fleeces to wool staplers on-farm or through specialist wool fairs such as those held annually at Thetford and Lewes. Wool stapling originated in the Middle Ages. Traditionally, a brogger (broker) bought wool from a farmer and sold it to a merchant or jobber. But by the 18th century, broggers had morphed into jobbers, and merchants had become wool staplers, acting as the middlemen between producer and manufacturer. Staplers bought fleeces directly from the farmer or through jobbers and then sold them to the manufacturer after sorting and grading the wool to suit the buyer's requirements. While broggers or jobbers also bought wool from farmers, unlike staplers, they did not own warehouses or hold sufficient capital to employ a workforce to sort and grade the wool.[22]

From the Middle Ages, merchant wool staplers (mercers) had been wealthy men, many of them controlling the export of wool and often diversifying into other products, becoming very rich in the process and gaining positions of high authority, such as lord mayor or alderman. The story of Richard (Dick) Whittington illustrates well how a wool stapler like Whittington could move up in society. Although in the late 18th century, the rewards for wool stapling were nowhere near as lucrative as their predecessors had enjoyed, there was still good money to be made for wool staplers who owned their own businesses. This made them a strong body of men in the farming world, and they could be very intransigent to deal with. However, as wool prices fell in the early 19th century, many were made bankrupt.

One influential stapler was William Oakley, who operated out of Southwark. He was one of several staplers who were always to be found at the early agricultural shows and sometimes judged at them. Honeyborn knew Oakley, as the two frequented the same shows for many years.[23] Another stapler Honeyborn likely met was Henry Lacocke, who often attended the London farming shows. He had been a foreman in John Buxton's London wool operation but became Sir Joseph Banks's right-hand man, helping Banks manage George III's Merino flock. Through Banks, Lacocke sorted the wool of many of Banks's eminent friends. He became a specialist with Merinos, and Banks considered him one of the first practical experts in Spanish sheep and one of the first to classify them.[24]

Although Lacocke was well respected for his prowess with the Merino breed, it was Thomas Walton who was considered the renowned livestock expert by the most eminent men interested in agricultural improvement during this period. Although dubbed 'the properest man in England with sheep' and called 'Walton, the famous tup man', his cattle expertise was also revered.[25]

Walton, the son of a draper, was born in 1746 in Castle Donington, a short distance from Dishley Grange.[26] It was at Dishley, under Bakewell's tutelage, where Walton honed his expertise in livestock husbandry. After he left Dishley, Walton's expertise, skill, and knowledge gained him an illustrious list of clients, including George III, the Dukes of Bedford, Coke of Norfolk and the 2nd Earl Spencer. Not only was Walton employed by the agricultural hierarchy of the country to advise on their livestock, but they also valued his companionship. So, despite his humble origins, he was invited to stay in many palatial residences while on business.[27]

Bakewell first took Walton on as a pupil and then as his assistant, the pair working alongside the Dishley shepherd John Breedon.[28] Walton left Dishley shortly before Honeyborn moved to Dishley Grange. He possibly thought his prospects were limited by the impending arrival of Bakewell's nephew. But although they never worked together, Walton and Honeyborn knew each other well, as he did the other Dishley Society members. Walton met up with them regularly at all the major agricultural meetings. Indeed, even in later years, Walton's connection with Bakewell and his close band of associates never waned, and he was even considered a Dishley man.[29]

Walton left Dishley Grange in 1786. In a letter to his friend and fellow livestock breeder George Culley, Bakewell divulged that Walton had gone to work for John Dutton, 1st Baron Sherborne, in Gloucestershire.[30] Sherborne was a keen New Leicester breeder for many years and a close friend and brother-in-law of Coke. Walton may have already met Coke at Dishley before he went to Gloucestershire. Coke and Bakewell were friends, and on one visit to Dishley, Bakewell taught Coke how to condition score sheep.[31]

Condition scoring sheep is an easy and accurate method of estimating the condition or 'nutritional well-being' of a sheep flock. It requires an assessment of the amount of muscle and fat covering the backbone and the short ribs of each sheep. This gives a picture of the sheep's store of energy. It is the best method for monitoring pregnant and lactating ewes but is also useful for monitoring the growth of weaners. It is done by feel on a scale of 1-5: a score of 1 indicates emaciation, whilst a score of 5 indicates an animal is obscenely overweight. A score of 3 is generally considered optimum.[32]

It was Walton rather than Bakewell who Coke later asserted taught him everything he knew about sheep.[33] Walton worked on and off for Coke for many years, and the pair inevitably became friends. The painting and subsequent print of *Coke among his Southdowns* shows Coke and Walton together with the Holkham shepherds.[34] Walton still auctioned the livestock at Coke's Holkham sheep shearings into his seventies.[35] Through Coke, Walton met Banks, Coke's good friend. After Lacocke retired, Walton became Banks's assistant with George III's merino flock, summoned from wherever he was working in the country if there were problems such as sheep scab in the royal flock.[36]

As Walton travelled the country advising on livestock matters, he kept Coke informed of anything he thought might interest the Norfolk landowner and passed messages between Coke and his friends. In return, Coke passed messages from Walton on to his tenant farmers.[37] This exchange of information between Walton and Coke was not unusual because these journeymen specialists like Walton were a conduit of news and gossip for the men who employed them. Richard Dovey, who castrated lambs in the Staffordshire area, passed information between his clients.[38] Lacocke was a mine of information for Banks, keeping his employer and friend updated with news and gossip. On one occasion, Lacocke conspiratorially told Banks how he wished he had never got involved in sorting and grading the wool from the agricultural writer Arthur Young because it had arrived in a disgraceful state![39]

George Garrard, the livestock artist, portrayed Walton and a handful of other specialists in *Wobourn Sheepshearing*, his seminal agricultural print from the early 19th century.[40] Along with eminent breeders, graziers and other livestock and arable men, Garrard referred to these specialists as 'Celebrated Agriculturalists', seeing them as integral in moving agriculture improvement forward in the early 19th century.[41] Still, the part these journeymen livestock specialists played within the farming world and their role as the promulgator of news and information within the farming community is an area that needs further research.

Endnotes

1 Chambers, J.D., & Mingay, G.E., The Agricultural Revolution: 1750-1880 (London: B.T. Batsford Ltd., 1984 ed.), p121.

2 Richard Parkinson discussed tying or cutting testicles off, whilst John Lawrence said shepherds routinely bit them off (all three methods are still in use). Richard Parkinson, Treatise on the Breeding and Management of Livestock, Vol I (London: Cadell & Davies, 1810), p287-9; Lawrence, J., A General Treatise on Cattle, the Ox, the Sheep and the Swine (London: Sherwood, Neely & Jones, 2nd ed. 1809), p296.

3 George Tollet to John Crewe, 21 April 1805, F/TWC2, Holkham Hall Archives.

4 Colyer, R., 'Welsh Cattle Drovers in the Nineteenth Century, Part I', in *National Library of Wales Journal*, Winter Vol XVII/4, (1972), https://www.genuki.org.uk/big/wal/Archives/NLWjournals/CattleDrovers1, (accessed 15 April 2024).

5 A Farmer and Breeder [John Lawrence], The New Farmer's Calendar (London: H.D. Symonds, 4th ed., 1802), p536.

6 Parkinson, R., Treatise on the Breeding and Management of Livestock, Vol II (London: Cadell & Davies, 1810), p255.

7 Foote, J., letter dated 13 December 1796 in *Annals of Agriculture and other Useful Arts*, Vol XXVII, p636-9.

8 Lawrence, R., An Inquiry into the Structure and Oeconomy of the Horse (Birmingham, 1801).

9 Richard Lawrence to Thomas William Coke, 21 June 1805, F/TWC 2, Holkham.

10 *Annals,* Vol XXXVI, p227-30.

11 B. Smith, whose father took over Dishley Grange after Honeyborn's death said Vickers 'was a constant guest at Bakewell's table'. Smith B., 'The Leicester Sheep', *The Gardeners' Chronicle and Agricultural Gazette for 1871,* Vol XXXI, Part 2, (1871), 1560, [online facsimile], https://books.google.co.uk/books?id=9I8SGsPl5JAC, (accessed 17 April 2024).

12 Robert Bakewell to George Culley, 22 October 1791, in Cecil Pawson, H., Robert Bakewell: Pioneer Livestock Breeder, (London: Crosby, Lockwood & Son, Ltd., 1957), p159.

13 *Annals,* Vol XXXVI, p241-2.

14 Rowlin, J., The Complete Cow-Doctor; or Farmer's Companion, (London, 2nd ed. 1799).

15 Navel or Joint ill is the result of infection entering through the umbilical cord of new-born animals. Unless treated quickly, the condition often results in death.

16 Frank Sitwell, 6th May 1806, *The Farmer's Magazine* 1806, Vol VII (Edinburgh: Arch. Constable & Co.), p285-6.

17 John Ellman to Thomas William Coke, 2 November 1803, F/TWC 2, Holkham.

18 For more information about subscriptions during this period, see 111432 | Matthews, H., (2022) 'Indeed amongst agriculturalists, there is something like free-masonry – we are all brethren': Exploring agricultural friendship in late Georgian England. PhD thesis, University of Reading. doi: https://doi.org/10.48683/1926.00111432, p358-67.

19 Richard Lawrence to Thomas William Coke, 21 June 1805, F/TWC2, Holkham.

20 Matthews, 'Indeed amongst agriculturalists…', PhD thesis, doi: https://doi.org/10.48683/1926.00111432, 359.

21 'Miscellaneous Information: Remedy for Scouring', *Annals,* Vol XLI, p478-9.

22 For more information on the history of wool stapling in England, see *Company of Merchants for the Staple of England,* [website], https://merchantstaplers.co.uk. For more information on the company and its archives, see http://archiveshub.jisc.ac.uk/data/gb193-staple (both websites accessed 17 April 2024).

23 For more information on the group of agriculturalists, including Honeyborn and Oakley, who frequented the agricultural shows in late Georgian England, see Matthews, 'Indeed amongst agriculturalists...', PhD thesis, doi: https://doi.org/10.48683/1926.00111432 For more information on William Oakley, see his biography in this thesis, 473.

24 For more information on Henry Lacocke and his relationship with Sir Joseph Banks, see Carter, H.B., His Majesty's Spanish Flock (Angus & Robertson, 1964), Regarding his expertise as a classifier, p429. For the various letters between Lacocke and Banks, see Carter, His Majesty's Spanish Flock; The Sheep and Wool Correspondence of Sir Joseph Banks 1781-1820 (The Library Council of New South Wales in association with the British Museum (Natural History), 1979).

25 Sir Joseph Banks to Richard Snart, 8 December 1807, letter 1298; Sir Joseph Banks to George John Spencer, second Earl Spencer, 31 July 1809, letter 1362, in Carter, The sheep and wool correspondence, p475 & 493-4.

26 For more information on Thomas Walton, see Matthews, 'Indeed amongst agriculturalists…', PhD thesis, doi: https://doi.org/10.48683/1926.00111432, 502.

27 For more information on Thomas Walton and his relationship with the agricultural elite, see Matthews, 'Indeed amongst agriculturalists…', PhD thesis, doi: https://doi.org/10.48683/1926.00111432, p191-202.

28 Parkinson, Treatise on the Breeding and Management of Livestock, Vol I, p281.

29 In *Wobourn Sheepshearing*, George Garrard portrayed Walton between two other staunch Bakewell disciples in his print, whilst reports of attendees at shows and meetings invariably listed him with the Dishley breeders.

30 Robert Bakewell to George Culley, 11 April 1786, in Pawson, *Robert Bakewell,* p103. Pawson mistakenly transcribed Lord Sherborne in Bakewell's letter as Lord Lisbon.

31 Bakewell taught Coke to condition score when he visited Holkham soon after Coke moved to Norfolk. Stirling, A.M.W., Coke of Norfolk and his Friends (London: John Lane The Bodley Head, 1912 edn.), p159.

32 'Condition Scoring of Sheep', *Agriculture and Food, Department of Primary Industries and Regional Development*, [website], https://www.agric.wa.gov.au/management-reproduction/condition-scoring-sheep, (accessed 16 May 2024).

33 'Monthly Register July 1809-Norfolk Agricultural Society', *The Agricultural Magazine*, Vol V (1809), p66-7.

34 Thomas Weaver, *Thomas William Coke and his Southdown Shee*p, Oil on Canvas (1807). Norfolk: Holkham Hall.

35 'Holkham Sheep Shearing', *The Agricultural Magazine*, Vol III (1809), p131; The Holkham Flock Book shows Walton's involvement, particularly in later years. His involvement is documented until 1819. 'Holkham Sheep Shearing Years 1804, 1805, 1806, 1807', E/X3, F/TWC2, Holkham.

36 Banks to Snart, 8 December 1807, letter 1298, Carter, The sheep and wool correspondence, p475.

37 Thomas William Coke to Sir Joseph Banks, 12 May 1806, letter 1238, in Carter, The sheep and wool correspondence, p455-6; Thomas Walton to Thomas William Coke, 12 August 1811, F/TWC2, Holkham.

38 George Tollet to John Crewe, 21 April 1805, F/TWC2, Holkham.

39 Henry Lacocke to Sir Joseph Banks 8 October 1799, letter 735, Carter, The sheep and wool correspondence, p317-8.

40 Garrard, *Wobourn Sheepshearing*, Uncoloured Mixed Method Engraving, 545 mm x 785 mm (1811), Painted and Published by G. Garrard, 11 May 1811. 'Stipled and Outline Etched by M.N. Bate, Figures and Landscapes by J.C. Stadler, Lined by T. Morris, the whole touched and arranged by the original painter'.

41 George Garrard, *Proposals for Publishing a Print of the Woburn Sheep Shearing from a Picture by Mr Garrard, Associate of the Royal Academy* (1811), front cover.

PART THREE

Bankruptcy and Related Matters

CHAPTER SEVENTEEN

Robert Bakewell's Bankruptcy: Background and Process

Janet Spavold

> Bankruptcy is perhaps the greatest and most humiliating calamity which can befall an innocent man.
>
> (*Adam Smith*)

The family history shows that the Bakewells had a successful farming record over several previous generations, gradually increasing the family's wealth. They owned some of their land and were reliable tenants of other land, which was leased over varying periods as and when they needed extra grazing. This record of good husbandry no doubt influenced the Phillipps family of Garendon Park when they were looking for a new tenant for Dishley Grange after the unexpected death of Benjamin Clark in 1707. It seems to have taken two years to sort out Clark's affairs, as he died intestate, before the Bakewells moved there in 1709. At this time, large estates were expanding as smaller ones sold up because of heavy taxation for the 1692-1715 wars; large landowners managed better because they had non-agricultural income to help. They actively sought good, reliable tenants for their farms and they were able to undercut smaller landlords by letting on favourable terms.[1] Robert Bakewell (2), took Dishley Grange 'at an easy rent', and the Phillipps rent rolls show that the landlord paid the land tax, not the tenant.[2] It was a 99-year repairing lease which ran until 1808. The farm covered 350 acres then, 'with the Privilege of Plowing without restraint, 250 acres or upwards'.[3] Three generations of Bakewells clearly observed the terms of the lease fully; when Robert (4) died in 1795 and the farm passed to his nephew Robert Honeyborn, the Garendon Estate records show it was then valued at £711 3s 5d,[4] and let to him at £735 p.a. It was described as 'House in good repair and the outbuildings, which are many, pretty generally so'.[5] A contributing factor in Robert's (4) bankruptcy would have been the cost of not only keeping existing buildings such as the medieval barn, which measured 50 by 15 yards,[6] in good repair, but building new ones for his experiments, for example, the one built to measure pig growth. The significance of a long, secure tenancy for an experimental farmer cannot be underestimated. Farming success brought social respect, with the tenant referred to as 'Mr Bakewell' in his landlord's rent books.

But long leaseholds were not common, a matter returned to in the following chapter.

When Robert (3) married Rebecca Brodhurst in 1722 her marriage settlement included land owned by the Bakewells at Swepstone: the three Butt Closes, the four Moor Closes, some other land and a tenanted messuage as well as land from her family. He was able to lease or mortgage the Swepstone lands or sell them if 'other sufficient estates' were substituted.[7]

Robert (3) rented extra parcels of land from his landlord during his tenancy. In June and August 1739, he paid £10 10s and £5 5s for clover grass.[8] Between 1740 and 1749, then 1751 and 1753, he rented extra land at Kettlebrook, part of Bishop's Meadow on the edge of the Dishley land for £20 per year; furthermore, at Michaelmas 1751 and Lady Day 1752 he paid £20 for 'part of Kettle Brook Acre', which adjoined his farm, with a repeat payment in 1752 and 1753. Between 1749 and 1751, however, the rent was £30 per year.[9] Like the other Garendon tenants, he was always in arrears, paying each rent about two weeks before the next was due.[10]

Robert (3) also owned a cottage and 52 acres of land including the Barn Closes (20 acres) in Swepstone, Leicestershire, which, unlike Dishley Grange, he could use as security to borrow money. He had bought most of the property in 1728 for £300,[11] and the Barn Closes for £250 in 1729.[12] In 1756 he borrowed £250 on a 21-year mortgage at £3 10s annual interest from Henworth Newton, a tanner from Ashby de la Zouch, secured on the Swepstone cottage and the Barn Closes.[13] It has been suggested that this was to fund his son's travels to further his farming education but, from the dates, it is more likely that the money was used to build a new farmhouse on post-enclosure lines and add extra farm buildings. From around this time, Robert (3) increasingly passed the management of Dishley Grange to Robert (4), who in 1765 won first prize at the Ashby Horse Show.[14]

In November 1767, Robert's (3) daughter Rebecca married Joseph Smith of Hayhall, Yardley, in Worcestershire. He was a toymaker and part of the Bakewell family's Unitarian and Nonconformist network. Toymakers then made small items such as buckles, hinges, buttons and hooks in steel or other materials, and the term also covered decorative small items in tortoiseshell, gold or silver. Although she was a widow, a marriage portion was provided for her. Her sister Hannah never married but stayed at home as housekeeper for her brother. As there was no inn in the area, visitors stayed at Dishley Grange, where Hannah would receive the visitors and look after them, sometimes for weeks. Some have suggested that his very generous hospitality was partly to blame for Robert's (4) later bankruptcy.

Robert's (3) next financial move in 1768 was to set up a family trust with his wife Rebecca to benefit their children Robert (4), William, Hannah and Rebecca. Their son Robert (4) was to manage it, helped by John Ashworth as their trustee. Two properties were put into trust: Rebecca's marriage portion of Flaxholme Close in Little Eaton, Derbyshire, worth £400, and 36 acres 3 roods 20 perches of Swepstone land then rented to John Woodward and John Clayton. The Swepstone property mortgaged to Newton was also included.[15] Ashworth was living at Dishley Grange as one of Robert's (4) agriculture pupils at the time. He came from a highly-respected Nonconformist family in Daventry, Northamptonshire, where the family owned the Wheat Sheaf Inn and ran a Nonconformist academy. He was described as a gentleman and later acted in Robert's (4) bankruptcy.

In 1769 Robert's (3) wife Rebecca died.

By this point the Bakewells, father and son, had some independent income from the rent on the bulk of their Swepstone property (now called Dishley Farm) and from Rebecca's marriage portion land, and the income from Robert's (4) farming. One steady and substantial element of income was the wool he sold each year from his large flocks of sheep. The farming income could be erratic from 1760 when Robert (4) first hired out his rams at modest prices. Payments from the ram and stallion hires might not appear until the animals returned to Dishley after some months away. Count Rochefoucauld later commented on the unreliability, particularly of his Scots customers who simply did not pay, while Bakewell (4) himself later commented that he went bankrupt because an Irish ram hirer had defaulted. However, Count Rochefoucauld confirmed that by 1784 the unnamed defaulters had paid their debts.[16] Rochefoucauld the younger, writing in 1785 after his visit to Dishley, also said that it was down to Irish defaulters.

While the valuable tups went out in horse-drawn carriages, any ewes sold, for example to Culley or other Borders farmers from 1767 were probably taken to Northumberland as a by-employment by the drovers returning from London using the established route through the Midlands: Leicestershire and Derbyshire to Doncaster and on to the Great North Road (the A1). As droving was seasonal, sheep collected on the return north in spring might not be paid for until the autumn drove south when the drovers could bring payment, in either cash or promissory notes drawn on specialist droving banks. There were costs for the drove: grazing for the animals and food and accommodation for the drovers. The drovers used green lanes to avoid toll roads where they could but some tolls had to be paid. While cattle droves could take eight or nine days to cover 100 miles, sheep needed longer.[17] It is easy to see how cashflow problems could build up.

Outgoings for the family were their rent on Dishley Grange and the costs of its maintenance, the interest on the Newton mortgage, generous hospitality costs for visitors to Dishley, wages for the farm staff and the self-imposed costs of Robert's (4) long-term experiments developing both his livestock and his land. George Culley, visiting in 1765, thought that Robert (4) 'employs more men than he has occasion for ... I suppose [he] pays more money in wages than rent'.[18] This would have been correct, as this was before Bakewell's rent was increased. It should be remembered that there were more people at Dishley than the Bakewells; some farm servants and female domestic servants lived in but there does not seem to be any record of exactly how many people were there. The employees with families lived out, in the immediately surrounding area. Visitors commented on the way their children, brought up to help, were able to handle the bulls safely because of Robert's (4) policy of kindness to animals. As well as the farm buildings, he stood the whole cost of land development such as the canal and his experiments in fencing and watering.

Financial problems were certainly building and in April 1770 Robert (4) mortgaged more of the Swepstone property to Catherine Orton of Stothill, Kingsbury, in Warwickshire for £500,[19] at 4% per annum for 500 years. It was a 33-acre messuage comprising the Great Bull Close and the Little Bull Close, and the four Thorn Closes.[20] While there is documentary evidence for this debt, it is clear from the way the bankruptcy proceedings developed that Robert (4) took out 15 or more other loans between about 1770 and 1776. Three may have been informal loans without written confirmation or maybe even interest to pay as they were loans from family and close friends in his Unitarian network, but the others certainly expected interest on their money. There is no suggestion that any of these debts were taken out by his father.

His problems were becoming known by 1771. An unnamed 'Scotch Farmer' visited that year and his 'Observations at Mr Bakewell's in 1771' noted that Robert (4) 'has commissions from all parts of the Kingdom and Ireland, and gets prices quite extraordinary; he has run into great expense, and many farmers in his neighbourhood treated him as a whimsical man, and gave out that he was bankrupt; and in fact, before he completed the improvement of his stock as it now [1797] is, he was run very low; but for some years past his returns have been great, though his losses are sometimes very considerable'.[21] That year he had about 60 horses, 150 cattle and 400 sheep on the farm.[22]

Also in 1771, Robert (4) was 'seriously embarrassed by a major drought', and having sold some bulls to Jamaica, had difficulty getting payment for them.[23] Another factor beyond his control could have been disease, such as the widespread 'most fatal pestilence' (probably anthrax) noted in the Annual Register for 1770,[24] or scrapie, first described in the 18th century. Its 'rapid expansion' was 'a result of the practice of inbreeding to improve the quality of wool'.[25] While Robert (4) was not primarily aiming to improve wool quality, he certainly was using inbreeding, and Young said that his wool quality was equal to any. While it is on the record that liver fluke affected Robert's (4) sheep, there is no written record for scrapie and, if it affected his farm, it may not have been recognised as such. The 1768-71 outbreak of cattle plague (rinderpest) however was not severe in Leicestershire. Later, he lost £500 through freak weather; having been offered this sum for one of his stallions, which he refused, 'a few days later it was struck dead by lightning'.[26] A similar event in 1788 occurred when, having let 'one of his finest rams' for the coming season for 400 guineas,[27] the animal fell ill and died.[28] It was a severe loss. Such events were unpredictable and there was no insurance. In addition, there was a British credit crisis in 1772-3.

In 1773, Henworth Newton died, and the mortgage transferred to his executors who were his son John Newton and his kinsman William Newton, a surgeon of Burton upon Trent. In May that year, Robert (3) died; it did not ease his son's financial problems. When Catherine Orton no longer wanted to continue the mortgage with Robert (4) in 1774, it was assigned to Elizabeth Aston of Stonehill, Lichfield, but the loan was increased to £1,000,[29] and the interest increased to £4 per cent.[30] Robert (4) was clearly not seen as totally financially reliable at the time.

There were more problems in 1774. Robert's (4) generous landlord Samuel Phillipps of Garendon died, and after probate was granted in April his widow Mary was left £7,000 and a life interest in, and life ownership of, Garendon Park. Mary came from the Allsopp family of lawyers and was related to the lawyer Caleb Lowdham (part of the Bakewell network), and distantly related by marriage to Robert (4). Mary immediately began to modernise the Garendon Estate, perhaps with Lowdham's advice, to maximise her income. The first thing she looked at was her rent book and the family tie to the Bakewells did not deter her from raising his easy rent to a commercial level. On Lady Day 1774 she must have given Robert (4) the necessary quarter's verbal notice of a rent rise, followed by the formal written year's notice at Michaelmas 1774. The term was reduced to 75 years ending in 1784, the rent was raised to £310 10s per year,[31] and the repairing clause maintained; the first half-yearly payment was due at Michaelmas 1775.[32] He had to raise an extra, unplanned sum at a time when his cashflow was already strained. This is probably the reason why he sold the four and a half acres of land in Little Eaton, Derbyshire, inherited through his mother, to his tenant Henry Peate early in 1775.[33] The £340 it brought gave him temporary security at Dishley, knowing he could pay his rent, which he duly did on Lady Day 1776. This meant that when he was declared bankrupt on 3rd June 1776, there was no need to invoke the deferment of rent which was customary when tenants were waiting for long-distance payments for stock, or similar monies due. Maybe the fact that the planned new buildings at Dishley Grange had by then been completed also helped.

A later namesake, quoted by Pawson on the rent rise, shows how family stories can distort events. While praising Robert (4) for his enormous financial success with letting his stock, this Robert Bakewell (born 1775) exaggerates the extent of the farm and mistakes the landlord as Thomas March Phillips Esquire, who did not inherit Garendon Park until after Mary's death. He claimed that the rent was raised to £400 per year and said that Robert's (4) response was 'that it was only taking half another sheep from him'.[34] If this is accurate it shows a remarkable degree of prescience on Robert's (4) part as his letting prices at this date were nothing like as high as in later years.

The outbreak of the American War of Independence on 19th April 1775 severely affected trade and caused a recession and high taxation at home. The impact on Bakewell's business was probably mixed. The trade in breeding animals was not affected as they were exempted from American import duties, and raw wool was also tax free.[35] Robert (4) did not export live animals there yet and it is unlikely that he exported his raw wool; this would have been sold to wool merchants here for further processing. The impact on his business would be through the immediate cessation of any American trade, and the inability of the wool merchants to buy on their usual scale. Marshall wrote that the New Leicester sheep 'was not deficient in wool, wethers' fleeces generally going at the rate of 4 to the tod (of 28lb), and those of ewes about 4½ – say, wethers' 6lb-8lb, ewes 5lb-7lb'.[36] At this date, 1lb of raw wool sold for about 10d,[37] but it fell to 3d or 4d. With 400 sheep, an unexpected fall of two-thirds in his wool income would indeed be a 'great and unavoidable Loss', as he put it in the petition. He discussed the wool trade, and his support for the Parliamentary Bill affecting it, in his letters to Young. The war ended on 3rd September 1776 so his finances would have been affected for two if not three years before trade was eventually back to previous levels.

Through the rest of 1775, Robert (4) continued his business even if it was difficult. In September, he asked John Holland, his only formally bound apprentice,[38] to write to the Duke of Buccleuch to acknowledge the receipt of £150 on his behalf and to say that Bakewell would write himself 'in a few days'.[39] This was probably for a ram, but it is not clear whether the animal was destined for the Boughton House Estate at Kettering, Northamptonshire, or for one of the Duke's Borders estates. At Michaelmas he paid his half-year's rent, and the other half at Lady Day 1776.[40] Early in 1776 he sent more bulls to Jamaica,[41] but payment would be delayed for a year or more until after they had arrived. Like the rams he let at home, the terms of payment were such that he did not receive the income until they had 'proved their worth'. His petition tells us that he was also sending animals to Germany.

While Bakewell was obviously a very good manager of his livestock, the same could not perhaps be said for the management of his financial affairs. Perhaps he could even be described as extravagant in his management of the farm.

> The expanded heart of this man demanded more capacious means for the gratification of its generous desires; and it is evident, from his conduct, that he was ambitious rather of the honour than the profit of his calling.[42]

Disaster struck in 1776, when a Commission of Bankruptcy was issued against Robert (4) 'on or about' 3rd June 1776.[43] On midsummer quarter day, when Robert (4) was due to pay interest on his loans and mortgages, he stopped the payments.[44] While some of his creditors were sympathetic, one especially was not; John Smellie of Nottingham, the son of a wealthy and well-connected hosier. He was also distantly related to Robert (4) by marriage. Smellie junior had a web of loans and borrowings and needed Robert's (4) interest payments to pay interest that he owed on his own loans. From the sequence of events, it was possibly Smellie who triggered the bankruptcy and forestalled Bakewell's friends who were arranging help, fomenting the acrimony and ill-feeling among the creditors that dogged the bankruptcy proceedings for so long. The total borrowed from the unnamed creditors is not known, so it is impossible to estimate the full extent of Robert's (4) indebtedness.

Some of the reasons contributing to his cash flow problems have been discussed above and elsewhere, but there would have been others that have left no record. Robert (4) himself did not write about it other than in his petition for help during the bankruptcy proceedings, and his statement there that it was due to his 'Great and unavoidable Losses to the Amount of many Thousand Pounds' must be given weight. His friends ascribed the problems especially to non-payments from the Scots and Irish breeders; they were probably the largest single factor in a wider problem and Bakewell himself blamed the default of an Irish ram lessee.[45] This could be Mr Thomas Johnston who, as described in Chapter 8, had been illegally smuggling livestock into Ireland 'for several years past' at 'vast expense' and very high risk prior to 1777. The cargo caught by the authorities in 1776 included 30 rams and ewes, which starved to death when Johnston could not provide proper papers for them. Following a petition to the Irish Parliament in 1777 he was granted financial support in recognition of his success in improving Irish livestock through his smuggling. However, debts incurred illegally, like gambling debts, were not enforceable in law so that while he could then pay his debts over the sheep, there was no mechanism to do so. If the unnamed Irish ram lessee was indeed Johnston this legal impasse could be a significant part of Bakewell's (4) problems. However it was managed, it seems that the Irish debt was paid sometime in the late 1780s. Count Rochefoucauld was specific in his accusation, writing in 1784, that it was due to 'the enormous number of males that Bakewell let, and the prodigious number of non-payers among his customers',[46] and in the next year his son also accused the Irish. The slump in wool prices probably came second as a contributory factor.

In July 1776, Mary Phillipps married Sir William Gordon, who then became Bakewell's landlord during Mary's life. He was something of an adventurer who saw the Garendon Estate as a pot of money to be enjoyed while it lasted. Nichols observed disapprovingly that thereafter he and his wife 'were active enclosers and, taking full profit from their liferent possession, cut down trees to the value of £9,500, which they invested for their own benefit'.[47] Nevertheless, they did not evict Bakewell from Dishley Grange during the long process of bankruptcy and must have agreed to the sale of the lease when that was proposed. After 1777, Robert (4) may have invoked the allowable deferment of his rent, at least for two or three years.

Bankruptcies then were governed by two Acts of Parliament, 4 & 5 Anne I c4 1705 and 6 Anne I c6 1708, the second being necessary amendments to the first. They were the first Acts to legally recognise a difference between a debtor, who could be fraudulent or in debt through his own fault, and a bankrupt who was unable to pay his debts through no fault of his own. Debt was a felony with a possible death penalty, but bankruptcy was not. Debts could be pursued to the full amount with no discharge, but bankruptcy offered the possibility to agree a lower payment and discharge from debt. Bankruptcy was initially only available to merchants and traders; practical experience showed that trading credit could be affected by circumstances beyond the trader's control. Receivers of taxes, drovers, graziers, farmers and innkeepers were excluded from the advantages of bankruptcy. However, as graziers, farmers and innkeepers included some forms of buying and selling independently as well as their main trade, most such bankrupts were able to claim, for example, that they were dealers and chapmen (if graziers or farmers) or wine merchants (if innkeepers), so there was a way round it. Bakewell was cited as a dealer and chapman (trader), and his friends recognised that his problems were not directly his fault. The younger Rochefoucauld made this point, 'he let many of his beasts to the Irish who have never paid up, so that he was ruined without its being his fault'.[48]

Under the Acts, a single creditor, or a group of them, could apply by letter to the Lord Chancellor for a Commission of Bankruptcy but they had to deposit a £200 bond to guard against a malicious application. If a single creditor's loan was at least £100, or two creditors were owed £150, or three or more creditors were owed £200 or more, a Commission of Bankruptcy must be issued. Three commissioners were appointed with power to summon anyone who had information about the bankruptcy to be examined upon oath. They first had to serve 30 days' notice on the bankrupt in the *London Gazette* and at his home, giving the time and place of a meeting when he had to disclose all his assets except 'necessary Wearing Apparel'. If he surrendered to the Commission and conformed to their requests, he would be allowed 5% of the net product of everything that was received by the commissioners' assignees to a limit of £200, providing the assets could pay the creditors at least 8s in the pound. The commissioners then granted a certificate of bankruptcy and decided on the time and place of at least three creditors' meetings, to be notified in the gazette; they were usually also advertised in the local papers. Creditors had to prove their debts. The assignees were responsible for collecting the bankrupt's assets, holding meetings and distributing dividends, and they were elected by a majority of the creditors. They had to keep account books which must be open to the creditors' inspection. They could be removed and replaced by a majority creditors' vote when all the books must be passed to the new assignee. When the commissioners were satisfied that the bankrupt had made a full disclosure of assets, they granted a certificate of conformity, but the creditors could appeal if they did not think all assets had been found. Under the amending Act, the certificate of conformity had to be signed by four-fifths of the creditors by number and value. When all the assets had been collected and the final distribution made, the commissioners issued a certificate of discharge to complete the procedure, and a certificate could be granted without the full debts being paid if four-fifths of the creditors agreed.[49]

For the first time, the law recognised that bankruptcy could be due to 'unavoidable misfortune' rather than deliberate fraud. It gave individuals the chance of release from bankruptcy without full repayment, and with protection from future claims on these unpaid debts. For this to work, creditors had to be prepared to settle for a lesser payment, perhaps in order to get any repayment, and this could be difficult. An old Public Records Office's *Bankruptcy Glossary of Terms* noted that 'creditors frequently withheld consent until they had been paid what they considered a reasonable proportion of their debts and many, it is said, refused from spite or because they thought bankrupts had acted fraudulently'. Discharge could not be taken for granted; Carlos, Kosack and Penarrieta estimated it at only about 50%,[50] so if Bakewell's creditors demanded full or almost full payment, this would have contributed to the long delays in granting discharge. Postles's analysis of Leicestershire bankruptcies over the century concluded that while most people were discharged after about a year, those whose bankruptcies extended for 10 years or more had complex business dealings covering an extensive geographical area.[51] This was certainly the case with Robert (4). In the petition, he wrote that he had sent his stock into 'the Counties of Bedford, Bucks, Cambridge, Chester, Cumberland, Derby, Devon, Dorset, Durham, Essex, Gloucester, Hereford, Herts, Huntingdon, Kent, Lancaster, Leicester, Lincoln, Norfolk, Northampton, Northumberland, Nottingham, Oxon, Rutland, Salop, Somerset, Southampton, Stafford, Suffolk, Sussex, Warwick, Westmoreland, Wilts, Worcester, Yorks, into North Britain, Wales, Ireland, Germany and Jamaica'.[52]

Robert (4) had considerable support over his bankruptcy from within the family circle. His third cousin Robert (8) was a respected Leicester lawyer and local politician, the son of another lawyer. He lived first at Gresley Old Hall in south Derbyshire, then at Tempe Farm in Swepstone. He was experienced in dealing with bankrupts and astute enough to understand how to preserve a bankrupt's property for him within the legal limits. For example, Thomas Brentnall of Derby owned the Town Arms public house in Lenton, Nottingham, and half of a block of cottage property and land there, inherited in three one-sixth parts. Brentnall claimed to be a wine merchant, dealer and chapman in his bankruptcy papers. Robert (8) had lent him £500 on interest in 1775 as a mortgage on the cottage property and land. When Brentnall was declared bankrupt in 1779 Robert (8) was one of the commissioners, and two Derby men were appointed the assignees; one was Thomas Evans, the cotton mill owner. The mortgaged lands were held to pay the creditors. The assignees put two-sixths up for an auction which was held by private bids, not in public, and they declared Brentnall the best bidder at £1,050. He was probably the only bidder. Robert (8) agreed that the £500 mortgage on the land would remain in place, albeit at a higher rate of interest. Evans's daughter was due to marry Charles Upton, and Brentnall had borrowed £400 from Upton towards the purchase money. Brentnall then borrowed a further £150 in cash from Upton to complete it, which he added to the mortgage. The security for the £400 went to Upton's marriage trustees; it was two-sixths of the property and the trustees paid Brentnall a peppercorn rent on it until the mortgages were redeemed.[53] So, although he was a bankrupt, Brentnall was able to borrow money from supportive creditors in order to buy his own property back. It was probably on similar advice from Robert (8) that Robert's (4) assignees dealt with his bankruptcy and succeeded in preventing him losing his stock and lease. A noticeable feature of the assignees' methods is the use of private sales such as Brentnall's rather than open auctions run by an independent auctioneer. The private sales could be spread out over long periods giving them more time to get payments due to Robert (4) and eventually pay off his creditors. For the discontented creditors, private sales made it more difficult for them to know how well the assignees were working.

A timeline of Bakewell's bankruptcy is given in Appendix 7; it stretches from 3rd June 1776, when a Commission of Bankruptcy was issued against him, to November/December 1789, when he was granted a Certificate of Release, which discharged his bankruptcy. It shows the complexities and dangers of the process for the bankrupt and all those caught up therein.

Endnotes

1 Jones, E.L., Agriculture and Economic Growth in England 1650-1815 (1957), p160-61.

2 Record Office of Leicester, Leicestershire and Rutland (ROLLR), DE7454/23-34, 1738-69, Phillipps Rent Rolls.

3 The British Newspaper Archive (BNA), *Aris's Birmingham Gazette*, 12 Feb 1781.

4 Now £78,170. Measuring Worth (2004), 'Five Ways to Compute the Relative Value of a UK Pound Amount, 1270 to Present'. URL: www.measuringworth.com/ukcompare/

5 ROLLR, DE7454/1, 1797.

6 White in his Directory of Leicestershire and Rutland, (1863), p511, says it was demolished in 1856.

7 Staffordshire Record Office (SRO), D661/1/7/Parcel 5, schedule No. 1.

8 ROLLR, DE 7454/23, 1739.

9 ROLLR, DE 7474/22, rent roll 1732-53, p70; DE 7454/25, 1751-2 & DE 7454/26, 1752-3.

10 ROLLR, DE 7454/4.

11 Now £43,170. Measuring Worth (2004), 'Five Ways to Compute the Relative Value of a UK Pound Amount, 1270 to Present'. URL: www.measuringworth.com/ukcompare/

12 SRO, D661/1/7/Parcel 5.

13 SRO, D661/1/7/Parcel 5.

14 Wykes, D.L., 'Robert Bakewell (1725-1795) of Dishley: farmer and livestock improver', *Agricultural History Review*, 52, (2004), p40.

15 SRO, D661/1/7/Parcel 6, schedule No. 2.

16 Pawson, H.C., Robert Bakewell Pioneer Livestock Breeder, (1957), p40-1.

17 Bonser, K.J., The Drovers. Who they were and how they went. An Epic of the English Countryside, (1970), p171 and 175.

18 Wykes, D.L., 'Robert Bakewell (1725-1795) of Dishley: farmer and livestock improver', *Agricultural History Review*, 52 (2004), p47.

19 Now £73,730. Measuring Worth, (2004), 'Five Ways to Compute the Relative Value of a UK Pound Amount, 1270 to Present'. URL: www.measuringworth.com/ukcompare/

20 SRO, D661/1/7/Parcel 6.

21 'A Scotch Farmer', 'Observations at Mr Bakewell's in 1771', Annals of Agriculture, Vol 28, (1797), p592.

22 Wykes, D.L., 'Robert Bakewell (1725-1795) of Dishley: farmer and livestock improver', *Agricultural History Review*, 52, (2004), p40.

23 Wykes, D.L., 'Robert Bakewell (1725-1795) of Dishley: farmer and livestock improver', *Agricultural History Review*, 52, (2004), p45.

24 The Annual Register, (1773), p10.

25 Brown, P., & Bradley R., '1755 and all that: a historical primer of transmissible spongiform encephalopathy', in the *British Medical Journal*, 1998, 19 December; 317 (7174), p1688-92.

26 'A Scotch Farmer', 'Observations at Mr Bakewell's in 1771', in The Annals of Agriculture, Vol 28, (1797). It is not clear whether this happened in 1771 or in the 1790s.

27 Now £56,390. Measuring Worth, (2004), 'Five Ways to Compute the Relative Value of a UK Pound Amount, 1270 to Present'. URL: www.measuringworth.com/ukcompare/

28 BNA, *Ipswich Journal*, 6 September 1788.

29 Now £133,900. Measuring Worth (2004), 'Five Ways to Compute the Relative Value of a UK Pound Amount, 1270 to Present'. URL: www.measuringworth.com/ukcompare/

30 SRO, D661/1/7/Parcel 6.

31 Now £41,500. Measuring Worth (2004), 'Five Ways to Compute the Relative Value of a UK Pound Amount, 1270 to Present'. URL: www.measuringworth.com/ukcompare/

32 ROLLR, DE7454/35, 1775-6.

33 Derbyshire Record Office (DRO), D2010/CT/10-11, 1775.

34 Pawson, H.C., Robert Bakewell Pioneer Livestock Breeder, (1957), p36-7. This Robert Bakewell was not related to the Dishley, Normanton-le-Heath or Swepstone families.

35 Cox, T., A View of the United States of America, (1795), p375 & 402.

36 Marshall, W., Georgical Essays, Vol XX, (1803), p386. Georgical is a reference to Virgil's Georgics which is a 4-part poem on the agricultural year.

37 Now £5.70. Measuring Worth, (2004), 'Five Ways to Compute the Relative Value of a UK Pound Amount, 1270 to Present'. URL: www.measuringworth.com/ukcompare/

38 The National Archives (TNA), IR1 58, 1773 National Register of Apprentices. He was indentured for 4 years from 4 June 1771. Bakewell's many other students were gentlemen pupils usually there for a shorter period.

39 National Archives of Scotland (NAS), GD224/1040/62.

40 ROLLR, DE7454/35, 1775-6.

41 Wykes, D.L., 'Robert Bakewell (1725-1795) of Dishley: farmer and livestock improver', *Agricultural History Review* 52, (2004), p45.

42 The Annual Necrology 1797-8, p203.

43 TNA, B1/74, p202, 1784. The Office of the Commissioners of Bankrupts & Successors: Order Books relating to Petitions against the Declaration of Bankruptcy.

44 BNA, *Leicester Journal*, 7 June 1783, Smellie's notice of complaint.

45 Pawson, H.C., Robert Bakewell Pioneer Livestock Breeder, (1957), p41.

46 Chivers, K., The Shire Horse. A History of the Breed, the Society, and the Men, (1976), p51-2.

47 Quoted in Namier, L. & Brooke J., (eds), The History of Parliament: The House of Commons 1754-1790, Member Biographies: Gordon, Sir William.

48 Chivers, K., The Shire Horse. A History of the Breed, the Society, and the Men (1976), p35.

49 Ruffhead, O., (compiler), The Statutes at Large. Bankrupts Act 4 & 5 Anne I c.4, 1705, p232-5 and Bankrupts Act 6 Anne I c.6 1708, p235-6.

50 Carlos, A.M., Kosack, E., & Penarrieta, L.C., 'Bankruptcy, Discharge and the Emergence of Debtor Rights in Eighteenth Century England', paper given at the Federal Reserve Bank of Atlanta, Georgia, Conference, 5 November 2015, p44.

51 Postles, D.F., 'Trading beyond his Capital: The Business of Bankruptcy and Local Economies in Leicestershire, 1706-1842', in *LAHS Transactions* Vol 96, (2022), p137-56.

52 Pawson, H.C., Robert Bakewell Pioneer Livestock Breeder, (1957), p181.

53 Nottingham University Manuscripts & Special Collections, Mi 4Da 2/8, Mortgage to the Trustees of Mr & Mrs Charles Upton's marriage, 28-29 September 1780.

CHAPTER EIGHTEEN

Landlords and Tenants

Tim Bacon

As we have seen above, Robert Bakewell was fortunate to have a landlord who appreciated his efforts and seems to have done everything he could to assist his tenant to remain on his farm at Dishley Grange. However, such an attitude was by no means universal in Britain. In a report to the Board of Agriculture, printed in 1794, Charles Vancouver bemoaned the usual practice in Cambridgeshire of granting only short agricultural tenures.

> GENERAL VIEW OF THE **AGRICULTURE** IN THE COUNTY OF **CAMBRIDGE;** WITH OBSERVATIONS ON THE MEANS OF ITS IMPROVEMENT. BY CHARLES VANCOUVER. DRAWN UP FOR THE CONSIDERATION OF THE BOARD OF AGRICULTURE AND INTERNAL IMPROVEMENT. LONDON: PRINTED BY W. SMITH. M.DCC.XCIV. [1794].
>
> TENURES
> There is no greater error in the whole oeconomy [sic] of a country business, than that which the gentlemen of Cambridgeshire, are too apt to fall into, respecting the tenures they grant of their estates. Few are inclined to give their tenants such assurances of the certain and quiet enjoyment of their improvements, as reason dictates and justice demands. Had the same jealousy prevailed in Norfolk, on the early improvement of that county, in vain would the landlords, in conjunction with their tenants, have expended such large sums in claying, marling, and otherwise improving their estates; if the tenant had not been assured of an eventual benefit, and rewarded for his expence [sic] and labour, under the protection and encouragement of a lease for twenty-one years. The general state of the husbandry in the county of Cambridge, demands the like assurance from the proprietor as to the means, and the same indulgence in point of possession, as a reward for the industry and labour of the tenant, who will otherwise be little inclined to bury his property in the earth, or improve the surface by the sweat of his brow. In all cases however, the most strict and rigid observance ought to be exacted, as

to the performance of the stipulated covenants, but the confidence of a lease, is indispensable to the advancement of agriculture in every county; and where it is found so backward as in the county of Cambridge, it is highly expedient, that the term should not be of shorter duration, than that of twenty-one years; though in the highly improved counties, such an extension of time, may not be found so essentially necessary. (p198-9 of the Report)

In contrast, the Bakewells had benefitted from long agricultural tenures and understanding landlords when they leased the farm at Dishley Grange, but set out below is the case relating to the same farm about 100 years later, which illustrates the nature of the problems that could be caused by an unsympathetic landlord.

The following are extracts from an article in the League of 5th April, in which this organ of the Free-trade party gives nearly the whole of the Report of the Dishley Farm case.

Leicester Chronicle 19th April, 1845
LANDLORDS AND TENANTS. YEARLY TENANCIES AND RESTRICTIVE RULES ILLUSTRATED.
We are indebted to the Leicester Chronicle for the report of two actions tried at the late assizes for that county, which demand the serious attention of every tenant-farmer in the United Kingdom. Now, it is pretty notorious that tenant-farmers have very little to do with the voluntary selection of the men who make the laws; and that, somehow or other, whenever disputes arise between landlords and tenants, the law always turns out to be in favour of the landlords. The fact that landlords alone are the makers of the law, will to some fully account for this result. The farmers, however, have not as yet made any efforts to be represented in Parliament by men whose interests and sympathies are in unison with their own. What they may do hereafter time will show. In the meantime let them ponder over the case we are about to state. It appears that Mr Smith occupied, as tenant to Mr C.M. Philipps, the celebrated Dishley Farm, in Leicestershire, on which the late Mr Bakewell made his well known improvements in the breeds of long-horned cattle and Leicester sheep, at the yearly rent of £989. Mr Smith had succeeded to the tenancy on the death of his father in 1831, and for several years managed it entirely to his landlord's satisfaction. It was the landlord's 'pet farm,' the object he constantly exhibited to his visitors. The farm was within half a mile of the landlord's residence, and if the farm was a show for the 'squire's amusement', of course he required the attendance of the tenant as showman. Of course, also, the 'pet farm' was let only from year

to year, and Mr Smith, being a good farmer and a man of enterprise, was scarcely contented to remain half employed for the purpose of acting as bailiff to his landlord. He therefore took another large farm in Derbyshire, besides occupying other land in Leicestershire. Thereupon, Mr Smith, in 1834 or 1835, received a notice to quit. Matters, however, were accommodated for a time: the notice was withdrawn, and Mr Smith continued to hold the farm down to 1843, when he again had notice to quit at Lady-day, 1844. It was then that the landlord or his agent, thought proper to attempt to enforce certain contested rights of entry by a distress for rent, and ultimately by an action against the tenant, for not complying with certain alleged customs, neglecting to repair the fences, crosscropping the land, and removing certain willows. Damages were laid at £500. As we have before mentioned, the law in cases of this sort is sufficiently favourable to the landlord-plaintiff, and that from obvious causes. But in this case Mr Philipps, or his agent, — we leave him that loophole, not satisfied with landlord-made law, determined to have a special jury of landlords to try the cause! Well might the tenant-farmers of the district regard the result of this action with interest, and the conduct of the landlord-plaintiff with indignation. There cannot be a farmer in Britain who, reading this case, will not so regard it. Most of the grounds of action were abandoned by the plaintiff's counsel in opening the case, and the rest was entirely disproved by his own witnesses. Let the farmer read this almost incredible case, and remember that it may be his own next week or next year. The first witness called to prove the plaintiff's case was Benjamin Kidger. [This man's evidence having been given, the article proceeds]. Here we have a yearly tenant whose wheat crop was forty-eight bushels to the acre, and barley in proportion, prosecuted, might we not have said persecuted? by his landlord for bad husbandry! Verily farmers may well say that landlords know little of agriculture, and their agents know less, when such actions as these are brought against ejected tenants. The above witness — the plaintiff's own witness — was quite enough to put the plaintiff out of court; nevertheless his counsel, relying possibly on the special jury of landlords, persevered, and called Mr Harris, a surveyor, who had been sent by the landlord to look over the farm. Now, Mr Harris seems to have been a sensible and honest man, and his testimony put the plaintiff's case quite beyond the aid of even a landlord jury. Mr Stokes, a farmer, was then called, and said there was no particular course of cropping followed in the district, and that some of the hedges wanted cutting. Mr Bosworth, the present tenant of the farm, was also examined, and admitted he had not cleaned the ditches, or cut the hedges, the non cutting and non cleaning of which formed the main ground of the action because 'the soil of the ditches would not pay for cleansing'. The defendant's counsel justly said he did not see what he was called upon to answer, and he called no witnesses. The result was a verdict for the plaintiff for the miserable sum of 44s 6d for damages on account of the fences, and a nominal sum for the willows; the legal question as to the tenant's right to cut them being reserved for the consideration of the judges. Let it be remembered that this pitiful verdict inflicted costs upon the tenant, perhaps, to the amount of many hundreds of pounds. What tenant farmer in England is there on whom an arbitrary or capricious landlord, or an ill-conditioned agent, might not inflict the same sort of injury? In a subsequent action Mr Smith recovered £1,311 and costs, for the illegal distress made upon his effects at the instance of the landlord. This sufficiently marks the whole character of the litigation. Now, one word to the farmers of Leicestershire. Do they mean quietly to see one of their own order thus persecuted? Let them follow the spirited example set by the tenant-farmers of Hertfordshire, in the case of Mr Horncastle. Let them instantly subscribe a sufficient sum to indemnify Mr Smith against every shilling of expense he has been put to in respect of this litigation with his landlord, and let it be presented to him at a tenant-farmers' public dinner. This will be a wholesome lesson to the litigious landlords of the county. And, further, let them follow up the movement by seeking some fit person from amongst their own body to represent them in Parliament at the next election for the county. Nothing is required for so doing but an understanding amongst the tenants. These are the only means by which tenant-farmers can emancipate themselves from the state of thraldom and vassalage in which they have hitherto been held by the landowners.

CHAPTER NINETEEN

Conclusion

Pat Stanley & Tim Bacon

As detailed earlier, Robert Bakewell was very innovative in his approach to farming generally, inventing new tools and methods, as well as adopting best practice wherever he saw it. But perhaps the best way to conclude this book about Robert Bakewell is to consider his impact on the various breeds of livestock that he strove to improve, particularly Longhorn cattle and the New Leicester sheep (see the painting of *Two Longhorns and Two Leicester Sheep* by Thomas Weaver in 1802 in the colour section).

The Longhorn

So what does the Longhorn offer having reached the 21st century? The answer is, very much indeed, although some recent writers may disagree. Animal welfare is under public scrutiny, more than at any other time in our history. With the call for less intensive and more extensive farming, an animal that can produce a large, lean carcase of high-quality beef, on a mainly roughage diet, and without the use of great quantities of high-cost concentrate, must surely have a great future. The Longhorn breed is renowned for its docility, ease of calving and mothering abilities, all absolute necessities for the modern suckler herd, keeping personnel management to a minimum, at complete variance to many of the modern continental breeds. The cows produce a high butterfat milk, ensuring speedy growth to their vigorous offspring, and their hardy nature means they can thrive outside all year round, cutting out housing costs and reducing the risk of respiratory and other ailments. The Longhorn bull used as a crossing sire on dairy heifers guarantees ease of calving, with resulting females making excellent sucklers.

The Longhorn has seen many changes in its long and chequered past. It has seen its rivals come and go! Its colour is now more uniform than ever before in its history. Once it ranged from black through to the palest roan, and even whole coloured animals, without the signature stripe or finch mark, made up quite a proportion of its number. Surely one of its greatest charms is that the whims and fancies of fashion will always be outwitted by this ancient race for, as all breeders know, no two animals mated repeatedly will ever produce consistent colouration in their offspring.

Will the Longhorn breed endure? On the basis of its history of coming into and going out of fashion, it may well do, providing it continues to find its own niche in our modern world. The growing awareness of its many wonderful qualities should ensure its survival in the current market place, enabling it to resist any future obstacles. May it constantly remind us of the debt we owe to that enlightened man Robert Bakewell and his fellow improvers, without whose legacy our country would not have held the lead in world agriculture at the time of its great awakening.

We will leave you with the thoughts of J. Neville Fitt from 1876, which are truly fitting for the present. 'The Longhorn has endured a long eclipse, but now there seems every indication that he is about once more to emerge from obscurity; and take his proper place amongst the magnificent breeds of cattle for which England is famous'.[1] Indeed, herds of Longhorns persist across the globe, with enthusiasts as strong in their belief in the qualities of the breed as their counterparts in the UK.

The New Leicester Sheep

The New Leicester continues as a breed, with a strong following in Leicestershire, but with fewer than 500 registered ewes, Leicester Longwools, as their Breed Society calls them, are one of the rarest native breeds left in the UK. While the 'original' breed may have dropped from popularity, as seen in Chapter 5, its impact on virtually every other sheep breed has been considerable and as Wrightson wrote,

> There is scarcely a breed which has not felt the influence of the Leicester. No race has been so largely employed as a means of improvement, and Southdowns, Cotswolds, Lincolns, Shropshires, Hampshire Downs, and probably every sort of sheep, directly or indirectly has benefited by it.[2]

Other livestock

We have tried to provide a good overview of Bakewell's impact on the breeds of horse, pigs and poultry above (see Chapter 7), where it is clear that his lasting success has been with horses rather than pigs or poultry. In relation to horses, once again, it is his crosses with other breeds where Bakewell leaves his legacy, as he turned his enormous skills to improving and differentiating the work-a-day animals rather than the 'elite' categories such as racers and hunters.

Finally, at a time when most people treated animals as merely a commodity to meet human requirements, Robert Bakewell really cared about his charges and how they were looked after. Bakewell was a humble but influential man of his times who changed farming practice for the better.

Endnotes

1 Fitt, Neville J., 'Longhorn Cattle: their History and Peculiarities' *Journal of the Royal Agricultural Society of England*, Second Series, Vol XII (London: John Murray) p487.

2 Wrightson, J., Sheep, Breeds and Management (1919), No. 1 of Sinclair, J., (ed.) Live Stock Handbooks, p21.

Appendix 1.1

The Bakewells of Normanton le Heath

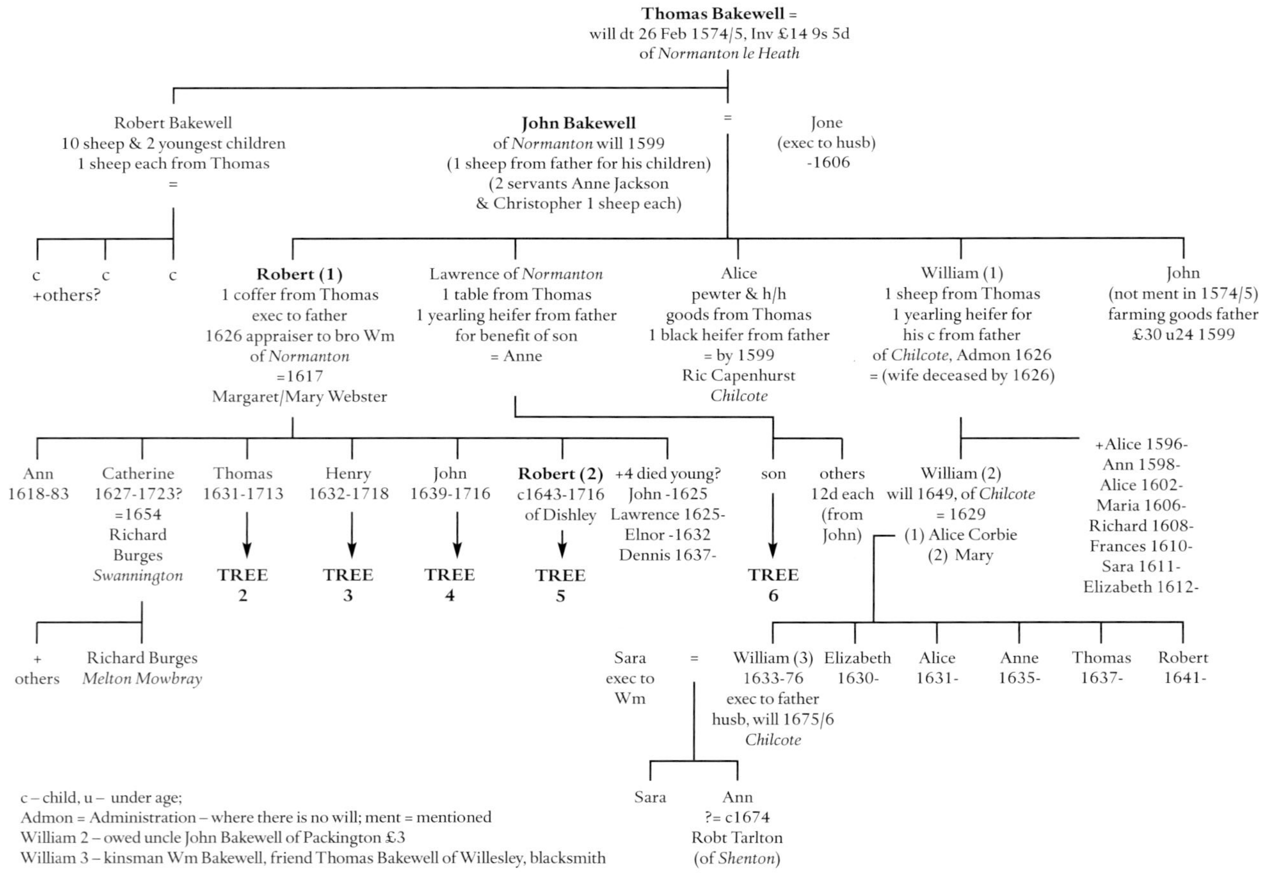

c – child, u – under age;
Admon = Administration – where there is no will; ment = mentioned
William 2 – owed uncle John Bakewell of Packington £3
William 3 – kinsman Wm Bakewell, friend Thomas Bakewell of Willesley, blacksmith

Appendix 1.2

Descendants of Thomas Bakewell 1631-1713

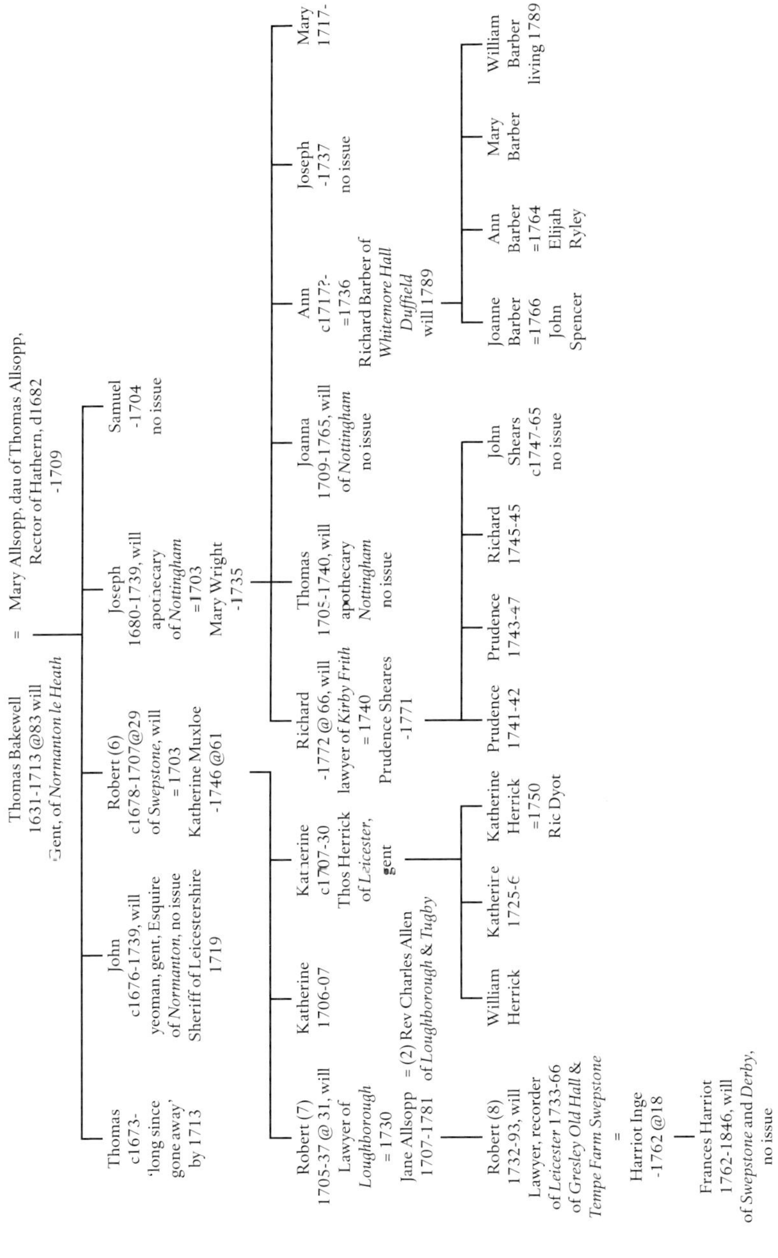

Appendix 1.3

Descendants of Henry Bakewell 1632-1718

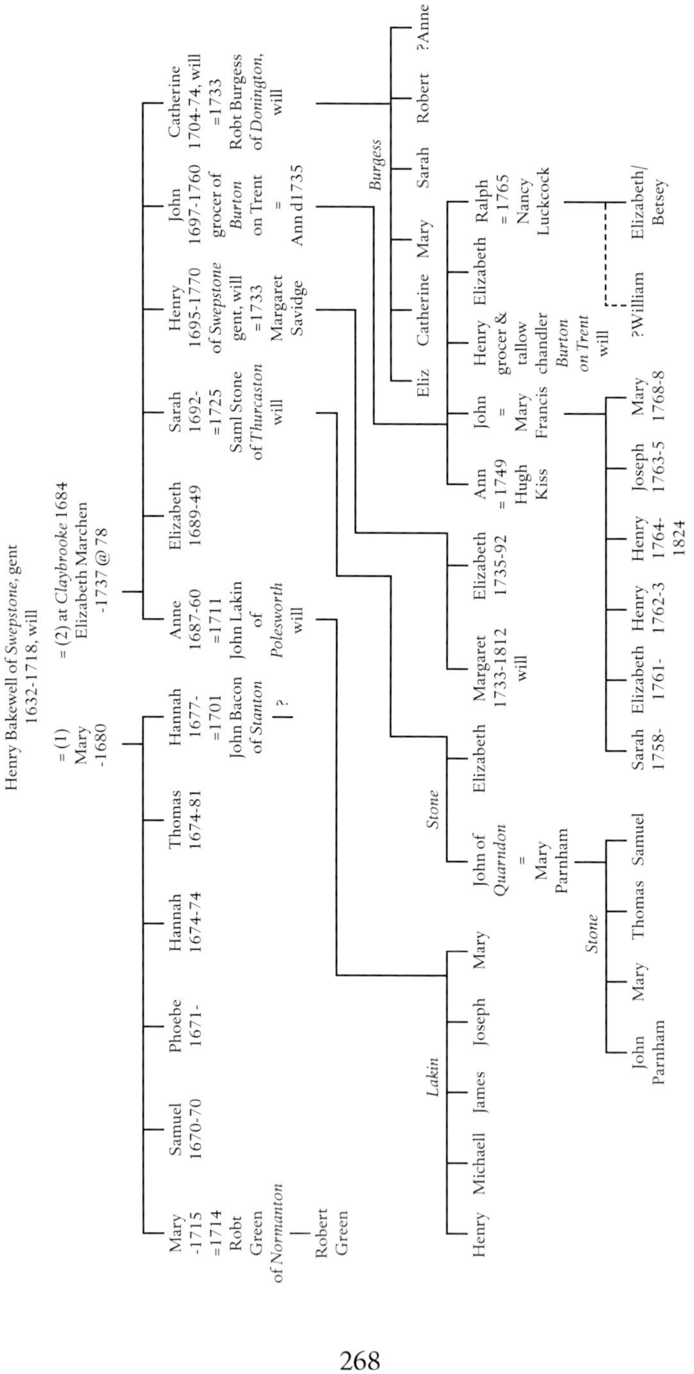

Appendix 1.4
Descendants of John Bakewell 1639-1716

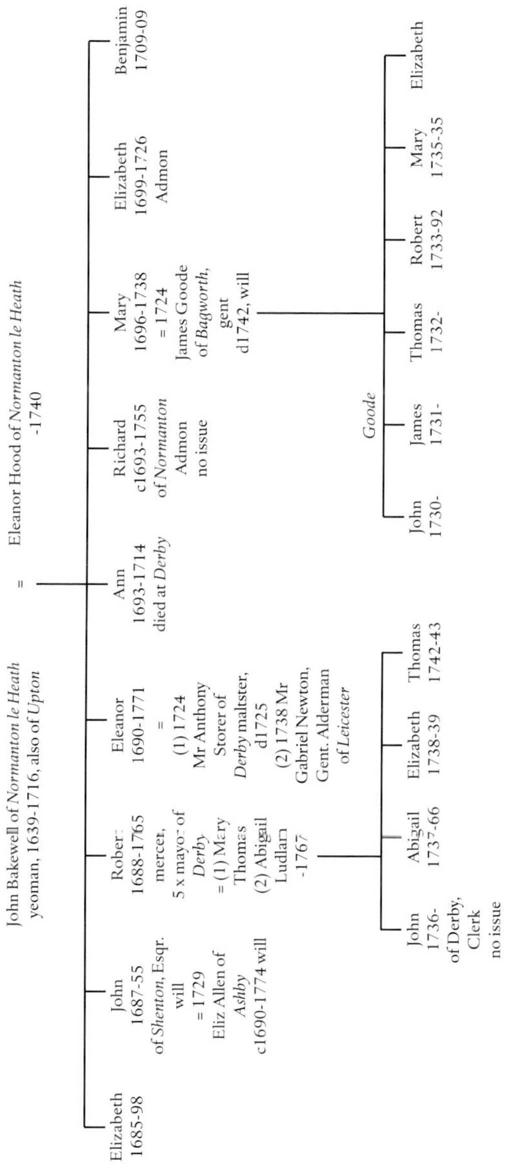

Appendix 1.5

Descendants of Robert Bakewell c1643-1716 of Dishley

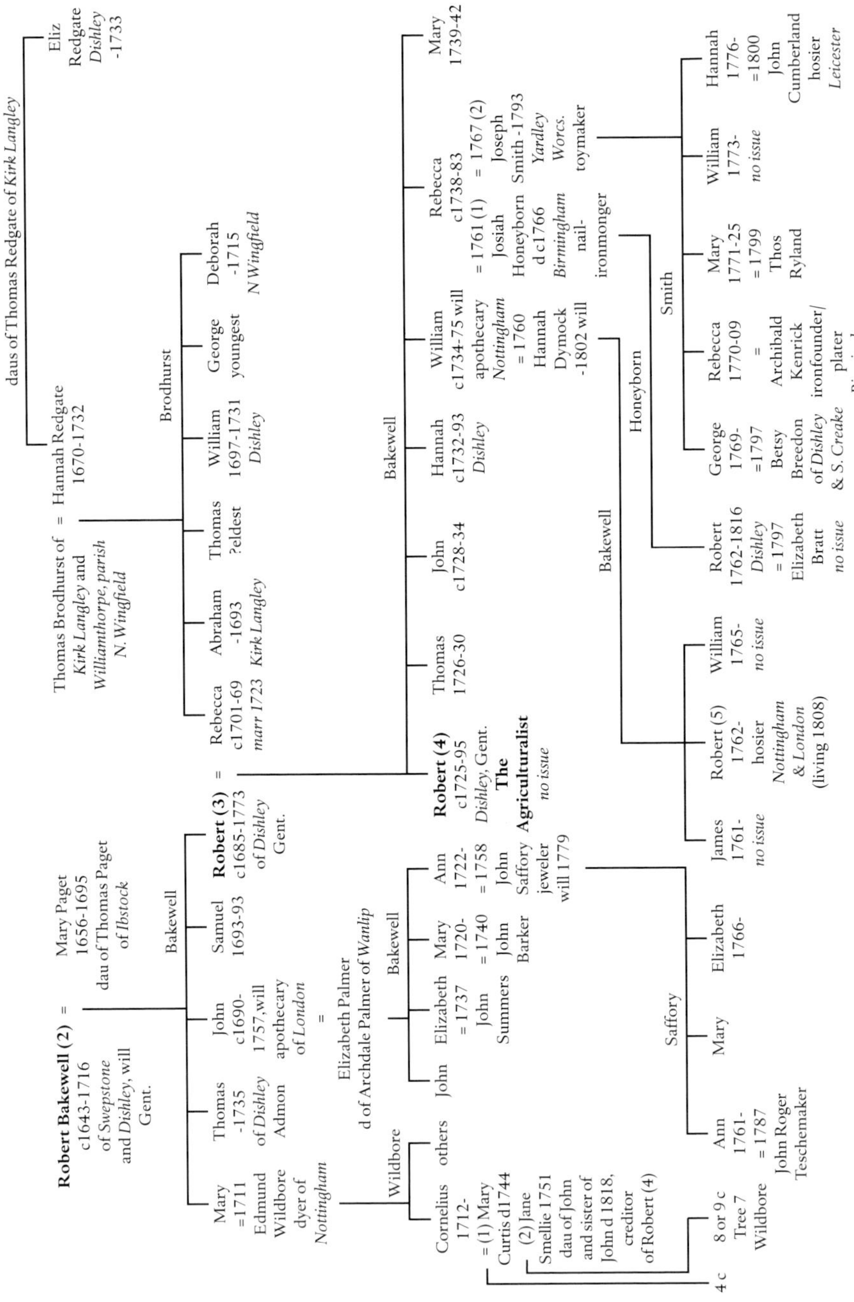

Appendix 1.6

Probable descendants of Lawrence Bakewell and others, Normanton le Heath

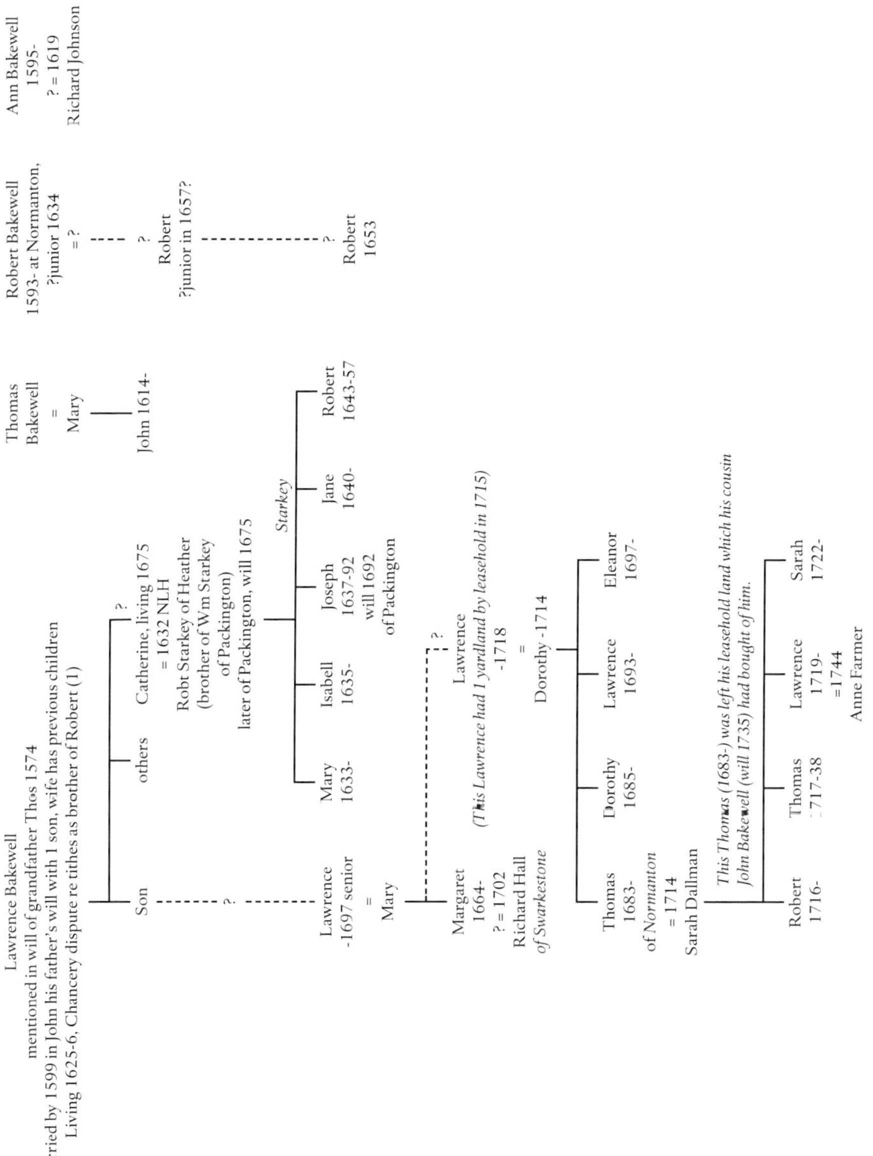

Appendix 1.7

The Wildbore Family, descent from Mary Bakewell

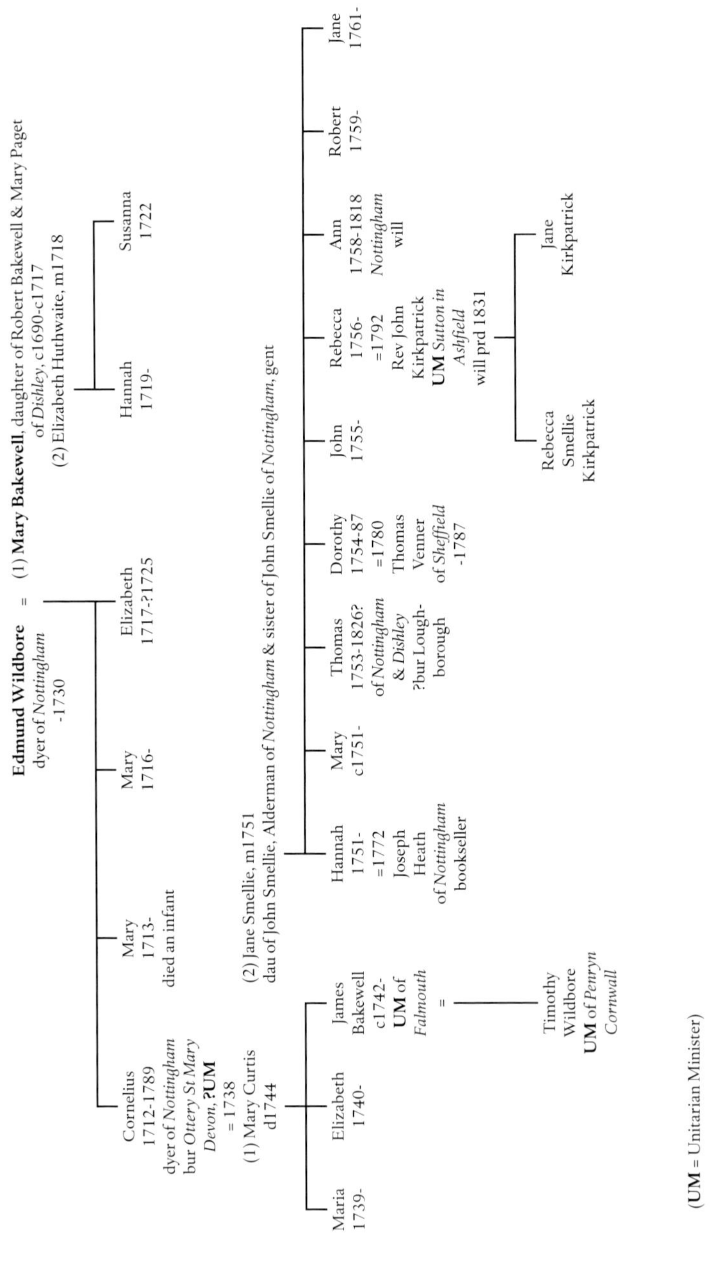

Appendix 2

The celebrated stock of Mr Robert Fowler, of Little Rollright, in the county of Oxford, which was submitted to the hammer 29, 30 and 31 March 1791.[1]

Lot	BULLS	£	s	d
1	GARRICK, 5 yrs, by Shakespeare, out of Broken Beauty. Mr Stone, Quorndon, Leics.	215	5	0
2	SULTAN, 2 yrs, by Shakespeare, dam Broken Horn Beauty. Messrs Freeman & Eden, of Norton, Glos.	220	10	0
3	WASHINGTON, 2 yrs, by Shakespeare, dam Bow Horn Red Cow. Mr Buckley, Normanton, Notts.	215	5	0
4	YOUNG TWOPENNY, 2 yrs, by Garrick, dam Long Horn Beauty. Messrs Cox, Harrison & Macey, Leics.	68	5	0
5	YOUNG GARRICK, 1 yr, by Garrick, dam Pillion Rump alias Old Skew Horns. Mr J. Smith, Warwicks. (Purchased for Mr Fowler)	49	7	0
	COWS			
6	YOUNG BRINDLED BEAUTY, 2 yrs, by Shakespeare, dam Brindled Beauty. Mr Russell, Warwicks. (for Messrs Knowles & Co.)	66	3	0
7	PILLION RUMP, alias OLD SKEW HORNS, by Hampshire Bull, dam Daughter of Ralph. Mr Millington, Wilts. (for Mr Fowler)	23	2	0
8	BULL CALF, by Garrick, dam Pillion Rump. Mr Baker, Farmcot, Glos.	25	4	0
9	GREAT BRINDLED COW, by D, dam Pillion Rump. Mr Millington, Wilts. (for Mr Fowler)	31	10	0
10	NELL, by Sampson, dam Old Nell. In calf to Sultan. Mr W. Freeman, Hitcot, Glos.	38	17	0

(Nell, a daughter of Old Nell (which was a daughter of Twopenny), for the time she has bred, which has not been more than 8 years, has made over 1,000 gns of herself and her stock, which is more than was ever made by any cow in the kingdom, and was highly esteemed with Mr Fowler as some of the best breeding on his farm.)

11	NELL'S BULL CALF, by Garrick. Mr Joseph Allcock, Longborough, Glos.	45	3	0
12	NELL'S WHITE BACK, 3 yrs, by Garrick, dam Old Nell. Lord Harborough, Leics.	89	5	0
13	YOUNG BRIND, 3 yrs, by Garrick, dam Old Nell's Daughter. Lot 41. In calf to Washington. Mr William, Huckvale, Over Norton, Oxon.	32	11	0
14	A BULL NAMED C, 1 yr, by Garrick, dam Short Tails Daughter, alias The White Back'd Cow. Lot 48. Mr James Moore, Charlcote, Warwicks. [This bull was supposed to be the best of the yearlings but unfortunately met with an accident before the sale]	52	10	0
15	LONG HORN'D BEAUTY, by her own brother, dam Long Horns out of Old Beauty. Mr Francis Robbins, Lillington, Warwicks.	44	2	0
16	LONG HORN'D BEAUTY'S cow calf, by Garrick. Mr Richard Robbins, Lillington, Warwicks.	22	1	0
17	NANCY, by Shakespeare, dam a daughter of D with bull calf by Garrick. Mr Brookes, Wolvers Hill, Warwicks.	52	10	0
18	BRINDLED SHORT TAIL, by Shakespeare, dam a daughter of Old Short Tail. In calf to Garrick. Mr Paget, Ibstock, Leics.	43	1	0
19	BEAUTY, by Shakespeare, dam old Long Horns Daughter. In calf to Garrick. Mr John Millington, Wilts. (for Mr Fowler)	52	10	0
20	A MOTTLED YRLING CALF, by Garrick, dam Long Horn'd Nancy. Mr John Zouch, Millcote, Warwicks.	27	6	0
21	YOUNG BEAUTY, of Lot 19, a yrling calf, by Garrick, dam Beauty, which is grand-daughter of Old Long Horned Beauty. (Mr Russell for Messrs Knowles & Co.)	34	13	0
22	A LIGHT COLOURED WELSH NURSE COW. Mr Richard Leythall (for Mr Fowler)	8	8	0*

*[This lot shows what the ordinary cow was fetching]

BULLS

23	YOUNG SHAKESPEARE, 1 yr, by Garrick, dam Daughter of the Short Tailed cow alias Douk Horn. Mr Leythall (for Mr Fowler)	29	8	0
24	A, 1 yr, by Garrick, dam Brindled Finch. Lot 46, daughter of the Great Brindled cow. Lord Harborough, Leics.	157	10	0
25	B, 1 yr, by Garrick, dam The Blue Heifer. Lot 45, daughter of the Great Brindled cow. Mr W. Seaton, Scasby, Yorks.	85	1	0
26	YOUNG SULTAN, 1 yr, by Garrick, dam Nell, Lot 10. Messrs Cox, Harrison, Macey, Leics.	210	0	0
27	D, 1 yr, by Garrick, dam a daughter of Short Tails. Mr Thomas Clarke, Lockington, Leics.	88	4	0
28	E, 1 yr, by Garrick, dam Nancy. Mr J. Zouch, Milcote, Warwicks.	152	5	0
29	F, 1 yr, by Garrick, dam bow horned red cow alias Washington's mother. Mr Francis Robbins, Lillington, Warwicks.	105	0	0

COWS

30	BRINDLED BEAUTY, by Shakespeare, dam Long Horned Beauty, she also being dam of Shakespeare (served by Garrick). Mr Russell (for Messrs Knowles & Co.)	273	0	0
31	GARRICK'S SISTER, by Shakespeare, dam Garrick's mother, alias Broken Horn Beauty (served by Garrick). Mr Russell (for Messrs Knowles & Co.)	120	15	0
32	WASHINGTON'S MOTHER, by a son of old D, brother of Shakespeare, dam Nell (served by Garrick). Mr Astley, Odstone, Leics.	194	5	0
33	LONG HORN'D NANCY, by Shakespeare, dam daughter of Old Nell (in calf to Garrick). Mr W. Freeman, Hitcot, Glos.	110	5	0
34	SPOTTED NANCY, by Shakespeare, dam daughter of Old Nell (served by Garrick). Mr Millington, Wilts. (for Mr Fowler)	84	0	0
35	BLACK HEIFER, 3 yrs, by Shakespeare, dam Brindled Beauty. Mr Russell (for Messrs Knowles & Co.)	141	15	0

36	GARRICK'S DAUGHTER, 3 yrs, dam Garrick's mother Broken Horn Beauty (served by Sultan). Mr Cox, Brailsford, Derbys.	47	5	0
37	YOUNG NELL, by a brother of Shakespeare, dam Old Nell (served by Garrick). Mr J Moore, Charlecote, Warwicks.	126	0	0
38	YOUNG NELL'S COW CALF, by Garrick. Lord Harborough, Leics.	31	10	0
39	BROKEN HORN'D BEAUTY, (dam of Garrick and Sultan) by the Hampshire Bull, dam Old Long Horn Beauty (dam of Shakespeare and Lot 30) (served by Garrick). Mr Eden, Norton, Glos.	46	4	0
40	RED COW, by Shakespeare, dam Old Skew Horns alias Pillion Rump (in calf to Garrick). Mr Cox, Brailsford, Derbys.	76	13	0
41	NELL'S DAUGHTER, by Shakespeare, Mr Russell, Warwicks. (for Messrs Knowles & Co.)	136	10	0
42	COW CALK, by Garrick dam above. Lord Harborough, Leics.	43	1	0
43	SALL, by Shakespeare, dam Old Sall (in calf to Garrick). Mr Richard Robbins, Lillington, Warwicks.	63	0	0
44	DOUK HORN, by Shakespeare, dam Old Short Tail (with cow calf by Garrick). Mr E. Higgins, Old Stratford, Warwicks.	81	18	0
45	BLUE HEIFER, 4 yrs, by Shakespeare, dam The Great Brindled Cow (with bull calf by Garrick). Mr Samuel Huckvale, Choice Hill, Oxon.	110	5	0
46	BRINDLED FINCH, (dam of Lot 24), by Shakespeare, dam The Great Brindled Cow. Mr Eden, Norton, Glos.	42	0	0
47	BRINDLED FINCH'S COW CALF, by Garrick. Mr Baker, Farmcote, Glos.	9	19	0
48	WHITE BACK'D COW, (dam of Lot 14), by Shakespeare's brother, dam a daughter of the Short tail cow. Mr Eden, Norton, Glos.	81	18	0

49	COW CALF of WHITE BACK, by Garrick. Mr S Huckvale, Choice Hill, Oxon.	32	11	0
50	WHITE BACK'D BROKEN HORN HEIFER, 2 yrs, by Shakespeare, dam a daughter of Old Nell. Mr Freeman, Hitcot, Glos.	52	10	0
51	BROKEN HORN, 2 yrs, by Garrick, dam The Great Brindled Cow. Mr Seaton, Scasby, Yorks.	69	6	0
52	A Welsh red cow.	8	18	6
53	A red and white nurse cow.	11	0	0

Endnotes

1 Young, A., *Annals of Agriculture*, Vol XVI (1791), p387-412. N.B., there are more animals here and some of the details differ from the Long-horn Herd Book.

Appendix 3

The sale of stock of Mr Thomas Paget of Ibstock, Leicestershire

First Day's Sale, 14 November 1793

Lot		Guineas
8,	Short Tail, by Shakespear. Bought at Mr Fowler's	38
9,	Eyebright, by a bull bred by Mr Varnam	51
14,	Strawberry, by a Dishley bull	31
16,	Brindled Eyebright	33
26,	Penn, [Dam by a son of the Hampshire Bull]*	35
[27,	Greg, Dam by a Dishley Bull	15]*
[28,	Nancy, Dam by a son of the Hampshire Bull	20]*
29,	Young Dandy	30
30,	Brindled-Finch-Tidy	29]
[31,	Young Lovely	20]
[33,	Young Eyebright – Dam by a Son of The Hampshire Bull	

Bulls and Bull Calves

34,	Shakespear, (bred by the late Mr Fowler) by Shakespear, off Young Nell. Whoever buys this lot, the seller makes it a condition that he shall have the privilege of having two cows bulled by him yearly.	400
35,	Bull-calf by lot 34	23
37,	A ditto, by ditto 34	31
38,	A ditto, by ditto 34	31

Second Day's Sale

Lot		**Guineas**
45,	One three years old heifer [Bought by Lord Harborough]	70
47,	,, ,, ,, ,, ,,	32
48,	,, ,, ,, ,, ,,	35
52,	,, ,, ,, ,, ,,	35
55,	One two years old heifer	25
57,	,, ,, ,, ,, ,,	60
58,	,, ,, ,, ,, ,,	84
60,	,, ,, ,, ,, ,,	29
61,	,, ,, ,, ,, ,,	25
64,	,, ,, ,, ,, ,,	27

Principal Sources:
Monk, J., General View of the Agriculture of the County of Leicester, (1794), p35.
Pitt, W., General View of the Agriculture of Leicestershire, including a survey of the County of Rutland, p222.
* additional material from research

Appendix 4

The sale of stock of Thomas Prinsep on 25th-26th September 1811

	Bulls	Guineas
Roscius	6 years	85
R	4	125
D	4	100
L	3	105
A	1	130
	Cows	
Dumpling	4 years	42
Mottled Rumps	4	42
Comely Lupin	6	64
Doll	5	78
Louk-horn Lupin	5	42
Mottle-back Lupin	6	200
White Lupin	7	220
Yellow Lupin	8	125
Beauty	6	200
Louk	8	45
Dowdy	7	80
Dishley Rumps	4	45
Cherry	8	43
Dishley Cherry	6	48
Dishley Daisy	7	68
Tidy	6	70
	Heifers	
Mottle Back Lupin II	3 years	58
Rumps	3	47
Yellow Lupin II	3	92
Dumpling	3	90
Lupin	3	60
Cherry II	3	90
White Lupin II	2	80
Strawberry	2	50
Louk II	2	40
Brindy	2	50
Beauty II	1	34

Lonk III	1	40
Cherry III	1	35
Brind	1	30
Strawberry II	1	35
Nell	1	60

Bull calves: 11 bull calves under six months made prices of 30, 34, 80, 55, 60, 37, 160, 32, 75, 32 and 45 guineas respectively.
Source: Lythall, J.B., Facsimile of The Long-horn Herd Book, Vol 1, p19-20.

Appendix 5

The sale of stock of R.H. Chapman on 16th December 1873

Lot	Guineas	Purchaser
1 Canley	29	Lord Bagot
2 Dairy Maid	27	Lord Bagot
3 Light of Other Days	30	Mr Parker
4 Oxford	24	Mr J.H. Burbery
5 Pallas, 4 years	44	W.T. Cox, Spondon Hall
6 Milkmaid	23	W.T. Cox
7 Moss Rose	24	Lord Bagot
8 Tulip	20	Mr J.H. Clay
9 Duchess	23	W.T. Cox
10 Kenilworth	29	Mr Shaw
11 Brindled Beauty	36	Sir John H. Crewe
12 Lady Calke, 2 years	50	Mr Parker
13 Upton, 2 years	40	Mr J.H. Burbery
14 Young Croperdy, 2 years	38	Mr Forrest, Kenilworth
15 Rose	38	W.T. Cox
16 Lady Stowe	36½	R. Warner
17 Young Dairymaid	42	Sir John H. Crewe
18 Young Fillpail	35	Mr T. Levett Prinsep
19 Ivanhoe	28	W.T. Cox
20 Nancy	40	W.T. Cox
21 Brailsford	35	Mr T.L. Prinsep
22 Lady Westmorland	30	Lord Bagot
23 Young Dishley	25	Lord Bagot
24 Cumberland Duchess	42	Sir John H. Crewe
25 Farewell	36	Mr Forrest
26 Croxall, 2 years	45	Sir John H. Crewe
27 Spondon	36	Mr Forrest
28 Fradley	25	Mr Shaw
29 Willscote	27	Mr Ball, Measham
30 Nellie	26	Mr Foster
31 Satchwell	25	Mr Foster
32 Countess	22	Mr Clarke
33 Fillpail	29½	Mr Degg
34 Sparkenhoe	24½	Mr Fox
35 Brailsford – (same name as 21)	29	W.T. Cox
36 Wry Neck	25	W.T. Cox

37 Cherry	30	Mr Clarke
38 Blossom	27½	Mr Clarke
39 Dishley	29 (fat cow)	
40 Milker	25 (fat cow)	
41 Croperdy	42	Sir John H. Crewe
42 Buttercup	22½ (fat cow)	
43 Woodcote	28 (fat cow)	
44 Hagley	30	Mr Shaw
45 Banbury	28	Mr Shaw
46 Lilly	27 (fat cow)	
47 Sally Dick	27 (fat cow)	
48 Ivanhoe	30 (fat cow)	
49 Lady Rollright (prize heifer at Birmingham)	46½	Mr Fox

Bulls

Twopenny, 3 years, bred by Mr J.C. Moore, sire of Mr Godfrey's Sampson	29 gns	Lord Bagot
Old Time, 1 year, bred by Mr Hewertson, Westmorland	40 gns	Sir John H. Crewe

Total of sale 1, 600 guineas
Average of 51 cows and bulls £32 18s 10d

Appendix 6

The sale of stock of Sir John Crewe on 19th May 1886

LONGHORN COWS AND HEIFERS

	Guineas
Trimmer 4th, calved 1876, by Peeping Tom	
Mr Morley, Derby	20
Brindle cow calf of above, by Cornet	
Mr Summerfield, Lichfield	8
[Lot here missing due to burn on the scrapbook]	
Red & white cow calf of above, by Cornet	
Mr Hunt, Morley	9½
Dairymaid calved 1879, by Hardy	
Mr Wilkinson, Hartshorn	24
Cow calf of above, by Cornet	
Mr W.E. Gladstone	6½
Tulip 16th, calved 1876, by Old Time	
Mr A.O. Worthington	20
Lofty 2nd, calved 1869, by York	
Mr W. Sandiland, Ashby	10½
Larkspur, 1878, by Earl of Upton 3rd	
Mr Morley, Derby	16½
Heroine, calved 1877, by Lord Hardendale	
Mr J. German	18
Tulip 24th, 1882, by Royal Duke of Upton 3rd	
Mr T.L. Prinsep	25
Tiny 5th, 1882, by Duke of Upton 3rd	
Mr R. Hall, Thulston	23½
Tulip 23rd, 1882, by Royal Duke of Upton 3rd	
Mr Thompson, Wilson	20
Cow calf of above, by Cornet	
Duke of Buckingham	8
Lofty 4th, calved 1879, by the Abbot of Calke	
Mr J. Summerfield	16½
Brindle cow calf of above, by Earl of Moira	
Mr Bates, Malvern	5½
Nymph, calved 1882, by May Duke	
Mr Watts, Melcombe, Dorset	22
Sparkenhoe 3rd, calved 1880, by Marshal Neil	
Mr J. German	26

Bull calf of above, by Cornet	
Mr J. German	10
Lofty 5th, calved 1881, by Marshal Neil	
Mr Watson, Cumberland	19½
Tulip 20th, 1881, by Marshal Neil	
Mr Eyke, Shrewsbury	24
Canley 10th, 1881, by Harlequin	
Major Burgess, Hendon	18½
Tulip 22nd, 1882, by Harlequin	
Mr Eyke, Shrewsbury	25
Cumberland Duchess 6th, calved 13 May 1879, by Farwell	
Mr A.O. Worthington	16½
Cumberland Duchess 5th, calved 6th May 1878, by Abbot of Calke	
Mr J. Summerfield	19
Bull calf of above	
Mr Bates	14
Tulip 18th, calved 5th May 1878, by the Abbot of Calke	
Mr Wragg, Ashby	17½
Tiny 4th, calved 1879, by the Abbot of Calke	
Mr J. German	17½
Coquette, calved 1882, by Spondon	
Mr Watson, Cumberland	18
Honesty, calved 1882, by Spondon	
Mr J. German	15½
Brindle Beauty 2nd, calved 1 May 1882, by Royal	
Mr Hudson, Castle Donnington	20½
Fanny 12th, calved 15 September 1881, by Marshal Neil	
Mr F.W. Symonds, Rugeley	19
Bull calf of above	
Mr Eardley, Market Drayton	4
Lovely 12th, calved 1883, by King of the Gypsies	
Duke of Buckingham	20
Brindle 7th, calved 1883, by Royal Duke of Upton 3rd	
Mr J. German	22
Lofty 6th, calved 1883, by Royal Duke of Upton 3rd	
Mr W.J. Leigh	23
Cumberland Duchess 8th, 1883, by Royal Duke of Upton 3rd	
Mr Prinsep	25

Upton Brindle Beauty, 1883, by May Duke
Mr Watson, Cumberland 21½
Upton Queen, 1883, by May Duke
Mr Bowley 27½
Speranza, 1883, by Royal Duke of Upton 3rd
Mr Matthews, Foremark 19
Queen of Spain, 1884, by King of the Gypsies
Mr Matthews 19
Tinted Venus, 1884, by King of the Gypsies
Mr W.J. Legh 19
Trifle, 1884, by Royal Duke of Upton 3rd
Mr Bates 16
Gem, 1884, by Earl of Moira
Duke of Buckingham 25
Diadem, 1885, by Earl of Moira
Mr Worthington 22
Diamond, 1884, by Earl of Moira
Mr Bates 16½
Castanette, 1885, by Exchange
Mr Preston, Coleorton 11½
Cow calf, by Cornet
Mr Spencer, Derby 8½
Cow calf, by Cornet
Mr Summerfield 7½

BULLS

Cornet, calved 1882, by The Captain
Mr Toone, Melbourne 35
Farmer, calved 1883, by Duke of Albany
Mr Toone 28
Nomad, calved 1884, by King of the Gypsies
Mr Moorcroft, Burton 25½
Robin Hood, calved 1885, by Earl of Huntingdon
Mr Watts, Melcombe 25
Little John, calved 1885, by Earl of Huntingdon
Mr Bates 24½
Friar Tuck, calved 1885, by Cornet
Mr Causer, Shenstone 10½
Blucher (property of Mr W.J. Matthews) calved 1883, by Prior of Ashby
Mr R. Hall 12½

Appendix 7

Timeline of Bakewell's Bankruptcy

(BNA is the British Newspaper Archive.)

3 June 1776. A Commission of Bankruptcy was issued against Robert (4).

24 June 1776. Robert (4) defaulted on creditors' interest payments, according to John Smellie's complaint in the *Leicester Journal*, 7 June 1783. (BNA). Over the next two months it is clear that Robert (4) and three of his closest farming associates and relatives tried to safeguard the Swepstone property from his creditors, as land could be seized under the bankruptcy laws. They were being advised by Robert (8) who was key to the second part of their plan, that Robert (8) would buy the property, thus keeping it in the family. First, Thomas Paget, Joseph Smith and John Ashworth, who were also creditors, planned to buy the property and various household goods from Robert (4), to hold as trustees to sell for him and use the money to pay the creditors (including themselves) a proportion of their debts, thereby avoiding bankruptcy.

27 September 1776. An indenture was drawn up between Robert (4) of the one part, and Paget, Smith and Ashworth of the other part, which recited 'that the said Robert Bakewell was indebted more than his effects were sufficient to pay and that the said Paget, Smith and Ashworth having been applied to be engaged for him in payment of 16s in the pound which they had agreed to accept'. Bakewell passed 'the Hereditaments & the premises comprised in the Indenture of 10 March 1768 and the Fine pursuant thereto' to the three as trustees to sell them, and 'his household goods, plate &c.' to pay his creditors.[1] The 1768 indenture was the family trust. They were willing to take only four-fifths of the money they were owed, and they must have hoped this would prevent a formal bankruptcy – if the other creditors agreed to take the same.

It was the creditors who started the bankruptcy process then, not the debtor, and four-fifths of the creditors must agree to do this and they must be owed at least £200 between them. There were therefore 15 creditors at least, with some of the other 12 having to act, i.e., being prepared to join those already mentioned to start bankruptcy proceedings, under the provisions of the Bankruptcy Acts. Paget, Smith and Ashworth were creditors, as were the assignees; these were William Bentley and William Hodges, both Leicester bankers; Henry Walker, a farmer of Thurmaston, Leicester; Samuel Huskison, a farmer of South

Croson (South Croxton), Leicestershire, and John Ashworth. The other creditors included John Smellie junior, who had a track record of aggressively pursuing his debts and mortgages in Nottingham, who had interest payments to make on his own borrowings as well. John Bowley, Matthew Steel, Thomas Flavell and William Bosworth were also creditors.[2] It has not been possible to trace these men with certainty, and Flavell not at all. Bowley may be the Hathern farmer who married in 1760. Steel may be the gentleman and farmer of Hoton who married in 1763. Bosworth may be the yeoman of Isley Walton who made his will in 1784 and had landholdings with large numbers of horses, cows and sheep. Their villages are all in Leicestershire.

27 September, and 2 & 4 October 1776. Clearly the money raised from the planned sale of the property would not have been enough on its own to make an adequate payment, so at the same time Robert (4) advertised the sale of some of his stallions, brood mares and fillies, bulls and cows, and rams and ewes. The sales description with it has a slight air of desperation in its fulsome praise.

> The extensive Connexions of Mr Bakewell, in various Parts of Great Britain and Ireland, render it superfluous to give a Description here of the above Stock; The repeated demands for different Parts of it, is a Circumstance from which may be inferred that great Satisfaction was afforded, and shews that the high Estimation it was held in, arise from the Advantage that was obtained.[3]

However much was raised from these sales, the creditors clearly thought that 16s in the pound was not sufficient. They might have done better to settle for 16s in 1776; they may have received less by dragging the process out to 1789, when dividends turned out to be 12s 9d plus an unknown final payment.

Before 19 November 1776. By this date three commissioners in Bankruptcy had been appointed. They were Thomas Pares the younger of Leicester, a lawyer and banker, a friend of the wider Bakewell family and a Nonconformist; Jonathan Foster, a cousin of Thomas Pares, a lawyer and Leicestershire landowner, and Caleb Lowdham, a lawyer who had been articled to Thomas Pares in 1761. He owned a lot of property in Leicestershire and was a relative of Mary Phillipps. They were all part of the Bakewell network and probably understood Robert's (4) difficulties. Bakewell is described as a dealer and chapman in the commission in order to qualify for bankruptcy not debt. 'Dealer and chapman' was the legal workaround for both farmers and innkeepers who were supposedly excluded from going into bankruptcy. A chapman was defined as an itinerant trader, and they often took assorted household goods round outlying villages and farms. Farmers counted as itinerant traders because they took stock to market to sell or sold via drovers. The difference really was that, formally, a trader had a place of trade, e.g., a shop or workshop, where the customers came to him, while a chapman took the goods to the customers.

19 November 1776. The required notice was placed in the *London Gazette*.

> Whereas a Commission of Bankrupt is awarded and issued forth against Robert Bakewell, of Dishley in the County of Leicester, Dealer and Chapman, and he being declared a Bankrupt, is hereby required to surrender himself to the Commissioners in the said Commission named, or the major Part of them, on the 6th Day of December next at Two in the Afternoon, on the 10th Day of December next at Ten in the Forenoon, and on the 4th Day of January following at Three in the Afternoon, at the House of Thomas Barry, called the Lion and Lamb, in Leicester, and make a full Discovery and Disclosure of his Estate and Effects; when and where the Creditors are to come prepared to prove their Debts, and at the Second Sitting to chuse Assignees, and at the last Sitting the said Bankrupt is required to finish his Examination, and the Creditors are to assent to or dissent from the Allowance of his Certificate. All persons indebted to the said Bankrupt, or that have any of his Effects, are not to pay or deliver the same but to whom the Commissioners shall appoint, but give Notice to Mr Thomas Pares, jun, Attorney, in Leicester; or Mr Coore, Attorney, Broad street, London.[4]

29 November 1776. The *Chester Chronicle* has Robert (4) in its bankrupts list. (BNA)

30 November 1776. The *Ipswich Journal* and the *Newcastle Courant* have the same. (BNA)

2 December 1776. *Aris's Birmingham Gazette* has the same. (BNA)

6 December 1776. The first commissioners' meeting was held where all the creditors had to prove their claims against Robert (4). He had to make a full disclosure of all his assets, and this would include the recent sale of the Swepstone lands to his trustees on 27 September. It also included the beneficial ownership of the lease on Dishley, and all his stock.

10 December 1776. The second commissioners' meeting was held. The creditors had to appoint the assignees from among their number. Their job was to examine Robert's (4) affairs, then manage the bankruptcy, pursue his debtors, keep accounts and records, and decide the dividends to be paid to the creditors. Money collected had to be deposited with the Bank of England, not with a local bank.

They appointed five assignees. William Bentley was a banker of Bentley & Buxton's bank, the second to be started in Leicester. (It would fail in 1803.) He was also the treasurer of the Unitarian Great Meeting in Leicester which had close links with the Loughborough and Mountsorrel Unitarians. He may have been Bakewell's personal banker and would have dealt with him over the Loughborough Chapel's finances. Bentley had loaned £3,000 to Henry Walker in September 1776, and when Walker went bankrupt he successfully petitioned to be added to his creditors in July 1782. William Hodges was also a Leicester banker, but not part of Robert's (4) close network. He was one of Walker's assignees. These two probably kept all the accounts, handled the money and ran the Bank of England account. Henry Walker was a farmer and grazier in Robert's (4) Unitarian friendship circle, initially farming at Thurmaston, then becoming the tenant of the large Beaumont Leys farms. He did well enough eventually to become a gentleman. When he did not repay Bentley's money with interest in September 1779, and was declared bankrupt in December 1781, two of his assignees were William Hodges and Thomas Paget. Like Robert (4), he was described as a dealer and chapman.[5] Although bankrupt himself after 1781, that did not disqualify Walker from acting as an assignee for Robert (4). Samuel Huskison was a friend of Robert (4) from youth and went on the European tours with him. He was a farmer and grazier at South Croson (South Croxton), Leicestershire. John Ashworth has been discussed above; he had the Wheat Sheaf to run in Daventry and would have been helped day-to-day to run Dishley Grange by the two local farmers. Thurmaston was about 14 miles from Dishley, South Croson (South Croxton) was about 16 miles and the Wheat Sheaf was 41 miles. Adding to these weaknesses in the system, by 1783 Bentley was 'very far advanced in years and unable to travel from home', Huskison was 'rather infirm and not Competent to the Sale and dispersal of such Effects as those of the Bankrupt' and Ashworth, because of the distance and of keeping a large inn, 'had not paid the Necessary Intention [attention] to the disposal of those Effects'.[6] Both Walker and Hodges were themselves subsequently declared bankrupt.[7]

Ashworth did most of the assignees' work, probably because, having lived at Dishley Grange as Robert's (4) pupil, he knew Robert, Hannah and both the indoor and outdoor staff well. They would have had confidence in him and his judgement. Robert (4) as a bankrupt was legally barred from taking further part in his business but as Dishley Grange was his home, he was still living there. This must have been a strain on his confidence and probably his health. He would have had the allowance from the assignees' money as directed by law. Ashworth was helped by Robert's (4) long-serving chief stockmen: John Bredon for the sheep, who was also Robert's (4) 'ground keeper' or farm manager;[8] William Peet, the head horseman, Edward Porter, a horseman and stallion leader, and William Arnold, who was in charge of the cattle. Having been there for years they knew how the farm ran; they had been running it on their own for years during Robert's (4) annual absences on his travels. Culley commented around 1765 that 'His whole business is carried on still with that surprising order and regularity, which I never saw anywhere else, and which is rather to be admired than imitated'.[9] These four all lived close to Dishley too and it stretches credibility that they did not discuss things with Robert (4) and then pass on his comments as advice to Ashworth and vice versa. Bredon's good stock sense came to the fore on at least one occasion.

> Young A was the sheep which Mr Ashworth would have sold to Mr Roby if Jno. Breedon [John Bredon)]had not convinced him of his folly. At that time they had two favorite Tups at Dishley – Viz – Charles and Young A. Mr Ashworth who had the principal direction of the business during the Bankruptcy, had used Charles two Seasons, contrary to Jno. Breedon's Superior Judgment and at the last would have sold A to Mr Roby if Breedon had not fortunately prevented him.[10]

Although we do get conflicting information, it would appear that Bakewell must have still been able to have some influence on how the business was run, especially the breeding side of it.

Subsequent events suggest that the assignees had such difficulty pursuing the unknown number of Robert's (4) debtors in England, Scotland, Ireland and (over such long distances) in Germany and Jamaica, that this, combined with their own limitations, was probably the cause of the inordinate delays to discharge. If there was only one debtor in each place, they had to pursue 40 people for money. They did persist, but the unavoidable delays trying to get the debtors to pay did not please the less sympathetic creditors.

11 December 1776. The next day, the commissioners met the assignees to pass on the assets they had collected or been notified of at the meeting on 6th December. An indenture was drawn up between Pares, Foster and Lowdham, the commissioners 'under the said Robert Bakewell's bankruptcy', and the five assignees, which conveyed 'a messuage and lands in Swepstone belonging to Robert Bakewell the Bankrupt and of all other Lands of a Trust he was possessed of' to the assignees.[11] This is the property he sold to his friends as trustees on 27 September 1776, but there is no mention of the household goods being taken into the commissioners' hands. There is no record of the meeting so it is not possible to say what other assets were seized.

Over the next month the assignees worked to trace debts, collect money and keep the required records. Ashworth would have had to spend time at Dishley, arranging how his direction of the farm would work. By the time of the third commissioners' meeting, they had not been able to complete a full account of Bakewell's debtors and assets, as the law intended; his affairs were too complicated.

4 January 1777. The third commissioners' meeting was held. By law they should have finished Robert's (4) examination of assets and have a full list to present to the creditors. If the creditors accepted that it was a full list, they would be asked to assent to the commissioners granting a Certificate of Conformity for Robert (4). The assets on the list would then be sold and the proceeds distributed to the creditors as dividends. No other assets could be added after the certificate without an appeal to the Lord Chancellor. But the creditors did not agree that it was a full list; they wanted more assets to be pursued. It may be that the assignees had got details of all the assets which could be easily obtained, and it was the long-distance ones, or monies owed by people in financial difficulties themselves, that were still outstanding. The creditors – possibly led by Smellie – wanted the assignees to continue to pursue those debts and obtain a better dividend. They therefore dissented, and the certificate was not granted. It may have been an unruly meeting.

1 March 1777. Some animals were offered for sale with the promise of a sale of all the rest later.

> To Be Sold by Private Contract, at Dishley, near Loughborough, Leicestershire. – Several Black Stallions, Horses, also Bulls of the long-horn kind, being Part of the Stock of Mr Bakewell. – The Remainder of the Stock, consisting of Mares, Cows, Sheep etc., will be advertised in this Paper at a future Period.[12]

The law did not state that asset sales had to be public, or that they had to be by auction to the highest bidder. There was some ill feeling among the creditors over this, as they did not know if the assignees were selling on favourable terms and not getting as much towards the dividends as they could when selling by private contract, or who they were selling to. Sales could be arranged without competition, which was what Robert (8) did in arranging the way Brentnall was deemed the best bidder for his own property. The stock could have been sold favourably to Robert's (4) friends so they could hold it until he could buy it back after his discharge. The private contract system would allow the assignees to delay the sales of most animals or sell selectively to keep the best animals at Dishley in anticipation of discharge. Given that Bakewell did have most of his stock by the time the process finished, this may be what happened. A comment following the auction sale at Dishley in March 1779 strongly suggests it, when a 10-year-old stallion was sold for 140 guineas and a bull for 130. While these prices seem high for the time 'many capital judges are of the opinion they were remarkably cheap'.[13] The assignees only had to enter the names and prices for the purchasers in the books that were open to inspection, not full details of anyone else who might have been involved. The advertised sales are private contracts combined with two public auctions, one of which may not have taken place. The results of private contract sales were never reported in the newspapers, where open auctions were. This may explain why so many of the sales advertised appeared not to have taken place, yet money was collected for dividends. Sales could work in Robert's (4) favour even with the auctions. Buyers at other such sales who were sympathetic towards the bankrupt might pay for an animal but not take it away.

26 & 27 June 1777. The second part of Paget, Smith and Ashworth's original plan for the Swepstone property was implemented, without any public offers for sale. The first indenture was between the five assignees (the first party, including Ashworth), Paget and Smith (the second party without Ashworth, and Robert's (4) trustees), Robert (4) of Dishley (the third party) and Robert (8) of Gresley (the fourth party). The assignees sold 'a messuage in Snarestone and Swepstone then in the possession and occupation of the Bankrupt' to Robert (8) for £1,450. This is the only reference found to the Bakewells owning property in Snarestone.

27 June 1777. The next indenture was between Henworth Newton's executors, his son John and brother William (the first party), the five assignees (the second party), Paget and Smith (the third party, trustees), Robert (4) of Dishley (the fourth party), Robert (8) of Gresley (the fifth party) and William Herrick of Westcotes, Leicester (the sixth party, who was Robert's (8) cousin). The Newton mortgage on the 'Messuage and the Barn Closes' continued and was assigned to Herrick for the rest of the term, 'In Trust for Robert Bakewell Esqr & to attend the Inheritance'. It would become Robert's (8) property when the mortgage ended; meanwhile he paid the interest.

27 June 1777. The third indenture was between Elizabeth Aston (the first party), the five assignees (the second party), Paget and Smith (the third party), Robert (4) of Dishley (the fourth party), Edward Abney (the fifth party, probably Elizabeth's trustee), Robert (8) of Gresley (the sixth party) and William Herrick (the seventh party). It assigned 'the several closes in mortgage to Elizabeth Aston and purchased by the said Robert Bakewell Esqr' for the rest of the mortgage term to Herrick, 'In Trust for the said Robert Bakewell Esqr and to attend the Inheritance'.[14]

Robert (8) had now bought all the property at Swepstone and Snarestone, and had taken over the existing mortgages, thus keeping it in the family. The £1,450 went to the assignees for creditors' dividends.

11 & 18 July 1777. The *Derby Mercury* (BNA) advertised that the assignees were 'Now Selling, by Private Contract … the intire Stock of Robert Bakewell, consisting of Horses, Mares, Bulls, Cows, Rams, Ewes etc.' at Dishley.

19 July 1777. The *Cambridge Chronicle and Journal* had the same advertisement. (BNA)

By now the assignees had started sales on an individual basis. During June and July buyers must have gone to Dishley to view the stock and decide which animals to treat for out of the whole of Bakewell's stock. The assignees may have hoped to sell enough this way that they did not have to hold public auctions. A first dividend of 10s was promised in September, presumably knowing they had the money from Robert (8) plus money they expected from these sales.

1 & 8 September 1777. An amended notice of sale was published in *Aris's Birmingham Gazette*. (BNA) The extra instructions strongly suggest that Dishley had been swamped by potential buyers, and maybe just by interested viewers.

> Now Selling by Private Contract, at Dishley, near Loughborough, Leicestershire, The entire Stock of Robert Bakewell; consisting of Horses, Mares, Bulls, Cows, Rams, Ewes etc. It is requested that whoever intends viewing any Part of the above Stock, would, if convenient, give a few Days Notice by Letter, directed to R.B. that proper Attendance may be given.

3 & 10 September 1777. The *Caledonian Mercury* had the same advertisement. (BNA)

The notice also suggests that Bakewell himself arranged the visits, though he would not have been able to settle any sales.

25 April 1778. Smellie put a notice in the *Leicester Journal* (BNA) complaining that the promised 10s dividend had not been paid. He seems to be the only creditor who so publicly pursued the assignees.

12 May 1778. The commissioners published the following notice.

> The Commissioners in a Commission of Bankrupt awarded and issued forth against Robert Bakewell, late of Dishley in the County of Leicester, Dealer and Chapman, intend to meet on the 2nd Day of June next, at Three of the Clock in the Afternoon, at the Lion and Lamb in Leicester, in order to make a Dividend of the said Bankrupt's Estate and Effects; when and where the Creditors, who have not already proved their Debts, are to come prepared to prove the same, or they will be excluded the Benefit of the said Dividend. And all Claims not then proved will be disallowed.[15]

16 May 1778. A notice in the *Kentish Gazette* stated that a dividend was to be paid to Robert's (4) creditors on 2 June 1778, at the Lion and Lamb in Leicester at 3p.m. (BNA)

18 May 1778. The same notice was placed in the *Northampton Mercury*. (BNA)

This should be the final warning for any of Robert's (4) creditors to register their claims.

2 June 1778. The dividend meeting was held.

3 June 1778. From Smellie's April complaint, the dividend seems to have been reduced to 6s at the May meeting but in fact only 5s was paid.[16] It is likely that the creditors objected to the reduction and then persuaded the assignees to pay the extra dividend. They pressed the assignees for more action in collecting the assets, demanding explanations for the slow progress in settling Robert's (4) bankruptcy. It was two years since the commission and most were settled after about a year.

21-23 July 1778. The following advertisement was placed in the *St James's Chronicle* or the *British Evening Post*.[17]

> To be SOLD by Mr CART, (without reserve) at Nottingham, on Wednesday the 5th of August, in the Race Week, in 20 Lots or upwards, agreeable to such Conditions as will then be produced. The entire STOCK of BROOD MARES and FOALS, two and Three-year-old

> Fillies, of Mr Robert Bakewell, of Dishley, being of the Coach and Cart Kind. The above Stock were bred by Mr Bakewell, and chiefly got by Blake and George, which Horses covered several Seasons at Five Guineas a Mare.
> N.B. The Stock at Dishley, consisting of Bulls, Cows, Rams and Ewes, are selling off by private Contract.[18]

27 July 1778. The *Northampton Mercury* carried the same notice but without reference to Mr Cart. (BNA)

1 August 1778. The notice appeared in the *Cambridge Journal and Chronicle*. (BNA)

3 August 1778. The notice appeared in *Aris's Birmingham Gazette*. (BNA)

The assignees seem to be making an effort to get a good crowd for this auction of one section of the stock, as it was race week. They had clearly been holding the private sales at Dishley over the past year, though without enough effort for the creditors. There is no report of the 5 August sale and it may not have been held, though the private contract sales continued. The financial crisis of 1778 would not have helped.

12 September 1778. Another promised dividend was not paid, according to Smellie's complaint in the *Leicester Journal*. It had not been paid by the time of the 1783 petition to the commissioners.

14 September 1778. Probably prompted by the creditors' anger, the assignees advertised another sale in the *Northampton Mercury*. (BNA)

> Peremptorily to be Sold, by Lady-Day, 1779, The remaining Part of the Stock upon the Farm at Dishley, near Loughborough, bred by Mr Bakewell; consisting of 12 Stallions, 10 Mares, 12 Bulls, 40 Cows, Heifers, &c., 100 Rams, and 120 Ewes; which Ewes are drawn in Lots of different Value, six in each Lot.

These numbers are much reduced from the stock numbers before his bankruptcy so a considerable number had been sold already. However, this advertisement is evasive about how they are to be sold and there is no record of a public auction sale.

1 December 1778. A meeting was advertised in the *London Gazette* to be held at the Lion and Lamb, when a dividend would be allocated to the creditors.[19]

The commissioners proposed to issue the Certificate of Conformity which confirmed that Robert (4) had disclosed all of his assets. This had to be accepted by the creditors. Any dividend had to be paid pro rata to the creditors. The assignees had the money from the Swepstone sale, and the money from the sales of stock, as well as any money they had been able to get from Robert's (4) own creditors by then. The Certificate of Conformity was proposed for Robert (4) confirming he had made a full disclosure; this time his creditors accepted it. No dividend was paid. The assignees continued to pursue Robert's (4) debtors.

5 December 1778. A further notice appeared in the *London Gazette* stating that his Certificate of Conformity would be allowed and confirmed.

> Whereas the acting Commissioners in the Commission of Bankrupt awarded against Robert Bakewell, of Dishley in the County of Leicester, Dealer and Chapman, have certified to the Right Hon. Edward Lord Thurlow, Lord High Chancellor of Great Britain, that the said Robert Bakewell hath conformed himself according to the Directions of the several Acts of Parliament made concerning Bankrupts. This is to give Notice, that, by virtue of an Act passed in the Fifth Year of His late Majesty's Reign, his certificate will be allowed and confirmed, as the said Act directs, unless Cause be shewn to the contrary on or before the 26th of December instant.[20]

15 December 1778. This was reported in the *Manchester Mercury*. (BNA)

6 March & 3 April 1779. More stock for sale was advertised in the *Leicester Journal,* to be sold on 31 March, at Dishley. (BNA)

> To be Sold by Auction, at Dishley, near Loughborough, Leicestershire, on Wednesday the 31st of March, and the two following days; agreeable to such Conditions as will then be produced, – The entire Stock, without Reserve, of Bulls and Stallions, bred by Mr Robert Bakewell. The other part of the Stock, consisting of Rams, Ewes, Cows, Heifers etc, are selling off by Private Contract. The Whole may be viewed at Dishley, any Time before the day of Sale.

13 March 1779. The *Oxford Journal* carried the same advertisement. (BNA)

6, 8, 15 & 22 March 1779. The *Caledonian Mercury* carried the same advertisement. (BNA)

If the sale on 5 August did take place, together with this one, almost all of the horses and the bulls would have been sold. It was at this sale that the comment quoted earlier that a stallion and a bull had sold remarkably cheaply was made. The two sales would have been major dispersals. It would have been interesting to know what the conditions of sale were but no information has been traced.

16, 23 & 30 August, 6 September 1779. *Aris's Birmingham Gazette* (BNA) advertised a further sale at the end of September.

> To be Sold, at Dishley, near Loughborough, Leicestershire, on Monday September the 27th, and the four following Days; All the Rams (consisting of more than One Hundred in Number) bred by Mr Robert Bakewell. Proper Persons will be appointed to treat with Customers for the same, or any other Part of the Stock; as it is the Determination of the Assignees, that the Whole shall be disposed of by Lady-Day next. Dishley, 12th August 1779.

21 & 28 August, 18 September 1779. The *Cambridge Chronicle and Journal* carried the same advertisement. (BNA)

28 August 1779. The *Oxford Journal* carried the same advertisement. (BNA)

18 September 1779. The *Leicester Journal* carried the same advertisement. (BNA)

With these sales, the assignees were certainly making sure they reached a wide pool of customers. They probably provided the money for the next dividend that was paid. Unlike the two previous sales, this one was not a public auction; the rams would all be sold by private treaty. With a deadline, the assignees are trying to complete their work on the bankruptcy. However, the stock was clearly not disposed of by Lady Day (25 March) 1780.

22 April 1780. A dividend meeting was advertised.

> The Commissioners in a Commission of Bankrupt awarded and issued against Robert Bakewell, of Dishley in the County of Leicester, Dealer and Chapman, intend to meet on the 12th Day of May next, at Four o'Clock in the Afternoon, at the House of Thomas Barry, known by the Sign of the Lion and Lamb, in Leicester, in order to make a Further Dividend of the Estate and Effects of the said Bankrupt; when and where the Creditors, who have not already proved their Debts, are to come prepared to prove the same, or they will be excluded the Benefit of the said Dividend, And all Claims not then proved will be disallowed.[21]

26 April 1780. The meeting was announced in the *Kentish Gazette*. (BNA)

1 May 1780. The meeting was announced in *Aris's Birmingham Gazette*. (BNA) It also added that the dividend payment would take place on 22 May at 4p.m. at the Lion & Lamb.

It is not known how many creditors there still were, but John Smellie was still one of them. This may have been intended to be the final dividend which had to be announced in the *Gazette*, following the previous dividend and the issuing of the certificate. There are no other formal announcements but that did not preclude further payments.

22 May 1780. A dividend of 5s in the pound was paid to the creditors who proved their debts when they went to the meeting, followed by a further dividend of 2s 9d.[22]

No further dividends were paid, according to Smellie's complaint of 7 June 1783, and it was over seven years since Robert's (4) bankruptcy had been issued. This strongly suggests that there was more disagreement between the assignees and the creditors at the meeting, and the creditors refused to accept both the level of dividend offered, and that this would be the last payment they would get. The assignees were still pursuing the long-overdue payments to Robert (4) from his Scottish and Irish clients, and there may well have been other debts still outstanding. The creditors insisted that the assignees must obtain more money; they demanded that the private contract sales should go more quickly. The assignees by then must have been hoping to obtain the certificate of discharge but they could not do that without the creditors' consent and the final dividend being accepted. They had been occupied with it for 4 years and they all had other businesses to run. It was dragging on well over the average year or so and continued to the end of this year without agreement. It would not have been an easy meeting. By then Robert (4) was moved to take action, perhaps with Robert (8)'s advice; in part, this may be because he knew by then of plans that not only his stock but the lease on Dishley Grange was going to be sold. The assignees had taken over the lease as beneficial owners, as part of Robert's (4) assets. Perhaps by this date his landlord was not prepared to carry non-payment of rent. Robert (4) would have been desperate to save his home, his farm and his life's work.

23 December 1780. He published his petition for support.

> *TO THE NOBILITY GENTRY AND OTHERS.*
> The Humble Petition of Robert Bakewell of Dishley in the County of LEICESTER

SHEWETH

That your Petitioner has for a Series of Years employed his Attention on a Plan for improving the Breed of Horses for Cavalry, Harness and Draught, as also of Meat Cattle and Sheep.

That your Petitioner in Pursuit of this Plan had many Difficulties to surmount having the Prejudices of other Breeders to combat and various Experiments to make in Order to ascertain which were the best kinds to breed from; and that such Experiments were attended with considerable Expense, and more trouble than he can well convey a Sense of.

That your Petitioner apprehends he has brought all the Different Kinds of Stock abovementioned to a greater Degree of Perfection than has been done by any other person and thereby rendered important Services to this Country, and in this Opinion he hopes he is justified, by the best Judges having purchased from his Stock at higher Prices than from any other, and having sent them into the Counties of Bedford, Bucks, Cambridge, Chester, Cumberland, Derby, Devon, Dorset, Durham, Essex, Gloucester, Hereford, Herts, Huntingdon, Kent, Lancaster, Leicester, Lincoln, Norfolk, Northampton, Northumberland, Nottingham, Oxon, Rutland, Salop, Somerset, Southampton, Stafford, Suffolk, Sussex, Warwick, Westmoreland, Wilts, Worcester, Yorks, into North-Britain, Wales, Ireland, Germany and Jamaica.

That your Petitioner has made considerable Improvements in Agriculture, Division of Lands, Watering of Meadows, &c.

That your Petitioner in Consequence of the aforesaid Difficulties and Expenses, as well as by many great and unavoidable Losses to the Amount of many Thousand Pounds, is rendered incapable of pursuing his Plan; and as considerable Part of the Stock is soon to be sold, and probably will fall into the Hands of those who for want of Experience or other Causes cannot be supposed to manage it to the same Advantage, consequently little if any further Improvement can be expected therefrom.

But if the Publick should take his Case into their Consideration and grant him such Assistance as would enable him to purchase the whole or the best part of his Stock, he is fully persuaded he could be highly instrumental to the general Good of this Nation, by continuing in his late Line of the Breeding Business, and carrying it forward in such a Manner as will be most conductive to the Public Service, and he apprehends he could make a great Improvement from the State the Stock is now in, as he has done from the State of Stock in general at the time he began this Business, an Object he thinks of great Importance to the Honour and Interest of the British empire; for if it be allowed that the Increase of Herbage by Improvement in agriculture is a real Advantage to the Public

> in general, he conceives the Improvement of Stock, so as to gain a greater Quantity and better Quality of Flesh from such Herbage to be of equal if not of greater Importance.
>
> Your Petitioner therefore most humbly Solicits &c.[23]

This petition had the desired effect as over £1,000 was raised. The exact sum is not known but if all the signatories paid up the appeal produced £1,531 10s.[24] The list of subscribers was long and included some very influential people of the day. The Duke of Richmond was said to have given 500 guineas and the Duke of Rutland 200 guineas. Lord Sheffield and the Earls of Hopetoun and Middleton were contributors as well as some of his close associates such as Thomas Paget and Matthew Culley.[25] Obviously Bakewell's influence on the agriculture of the day, as is stated in the petition, was such that his friends and acquaintances felt that they could not stand by and see him fail. This money would not have been counted in as part of his assets so, as he stated here, it 'would enable him to purchase the whole or the best part of his Stock'. The money from the purchase would go to the assignees, who could then pay the creditors. It is significant that he still had his lease and most of his stock on the farm, despite the apparently regular sales by private contract. It would be the biggest private contract purchase of them all and it enabled him to continue with his carefully bred stock and improve it further. The valuation of his stock and the lease in preparation for the advertised sale showed a balance in his favour of from £1,500 to £2,000. He could buy himself out of bankruptcy, as Thomas Brentnall had.

3 February 1781. The rest of Robert's (4) stock and the remainder of the lease of Dishley Grange were advertised for sale in the *Leicester Journal.* (BNA)

> To be Sold by Private Contract, on or before the 5th Day of March next, All the Stock of Mr Robert Bakewell of Dishley, Leicester, together with the Lease of the Farm, in which there will be three Years unexpired at Lady-Day next. The Farm consists of about 450 Acres of Land, with a Privilege of Plowing without restraint, 250 Acres or upwards. The Stock, &c., may be viewed at Dishley, and to Treat for the same, or further Particulars, apply to Mr J. Ashworth of Daventry, H. Walker of Thurmaston, S. Huskinson of Croson, or W. Hodges of Leicester.

3 &12 February 1781. The same advertisement appeared in *Aris's Birmingham Gazette.* (BNA)

5 March 1781. It seems certain that Robert (4) bought both stock and lease using the appeal money, and Robert's (8) technique to declare Robert (4) the best bidder as agreed by the assignees. The creditors could reasonably expect to see their dividends finally paid, but the assignees still seemed to be pursuing Robert's (4) debtors and there was no immediate notice of a meeting.

14 July 1782. The creditors presented a petition to the Lords Commissioners to remove the assignees. It was heard on 2 August 1783.

24 July 1782. One reason for their slowness was that there were unrelated disputes between two of the assignees, William Bentley and Henry Walker, whose 1781 bankruptcy had not been discharged. Walker had tried to exclude the £3,000 loan from Bentley from his debts, claiming it was a joint liability with his brother. Bentley disproved this, and then successfully petitioned the Lord Chancellor to be added to Walker's list of creditors, and to be paid dividends.[26] Although Walker was in debt to Bentley, and bankrupt, there was nothing in the law which prevented him still acting as an assignee for Robert (4), who was in debt to them both. Walker's assignees were William Hodges, Thomas Paget and Joseph Heycock.

When Hodges then went bankrupt, two commissions were issued against him for different partnerships. One was for William, his brother Edward and their partner Edmund Carr jointly, issued on 24 July 1782 and by August 1783 no dividends had been paid.

1783. Sometime this year Robert (4) founded the Dishley Society. Having bought the stock and lease, even though he had not been formally discharged from bankruptcy, he was clearly confident enough to resume his farming. The final dividends had not yet been paid but he must have known the creditors could be paid. Though often asked, he refused to be the Society's president until he was formally discharged.

7 June 1783. Smellie was becoming more dissatisfied with the assignees' work, and he published his second complaint in the *Leicester Journal.* (BNA) He cited the assignees' intention to dispose of all the stock by Lady Day 1780, which they had not done. The creditors had thought they would then be paid their dividends,

> yet in a patient expectation of an equal distribution, the creditors have waited above seven years, since the interest of their money, lent on Mr Bakewell's simple securities, terminated; but to this day they have not received eight shillings in the pound, towards the payment of their principal.

They had been paid 12s 9d, according to the August 1783 petition so it is uncertain why Smellie quotes a payment of 8s. It is interesting here that Smellie seems unaware of the sale of the stock and lease on 5 March 1781. If he knew about it he might have been even more dissatisfied with the assignees.

2 August 1783. Four of Robert's (4) creditors petitioned the Lords Commissioners on 14 July 1782 to remove the assignees, 'on behalf of themselves and all other the Creditors of the said Bankrupt'. They had all proved their debts. They were John Bowley, Matthew Steel, Thomas Flavell and William Bosworth. Their petition was heard on this date. It recited some of the history of the bankruptcy, including Walker's and Hodges' bankruptcies, and stated that,

> the Petitioners and the said Creditors of the said Robert Bakewell the Bankrupt sustained a great Injury by the delay and being kept out of a Final Dividend. And therefore praying That the said William Bentley William Hodges Henry Walker John Ashworth and Samuel Huskison might be removed from being Assignees of the said Robert Bakewell and other Assignees chosen in their stead and such of the said Robert Bakewell's Estate and Effects as remain uncollected might be duly transferred to such New Assignees …

They then requested that the old assignees should be ordered to hand over any records of Robert's (4) bankruptcy, and any monies they had in hand, to the new ones; they would be admitted as creditors. A further request was that they would be admitted as creditors under Henry Walker's bankruptcy and eligible for dividends, and the same under William Hodges and his brother's bankruptcies, for the time they were acting as assignees for Robert (4). The commissioners had ordered this meeting and those involved could attend with legal representation; it seems they all did except Ashworth who did not attend or send a representative. The commissioners ordered that the assignees 'be forthwith discharged and removed from being assignees' for Robert (4). His commissioners were ordered to give notice in the *London Gazette* of the time and place of a meeting of Robert's (4) creditors 'to proceed to the Choice of a New Assignee or Assignees'. The commissioners and the creditors then had to formally assign all the estate and effects to the new assignees. The old assignees had to personally appear before the commissioners on oath to account for all or any of Robert's (4) effects that they held. The new assignees were to be admitted as creditors for both Walker and Hodges and entitled to dividends.[27]

The creditors' meeting was held in Loughborough in 1783 to replace all the assignees. There is no record of the event, nor of who took over. The problem for the dissatisfied creditors was that they would have had to agree new assignees from their own number. Perhaps the four petitioners Bowley, Steel, Flavell and Bosworth, or some of them, are the most likely new ones. Whoever it was, they did work to get a final dividend paid.

22 December 1783. A notice was placed in *Aris's Birmingham Gazette.* (BNA) Dividends were to be paid to the creditors of Robert Bakewell of Dishley, Leicestershire, a dealer, at the Lion and Lamb, on 5 January 1784.

5 January 1784. The dividends were finally paid; there are no further notices or complaints.

This payment seems to have been sufficient to clear all Robert's (4) debts, so either they had been paid in full, or four-fifths of his creditors by number and value had agreed to accept this dividend as their final payment. This relieved Robert (4) from the limitations of bankruptcy and the normal procedure would have been to apply for the Certificate of Discharge, but this did not happen until late in 1789. Any such delay must have been at Robert's (4) insistence. This suggests that he intended still to pursue the debts due to him, maybe from a sense of the unfairness of the procedure, maybe because he resented the slur on his reputation or maybe because he simply wanted what was due to him. Once his own debts were cleared, Robert (4) could legally resume his normal life and business, and he clearly did this. It is possible that when he was declared bankrupt his long-distance debtors thought they could evade payment simply through the inherent delays in communications and general deviousness in dealing with the assignees. When Robert (4) was free he could pursue them as common debtors and he may have done this. Unlike bankruptcy, debt was a felony and punishable by imprisonment until the debt was paid, or death.

After his successful petition for financial help, and the eventual payment of his overdue accounts which enabled him to clear his debts, he was given back the control of the business even though he was still officially bankrupt. It is in the period following his final dividend payments that his stud animals reached their highest prices. From his correspondence, and the founding of the Dishley Society, it is clear that he was actively taking charge again at Dishley while pursuing the money owed to him. After he got the money, the certificate of discharge was granted.

Michaelmas 1784. The lease on Dishley Grange ran out, and a new one was negotiated with an increased rent of £552. This lease was still running when Robert Honeyborn took over at Dishley, but it is not known how long the lease term was. The repairing clauses carried over.[28]

Sir William and his wife Mary must have had faith in Robert's (4) eventual recovery to be willing to continue his existing lease before the final dividend payments were made, then grant a new one. He was probably their best tenant and they would not want to lose him.

1 April 1786. The series of letters to Culley, quoted in Pawson starts; they show how active Robert (4) was in his business despite not yet being formally discharged. In this first one, Robert (4) had already sent a stallion to Culley's subscription group of local farmers and hoped it had arrived.[29]

1786. On an unknown date, he sold more bulls to Jamaica,[30] after the final dividend had been paid but before his discharge.

1786. The Mill lands adjoining Dishley Grange were rented from the Garendon Estate to provide extra land to irrigate.[31]

8 December 1786. Another letter to Culley. The horse mentioned in the April letter to Culley, and Edward Porter the stallion leader, were back having given satisfaction. Robert (4) said he had sold 'great numbers' of foals this year, 'what more useful to pay rent with than this kind?' He offered the same horse next year on the same terms, 80 guineas, with Robert (4) paying Porter's wages and Culley paying all other expenses. Bakewell noted that the ram trade 'never was so good in this Neighbourhood and in other Places'.[32] Robert (4) was setting his own terms.

10 March 1788. This letter to Young shows that as soon as he could, Robert (4) resumed his visits and his observations on agriculture. He kept up with Young's *Annals of Agriculture* and collaborated with him on current issues. He began to send cattle to America and took one of his stallions for inspection by King George. The king's conversation shows that despite the bankruptcy, Robert's reputation continued to grow and he was treated with considerable respect.

> I left Dishley about a fortnight past, have brought a Black Stallion to London which has been seen by the King, Prince of Wales and many other great personages to whom he gives more general satisfaction than others I have yet shown. …
>
> I have sold a Bull and Two Heifers which are to put on board this week to go on to Mary land N. America and if these please I hope to have further orders. …
>
> I had near an hours conversation with the King on the subject of breeding which he seemed much pleased with and listened to what I said with great attention. I expect see again in a few days. …[33] (Apparently,

and understandably, he was in a hurry to get his letter to the next post and missed out 'to'.)

12 June 1788. There was a notice in the *Derby Mercury* (BNA) for the first meeting of the trustees for the proposed bridge over the River Trent at Sawley Ferry. Robert (8) of Swepstone and Robert (4) of Dishley were listed as investors. This was something Robert (4) could not have done while his affairs were under the assignees' management.

6 September 1788. There was a report in the *Ipswich Journal.* (BNA)

> That celebrated grazier Mr Bakewell has lately sustained a very considerable loss by the death of one of his finest rams, which he had actually hired for the approaching season to a neighbouring farmer for 400 guineas. The advice of the most eminent of the faculty was had to it, but the disorder being fixed upon the animal's liver, all medical skill was in vain.

The report suggests the ram had picked up 'the rot', alias liver fluke, for which there was no cure. Ironically Robert (4) knew a good deal about the disease and had used it to his advantage as a means of ensuring that sheep he fattened for the butchers could not be sold on for breeding: 'to be sure that they shall be killed, and not go into other hands, he rots them before he sells, which, from long experience, he can do at pleasure. It is only to flow a pasture or meadow in summer, and invariably it rots all the sheep that feed on it in the following autumn'.[34]

18 August 1789. Robert (4) replied later in August to a letter from Arthur Young asking for Robert's (4) opinions on one of Young's forthcoming publications. Despite his reception by the king, Robert (4) was well aware that his bankruptcy did affect how people judged him. We are not sure exactly when Bakewell (4) was able to clear his bankruptcy but it was still in effect in August 1789 as he himself states in his reply. However, his discharge is in sight.

> Your favour of the 18th instant I rec'd but of late have been so much engaged on the Ram business and other matters relative to my Bankruptcy which I hope will soon be settled that I have not had time to attend to anything else, and on account of the state my affairs have lately been in I should think any opinion as coming from me will rather injure than benefit any publication, and probably may have a tendency to make the whole disregarded, as most people think well or ill of any scheme by the success with which it is attended but this I leave to you.[35]

13 November 1789. By this date Robert (4) was confident enough to agree to be president of the Dishley Society, a post which he had previously refused because of his bankruptcy. The minutes of this Dishley Society meeting, held at the George Inn, Loughborough, proposed a 10 guinea subscription payable to the Treasurer, Robert Honeyborn. The first resolution was 'That Mr Bakewell be President'.[36] It was accepted.

18 November 1789. In a letter to Culley, Robert (4) discussed the difficulties of the Dishley Society's meetings over the past year, which had been held in different places. He had been frequently offered the chair but had declined 'on account of the state in which my affairs have been', but because of 'things now taking a favourable turn', he had accepted the position and intended to 'do every thing in my power to serve the cause I have engaged in'.[37] He and his assignees must have obtained the outstanding payments somewhere between 18 August and 13 November. They were then able to apply for discharge.

Late November or December 1789. Robert (4) was granted a Certificate of Release which discharged his bankruptcy.

9 September 1791. The *Stamford Mercury* (BNA) reported that 'Mr Bakewell lets three rams this season for three thousand guineas'.[38] Robert (4) was back on a sound financial footing.

17 October 1793. A letter from Robert Hepburn to the 4th Duke of Buccleuch notes the arrival of two rams from Mr Bakewell, one for each of them.[39] They may have been delivered to either Boughton House, Kettering, or the Borders estates, as with the previous rams.

18 November 1793. *Aris's Birmingham Gazette* (BNA) reported the death of Robert (8) Bakewell Esquire, Barrister at Law, at Tempe Farm, Swepstone. It was a late 17th-century two-storey farmhouse which he had then extended and gentrified with a three-storey Georgian wing and renamed.

1793. By this year both Robert's (4) sisters Rebecca and Hannah, both beneficiaries of the family trust, were dead and Robert (4) was the only surviving beneficiary. He could wind up the trust.

1 October 1795. Robert (4) Bakewell died. The lease passed to his nephew Robert Honeyborn and the Garendon Estate records show it was then valued at £711 3s 5d and let to him at £735 per year. This was an increase in rent from the £552 of the 1784 new lease. It was described as a 'House in good repair and the outbuildings, which are many, pretty generally so'.[40] Robert (4) had clearly observed the repairing lease fully, which is more than most of the Garendon tenants did. He not only kept the existing ones repaired but built new ones at his own expense.

22 October 1795. The obituary of Robert (4) was printed in the *Bath Chronicle & Weekly Gazette*. (BNA)

23 October 1795. It was printed in the *Kentish Weekly Post*. (BNA)

24 October 1795. It was printed in the *Staffordshire Advertiser*, the *Norfolk Chronicle*, the *Oxford Journal* and the *Hampshire Chronicle*. (BNA)

26 October 1795. It was printed in the *Gloucester Journal*. (BNA)

Endnotes

1 SRO, D661/1/71 Parcel 7.

2 TNA, B1/74, 1784, p202. The Office of the Commissioners of Bankrupts & Successors: Order Books relating to Petitions against the Declaration of Bankruptcy.

3 BNA, *Saunders's Newsletter* (Dublin), 27 September, 2 & 4 October 1776.

4 Recorded in *The Gazette*, issue 11720, 19-23 November 1776; *Derby Mercury*, 22 November 1776.

5 TNA, B1/73, 1773, p5-6.

6 TNA, B1/74, 1784, p203. The Office of the Commissioners of Bankrupts & Successors: Order Books relating to Petitions against the Declaration of Bankruptcy.

7 Wykes, D.L., 'Robert Bakewell (1725-1795) of Dishley: farmer and livestock improver', *Agricultural History Review* 52, (2004), p48.

8 Will of John Bredon, 8 January 1795, probate 2 March 1797.

9 Orde, A., (ed.) Matthew and George Culley Travel Journals and Letters 1765-1798, p67.

10 Pawson, H.C., Robert Bakewell Pioneer Livestock Breeder, (1957), p195.

11 SRO, D661/1/71 Parcel 7.

12 BNA, *Oxford Journal*, 1 March 1777.

13 Wykes, D.L., 'Robert Bakewell (1725-1795) of Dishley: farmer and livestock improver', *Agricultural History Review* 52, (2004), p47.

14 SRO, D661/1/71 Parcel 7.

15 Recorded in *The Gazette*, issue 11873, 12 May 1778.

16 TNA, B1/74, 1784, p202. The Office of the Commissioners of Bankrupts & Successors: Order Books relating to Petitions against the Declaration of Bankruptcy.

17 From the 17th and 18th century Burney Collection.

18 'Advertisements and Notices.' *St James's Chronicle* or the *British Evening Post*, July 21-23 July, 1778, 17th and 18th century Burney Collection.

19 Recorded in *The Gazette*, issue 11873, 1 December 1778.

20 Recorded in *The Gazette*, issue 11932, 5 December 1778.

21 Recorded in *The Gazette*, issue 12076, 22 April 1780.

22 TNA, B1/74, 1784, p203. The Office of the Commissioners of Bankrupts & Successors: Order Books relating to Petitions against the Declaration of Bankruptcy.

23 Pawson, H.C., Robert Bakewell Pioneer Livestock Breeder, (1957), p181-2

24 Now £216,300. Measuring Worth (2004), 'Five Ways to Compute the Relative Value of a UK Pound Amount, 1270 to Present'. URL: www.measuringworth.com/ukcompare/

25 Pawson, H.C., Robert Bakewell Pioneer Livestock Breeder, (1957), p181-2. Wykes, D.L., 'Robert Bakewell (1725-1795) of Dishley: farmer and livestock improver', *Agricultural History Review* 52, (2004), p48.

26 TNA, B1/73, 1782-3, p5-6, The Office of the Commissioners of Bankrupts & Successors: Order Books relating to Petitions against the Declaration of Bankruptcy. Entry Book Orders, p5 and 6.

27 TNA, B1/74, 1784, p202-5, The Office of the Commissioners of Bankrupts & Successors: Order Books relating to Petitions against the Declaration of Bankruptcy.

28 ROLLR, DE7454/1/Estimate or Survey of the Estates of Thomas March Phillipps, n.d. but 1796-1817. DE7454/3 has a pencil note that the rent was £1,100 in 1810.

29 Pawson, H.C., Robert Bakewell Pioneer Livestock Breeder, (1957), p102.

30 Wykes, D.L., 'Robert Bakewell (1725-1795) of Dishley: farmer and livestock improver', *Agricultural History Review* 52, (2004), p45.

31 Wykes, D.L., 'Robert Bakewell (1725-1795) of Dishley: farmer and livestock improver', *Agricultural History Review* 52, (2004), p40.

32 Pawson, H.C., Robert Bakewell Pioneer Livestock Breeder, (1957), p105.

33 Pawson, H.C., Robert Bakewell Pioneer Livestock Breeder, (1957), p172

34 Arthur Young, quoted in Pawson, H.C., Robert Bakewell Pioneer Livestock Breeder, (1957), p84.

35 Pawson, H.C., Robert Bakewell Pioneer Livestock Breeder, (1957), p175.

36 Pawson, H.C., Robert Bakewell Pioneer Livestock Breeder, (1957), p148.

37 Pawson, H.C., Robert Bakewell Pioneer Livestock Breeder, (1957), p147.

38 Now £403,100. Measuring Worth (2004), 'Five Ways to Compute the Relative Value of a UK Pound Amount, 1270 to Present'. URL: www.measuringworth.com/ukcompare/

39 NAS, GD224/31/10 item 24.

40 ROLLR, DE7454/1.

Appendix 8

Birmingham Royal Show, 1876

There were no separate classes for Longhorns at the Royal Show until the meeting at Windsor in 1851. Following is a complete list of entries in the Longhorn classes for 1876.

Class 81: Longhorn Bulls over two years

825. Thomas Taverner, Upton, Nuneaton, Leics. (now in Warwickshire) Royal Duke of Upton 1st, red, 4 years 3 months, bred by Mr Nicholas Taverner, Upton. s. Son of Earl of Upton. d. Damsel.

826. William Thomas Cox, Spondon Hall, Derby. Earl of Upton 6th, brindle and white, 4 years 3 months, bred by R.H. Chapman, St Asaph, Flint. s. Earl of Warwick. d. Countess of Warwickshire by Old Sparkenhoe.

827. W.T. Cox. Earl of Upton 5th, roan and white, 4 years 2 months, bred by R.H. Chapman, St Asaph. s. Earl of Warwick. d. Lady Willscote by Old Cropredy.

1st Prize 828. Duke of Buckingham and Chandos, Stowe Park, Bucking-ham. Conqueror 3rd, brindle and white, 4 years 11 months, bred by exhibitor. s. Young Conqueror. d. Lady.

3rd Prize 829. William Smith Shaw, Fradley Old Hall, Lichfield, Staffordshire. Earl of Upton 7th, light brindle, 4 years 3 months, bred by R.H. Chapman, St Asaph. s. Earl of Warwick. d. L Lady Upton 73rd by Old Sparkenhoe.

2nd Prize 830. Frederick Tomlinson, Southwood, Ticknall, Derbyshire. Peeping Tom, brindled, 3 years 1 month, bred by Mr Whitacre, Warwickshire.

831. Thomas Satchwell, Hernfield, Knowle, Warwick. Forester, brindled red, 2 years 3 months, bred by exhibitor. s. Twist. d. Lady Arden by Rockingham.

832. Samuel Forest, The Chase, Kenilworth, Warwickshire. King Lupin, red and white, 3 years 3 months, bred by exhibitor. s. Crown Prince. d. Lupin.

833. Samuel Forest. Crown Prince, red and white, 6 years 3 months, bred by Mr J.H. Burbery, Montague House, Kenilworth. d. Daisy.

Class 82: Longhorn bull over one year, not over two years

834. William Thomas Cox, Spondon Hall, Derbyshire. Lorne 1st, brindle and white, 1 year 3 months, bred by exhibitor. s. Lord Lorne. d. Pallas by Curzon.

835. William Thomas Cox. Lorne 2nd, brindle and white, twin to above.

837. Duke of Buckingham and Chandos. Earl of Dadford, brindle, 1 year 2 months, bred by exhibitor. s. Earl of Upton 3rd. d. Rose by Earl of Warwick.

1st Prize 838. Duke of Buckingham. The Marquis, brindle and white, 1 year 3 months, bred by exhibitor. s. Conqueror 3rd. d. Wild Fire.

2nd Prize 839. John Godfrey, Wigston Parva, Hinckley, Leicestershire. Tiger, red and white, 1 year, bred by exhibitor. s. Blue Knight. d. Daisy by Shakespeare.

840. Thomas Satchwell, Hernfield, Knowle, Warwickshire. Warrior, brindle and white, 1 year 2 months, bred by exhibitor. s. Ranger. d. Cherry by Peeping Tom.

Class 83: Cows over three years

841. Richard Hall, Walton-on-Trent, Burton on Trent, Staffordshire. Longhorns 4th, red and white, 6 years 2 months, in milk, calved 15 May, bred by Mr John Godfrey, Wigston Parva, Hinckley, Leics. s. The Stranger. d. Longhorns 3rd.

842. Henry Parsons, Haselbury, Crewkerne, Somerset. Prolific, brindle, 5 years 4 months, in milk, in calf, calved 24 January 1876, bred by Mr V.B. Watts, Melcombe Horsey, Dorchester, Dorset.

843. John Spencer, Villers Hill, Kenilworth, Warwickshire. Beauty, red and white, 4 years 1 month, in milk, bred by the late William Bolton, Finwood, Rowington, Henley-in-Arden, Warwickshire.

3rd Prize 844. Thomas Satchwell. Rosemary, red and white, 3 years 2 months, in milk, calved 20 May 1876, bred by exhibitor. s. Red Lion. d. Valentine by Monarch.

845. Thomas Satchwell. Lady Forest, red, 3 years 2 months, in milk, calved 26 March 1876, bred by exhibitor. s. Red Lion. d. Brindle Beauty by Peeping Tom.

846. Thomas Satchwell. Lady Arden, dark brindle and white, 9 years 3 months, in milk, in calf, calved 27 July 1875, bred by exhibitor. s. Rockingham. d. Crump by Blue Bull.

2nd Prize 847. Samuel Forest. Bluebell, brindle and white, 6 years 2 months, in milk, calved 25 February 1876, bred by Mr J.H. Burbery, d. Bluebell.

1st Prize 848. Samuel Forest. Lady, brindle and white, 6 years 3 months, in milk, calved 22 October 1875, bred by Mr J.H. Burbery. d. Beauty.

849. William Wheeler, Long Compton, Shipston-on-Stour, Warwickshire. (No name given), 6 years 3 months, in milk, calved 25 February 1876, bred by the late J. Wheeler, Cropredy Lawn, Banbury, Oxfordshire.

Class 84: Heifer in milk or in calf, over two years, not exceeding three years

850. William Thomas Cox, Spondon Hall, Derby. Mayflower, brindle and white, 2 years 1 month, bred by exhibitor. s. Earl of Upton 5th. d. Pallas by Curzon.

851. Richard Hemming Chapman, The Roe, St Asaph, Flintshire. The Bulkington Rose, white and coloured, 2 years 4 months, in calf, bred by Mr T. Warner, Weston Hill, Bulkington, Rugby, Warwickshire. s. Hercules. d. Fairy by Messenger.

852. R.H. Chapman, Lady Weston, white and coloured, 2 years 2 months, in calf, bred by Mr T. Warner, Weston Hill. s. Hercules. d. Lupin by Messenger.

2nd Prize 853. Duke of Buckingham and Chandos, 'Barmaid', brindle, 2 years 9 months, in calf, bred by exhibitor. s. Conqueror 3rd. d. Negress 2nd.

1st Prize 854. Duke of Buckingham and Chandos, Lady Twycross, red and white, 2 years 10 months, in calf, bred by exhibitor. s. Conqueror 3rd. d. Wildfire.

3rd Prize 855. Frederick Tomlinson, of Southwood, Ticknall. Lady Weston, red and white, 2 years 4 months, in calf, bred by Mr Weston.

856. Frederick Tomlinson, of Southwood, Loophorn 4th, strawberry, 2 years 2 months, in calf, bred by exhibitor. s. Conqueror, d. Loophorn by Duke.

857. John Godfrey, of Wigston Parva, Hinckley, Leics. Beauty, red and white, 2 years 10 months, in calf, bred by exhibitor. s. Shakespeare, d. Primrose by The Stranger.

858. John Godfrey, of Wigston Parva. Bashful, red and white, 2 years 10 months, in calf, bred by exhibitor. s. Shakespeare, d. The Plum by The Stranger.

859. Thomas Satchwell, Hernfield. Minnie, light red, 2 years 3 months, in calf, bred by exhibitor. s. Twist, d. Lady Bountiful by Rockingham.

860. Thomas Satchwell, Carrie, red, 2 years 2 months, in calf, bred by exhibitor. s. Twist, d. Cherry by Peeping Tom.

861. Samuel Forrest, (previously spelt Forest) The Chase, Kenilworth, 2 years 2 months, in calf, bred by exhibitor. s. Crown Prince, d. Cowslip.

Class 85: Heifers not exceeding Two years old

862. Henry Parsons of Haselbury, Crewkerne, Somerset. Prodigy, brindle, 1 year 2 months.

863. Henry Parsons. Promise, light brindle, 1 year 1 month, bred by exhibitor. s. Earl of Upton 3rd, d. Rollright by Earl of Warwick.

2nd Prize 864. William Neale Berry, of Stoke Golding, Hinckley, Leics. Pride of the Park, brindle, 1 year 2 months, bred by Mr Nicholas Taverner, Upton Park, Nuneaton. s. Brindled Boy, d. Beauty by Mr Chapman's Wild Bull.

865. William N. Berry. Young Strawberry's Perfection, red and white, 1 year 1 month, bred by exhibitor. s. Brindled Boy, d. Young Strawberry 2nd by Upton Hero.

3rd Prize 866. William N, Berry. Weston's Farewell, red and white, 1 year 2 months, bred by exhibitor. s. Royal Duke of Upton 1st, d. Damsel by Upton Hero.

867. William Thomas Cox, Spondon Hall, Derby. Ellen, roan and white, 1 year 2 months, bred by exhibitor. s. Tippo, d. Beauty by Emperor.

868. Duke of Buckingham and Chandos. Baroness, red and white, 1 year 7 months, bred by exhibitor. s. Conqueror 3rd, d. Lady Caroline by Conqueror.

1st Prize 869. Duke of Buckingham and Chandos. Countess of Temple, brindle and white, 1 year 10 months, bred by exhibitor. s. Conqueror 3rd, d. Lady Mary by Young Conqueror.

870. William Peyton Burbery, The Crofts, Stratford-on-Avon. Dumpling 4th, brindle and white, 1 year 1 month, bred by exhibitor. s. Crown Prince, d. Dumpling 3rd.

871. Frederick Tomlinson, Southwood. Pet 5th, strawberry, 1 year, bred by exhibitor. s. Peeping Tom, d. Pet 2nd by Conqueror.

872. Frederick Tomlinson. Pet 7th, brindle, 1 year 1 month, bred by exhibitor. s. Peeping Tom, d. Pet 3rd, by Duke 3rd.

873. Thomas Satchwell, Hernfield. Forest Gem, red and little brindle, 11 months, bred by exhibitor. s. Romeo, d. Lady Arden by Rockingham.
874. Thomas Satchwell. Rosalind, red and white, 1 year 1 month, bred by exhibitor. s. Ranger, d. Rosamund by Bulky.
875. Thomas Satchwell. Bloom, dark brindle and white, 1 year 3 months, bred by exhibitor. s. Ranger, d. Blossom by Peeping Tom.
876. Samuel Forrest. Brindle and white, 1 year 4 months, bred by exhibitor. s. Crown Prince, d. Sleepy.
877. Samuel Forrest. Red and white, 1 year 3 months, bred by exhibitor. s. Crown Prince, d. Beauty.

As well as the entries listed, Longhorns also competed in the mixed dairy classes.

Appendix 9

Stilton Cheese: History and Recipe

Pitt reported that although the Longhorn's chief use was as a beef animal, Leicestershire produced many excellent milkers. He considered Leicestershire a significant dairy county, and reported that around Hinckley, Bosworth, Appleby and Snareston many dairy herds, of approximately 12-25 Longhorns, could be found. The Longhorn's rich milk, 6 per cent butterfat, was ideal for cheese and butter manufacture. Indeed, it was said that 'Though you may skim a Shorthorn's milk with a feather, a mouse may run over the cream of a Longhorn without fear of falling through'.

Leicestershire in the late 1700s was celebrated for its Stilton cheese. The first person to produce this renowned cheese was Mrs Paulet of Wymondham, near Melton Mowbray, who was still living when Marshall toured the district in 1789. Mrs Paulet was related to an innkeeper who kept the Bell at Stilton in Huntingdonshire, on the Great North Road from London to Edinburgh. She supplied him with her quality cheese, which was well received and customers paid half-a-crown per lb. With unclear origins, it became 'Stilton' cheese.

Its origins were eventually discovered, and the art of production learnt by other women of the district, until production occurred in almost every village in that part of Leicestershire and Rutland. A recipe of the time, given by Pitt was:

> Take the milk of seven cows, and the cream of the same number; heat a gallon of water to scalding hot, and pour it upon three or four handfuls of marigold flowers, that have been bruised a little; then strain it into a tub to your milk, and put some rennet to it, but not too much, to make it hard: put the curd into a sieve to drain, it must not be broke at all, but as the whey runs from it tie it up in a cloth, and let it stand for half an hour or more; then pour cold water upon it, enough to cover it, and let it stand half an hour more; then put half of it into a vat, six inches deep, and break the top of it a little to make it join with the other; then put the other half to it, and lay a half hundred-weight upon it, and let it stand half an hour; then turn it and put it in the press, and turn it into clean cloths every hour the day it is made; the next morning, salt it and let it lie in salt a night and a day; keep it swathed tight till it begins to dry and coat, and keep it covered with a dry cloth a great while. The best time to make it is in August.

Source: Pitt, W., General View of the County of Leicestershire, including a survey of the County of Rutland, (1809), p158-160.

Appendix 10

Horn Shapes and Uses

It is said that the age of a Longhorn can be determined by the circular ridges which are present on its horns, like a tree with its annual growth rings. Until the age of two, the horn is smooth, but at three years a circular ridge appears at the base of the horn and another ridge is added in each succeeding year. Hence, by counting the ridges and adding two, the age of an older animal can be judged with tolerable precision, unless the rings have been defaced or artificially removed by scraping or filing.

These main annual rings are not to be confused, however, with other smaller rings that are sometimes found at the root of the horn and which are an indication that an animal has been poorly fed at some stage of its growth. Poor nutrition or accidents during the formative stages are also frequently responsible for horns that are crooked or unsightly.

By the age of three, there is often a distinct cone of horn at the extremity. This will be shed eventually, leaving a perfect horn tip underneath, and is no cause for concern. The horn should be fine in texture and of a yellowish colour, often with traces of pink towards the base, almost as if it had been bruised. Any trace of black in the horns is discouraged as showing evidence of an outcross.

The preferred shape for the horns, from the handler's point of view, is what are termed bonnet or wheel horns, that is, those that turn with an incurve towards the mouth, rather like the strings of a lady's bonnet. Horns which come from the head with a forward sweep are also quite acceptable.

Of course that is the theory; in practice, horns may take many directions. Sometimes animals can be seen exhibiting one bonnet and one straight horn. The only undesirable direction is horns that grow backwards from the head, which give the animal a curiously windswept appearance.

During the golden age of the Longhorn in the 18th century, it was thought that climate and region had a great effect on the direction that the horns took, but, as the fashion for horned cattle declined, so did the need to explain what affected their shape. The reader can ponder the possible validity of these suggestions for horn size and shape as visits are made to various herds in different parts of the country.

It is now generally acknowledged that Robert Bakewell fixed the style of the bonnet horn. His need was almost certainly one of economy, for he housed up to 170 cattle during the winter and animals sporting the bonnet type of horn took up less room than those with straight horns, and caused less injury to their neighbours.

Bakewell was certainly wise in back-fencing his hedge, a practice which immensely irritated the agricultural writer Arthur Young because he considered it to be uneconomic. But to anyone who has experienced Longhorns and knows the immense pleasure they gain from thrashing their heads around in a good thick stretch of hawthorn, back-fencing is an essential part of Longhorn husbandry.

In Bakewell's time there was a fashion for training horns, either by an attachment on the top of the animal's head or by using hot irons. By the latter method, incredible shapes could be achieved as, for example, in the case of the Fat Longhorned Heifer called Bandy, bred by Mr Paget of Ibstock, Leicestershire, and drawn by George Garrard in 1806.

It is usual to find with ruminants that the male carries the grandest horns, but this is not the case with Longhorn cattle. The bull carries thick, strong but comparatively short horns while the cow has finer but much longer horns. The best horns of all are to be seen on the heads of castrated males or oxen; they are of incredible length and almost invariably curve in the bonnet fashion. The length of horn is no indication of the temper of the animal. The Longhorn is an exceptionally docile breed (although there can, of course, be difficulties with individual animals in particular situations). Overall, however, it may be the case that, in cattle those with the longest horns are those with the most placid natures.

The entrance hall of Calke Abbey is hung with the mounted heads of prize oxen, the size of which is quite breathtaking. The horns are of great interest being both longer and thicker than are normally to be seen today. Admittedly they are the heads of oxen, which do grow horns far larger than both bulls and cows, but might it be that during this period, when horn was used extensively in the comb, button and lantern industries, cattle were bred as much for their horn as for their meat? Alternatively, over the years, when numbers of the breed have fallen to perilous levels, and outside blood has been most surely used, have we lost a little of what were once truly immense horns?

The horn industry was based in the south Midlands, which was well placed to take advantage of products from the cattle for which the area was famous, the Longhorn.

The horn consists of two separate parts – an outer horny case and an inner conical-shaped substance which is something between hair and bone. In the manufacturing process, the two were separated by a sharp blow. The outer case was then cut into three sections: the part from the root of the horn being flattened and used for combs; the middle section, after flattening by heat, was oiled to improve its transparency and then split into thin wafers to form a substitute for glass in the lanterns of the poor; and the tip of the horn was used by manufacturers of knife handles, whips and buttons, amongst other things. The core of the horn was boiled down to provide fat for the soap industry and the unusable bony substance left behind was ground down and sold back to the farmer as fertiliser. Drinking horns were also widely used at one time and little salt cellars of plugged horn, perhaps from the tip of the horn, have been used well into the present century by at least some workmen who have expected meat on some days with their midday bread.

Horn was thus an important raw material for a number of growing industries at one stage in our history. Some of the 'buttons and baubles' which our early explorers are said to have used as payment for vast tracts of territory in the New World are quite likely to have come from Longhorn cattle. And hoof and horn meal was that vital product of early industry which benefited the land, and led to the development of the superphosphate industry.

Appendix 11

A Chronology of Key Dates in Longhorn History

1725	Robert Bakewell born on 23 May
1760	First ram let by Bakewell for 16s
1783	Dishley Society meets informally
1788	Leicester Agricultural Society formed
1789	Dishley Society formed officially with minutes and subscriptions
1795	Bakewell dics on 1 October
1811	Sale of Thomas Prinsep's herd at Croxall, Derbyshire
1819	Sir George Crewe succeeded his inheritance
1844	Sir John Crewe succeeded his father
1873	Sale of Mr Chapman's herd at Upton
1876	Sale of Mr R. Brown's herd, Farewell, at Lichfield
1878	First volume of the Longhorn Herd Book published
1886	Sale of Calke herd on 19 May
1896	Bakewell's chair and portrait given to the R.A.S.E.

Appendix 12

Livestock Artists of the 18th and 19th Centuries

Paintings of animals are considered to have been fairly faithful representations in the 18th century, before artists succumbed to the financial inducements of their patrons in the 19th century and produced 'likenesses' of animals that can hardly be accepted today. The development of photography from around 1850 has brought more recent paintings back into the believable world for the modern viewer. Several artists are of relevance to the Bakewell and Longhorn story.

John Boultbee (1753-1812)

Sporting painter who is sometimes claimed to have entered the Royal Academy Schools in 1775, although there is no record there of his enrolment. The Boultbee family records show that he was a pupil of Sir Joshua Reynolds. Boultbee's patrons included the Lords Stamford and Egremont, Robert Bakewell himself, Richard Tattersall and Thomas Coke, later first Earl of Leicester.

John Fernely, Senior (1782-1860)

Leicestershire-born sporting artist who showed great talent for painting when apprenticed as a wheelwright to his father. In 1801, the Duke of Rutland arranged for him to study under Ben Marshall for three years in London. Throughout the rest of Fernely's life, Marshall retained a strong influence on his work. He had many patrons, including Robert, first Earl of Grosvenor, later first Marquis of Westminster. Works by Fernely can be seen at Calke Abbey.

George Garrard (1760-1826)

A sporting artist, who studied at the Royal Academy and also with Sawrey Gilpin, who later became his father-in-law. He became famous for his anatomically accurate models of livestock and for his engravings.

Thomas Weaver (1774-1843)

The son of a Shropshire farmer, he was said to have been a pupil of John Boultbee, and was famous for his livestock paintings. His patrons included Thomas Coke, first Earl of Leicester, and Viscount Anson. He painted the famous *Mr Morris's Sheep Show at Barton-on-Humber, Lincolnshire.*

Bibliography

Books and Articles

'A Farmer and Breeder' [John Lawrence], The New Farmer's Calendar, (1801 & 4th ed., 1802).

'Albion', 'Barney's Bakewell Sheep', in *The American Farmer*, 2 October 1829.

Anderson, J., Recreations in Agriculture, Natural-History, Arts, and Miscellaneous Literature, Vol 1, (1799).

Andrews, G.H., Modern Husbandry, (1853).

Anon., 'The Obituary of Robert Bakewell', in *The Gentleman's Magazine*, November 1795.

Baldwin, T., in Curran, P.L., The Native Lowland Sheep of Galway and Roscommon: A History, (2001).

Betham-Edwards, M., (ed.) The Autobiography of Arthur Young, (1898).

Bewick, T., A General History of Quadrupeds, 3rd edn., (1792).

Bonser, K.J., The Drovers: Who they were and how they went. An Epic of the English Countryside, (1970).

Boswell's Treatise on Watering Meadows, 4th edn., (1801).

Brown, P., & Bradley, R., '1755 and all that: a historical primer of transmissible spongiform encephalopathy', in the *British Medical Journal*, 1998, December 19, 317(7174).

Brown, T., General View of the Agriculture of the County of Derby (1794)

Burgess, Revd W.H., A History of the Loughborough Unitarian Congregation, (1908).

Caddy, A., Robert Bakewell and His Times, (1925).

Caird, J., English Agriculture in 1850-51, (1852).

Carlos, A.M., Kosack, E., & Penarrieta, L.C., 'Bankruptcy, Discharge and the Emergence of Debtor Rights in Eighteenth Century England', paper given at the Federal Reserve Bank of Atlanta, Georgia, Conference, 5 November 2015.

Carter, H.B., His Majesty's Spanish Flock, (1964).

Carter, H.B., His Majesty's Spanish Flock; The Sheep and Wool Correspondence of Sir Joseph Banks 1781-1820, (1979).

Chambers, J.D., & Mingay, G.E., The Agricultural Revolution: 1750-1880, (1984).

Chivers, K., The Shire Horse: A History of the Breed, the Society, and the Men, (1976).

Coleman, J., (ed.) The Cattle of Britain, a series of articles, (1875).

Coleman, J., (ed.) The Sheep and Pigs of Great Britain, (1877).
Collins, D., An Account of the English Colony in New South Wales, (1798).
Colyer, R., 'Welsh Cattle Drovers in the Nineteenth Century, Part I', in *National Library of Wales Journal*, Winter Vol XVII/4, (1972).
Copland, S., (The Old Norfolk Farmer). Agriculture, Ancient and Modern, (c1861).
Cox, J., Wills, Inventories and Death Duties, A Provisional Guide, (1988).
Cox, T., A View of the United States of America, (1795).
Culley, G., Observations on Livestock, (1786), 4th edn., (1807).
Copus, A.K., Changing markets and the development of sheep breeds in Southern England 1750-1900 , *Agricultural History Review*, Part 1, (1989).
Dalton, R., 'John Farey's Derbyshire: Sheep Farming in the Early Nineteenth Century', Derbyshire Miscellany, Vol 15, Part 4, (Autumn 1999).
Daniell, T. & W., 'British Farm Animals', *Walkers Quarterly*, (1932).
Dickson, R.W., Practical Agriculture; or, A Complete System of Modern Husbandry, (1805).
Dickson, R.W., Practical Agriculture, Vol 1, (1805).
Dixon, H.H., in the *Royal Agricultural Society's Journal*, Vol 4, (New Series).
Druid, The (Henry Hall Dixon 1822-70), *Saddle and Sirloin* n.d.
Edwards, P., Horse and Man in Early Modern England, (2007).
Encyclopaedia Britannica, 3rd edn., Vol VI, (1797), Vol VIII, (1797).
Ernle, Lord, The Land and its People, (c1924).
Ernle, Lord, English Farming Past and Present, 4th edn., (1927) 1st edn., (1912).
Farnham, G.F., Leicestershire Medieval Village Notes, Vol II, (1929-33).
Fitt, J.N., 'Longhorn Cattle: Their History and Peculiarities', *Journal of the Royal Agricultural Society of England*, Vol XII, Part II (1876).
Fream, W., The Complete Grazier, 14th edn., (1900).
Fream, W., Elements of Agriculture, 8th edn., (1911).
Fussell, G.E., James Ward R.A., Animal Painter 1769-1859 and his England, (1974).
Garrard, G., *Proposals for Publishing a Print of the Woburn Sheep Shearing from a Picture by Mr Garrard*, Associate of the Royal Academy, (1811).
Garrard, G., *Woburn Sheepshearing*, Uncoloured Mixed Method Engraving, Painted and Published by G. Garrard, (1811).
Gordon, Revd A., 'Robert Bakewell' in the Dictionary of National Biography.
Gordon, Revd A., 'John Paget' in the Dictionary of National Biography.
Hartopp, H., Register of the Freemen of Leicester 1196-1770, Vol 1, (1927).
Heazlewood, B., The New Leicester and its Fat.
Heazelwood, B., The English Leicester Sunday Roast.
Heazlewood, B., Should the English Leicester have Modernised?

Heazlewood, B., Longwool Origin.
Heazlewood, B., Bakewell's Legacy in Australia.
Housman, W., 'Robert Bakewell', *Journal of the Royal Agricultural Society of England*, Vol V, Part 1, (1894).
Housman, W., (ed.) Live Stock Handbooks, Vol IV, Cattle, Breeds and Their Management (1915 & 1919).
J.H. of Derby, 'A Visit to Robert Bakewell's Farm, July 9th 1793' in *The Gentleman's Magazine*, September 1793.
Jones, Prof. C.B., Livestock of the Farm, Vols I & II, (1920).
Jones, E.L., Agriculture and Economic Growth in England 1650-1815, (1957).
Lawrence, J., A General Treatise on Cattle, The Ox, The Sheep, and The Swine, (1809).
Lawrence, R., An Inquiry into the Structure and Oeconomy of the Horse, (1801).
Library of Useful Knowledge, The Horse, (1831).
Library of Useful Knowledge, British Husbandry, Volume the First (1834) & Second, (1837).
Lisle, E., Observations on Husbandry, (1757).
Long, J., The Book of the Pig, (1886).
Low, Prof., 'An Evaluation of Robert Bakewell's New Leicester Sheep', from Illustrations of the Breeds of Domestic Animals of the British Islands No. XII, in *The Sporting Review*, Vol VII, January-June, (1842).
Lowe, D., On the Domesticated Animals of the British Isles, (1842).
Lythall, J.B., The Long-Horn Herd Book, containing the Pedigrees of Long-Horned Cattle, and a Short History of the Breed, Vol 1 Facsimile.
Marshall, W., The Rural Economy of the Midland Counties, Vols I & II, (1790), 2nd edn., (1796).
Marshall, W., Georgical Essays, Vol XX, (1803).
Matthews, H., 'Indeed amongst agriculturalists, there is something like free-masonry we are all brethren': Exploring agricultural friendship in late Georgian England. PhD thesis, University of Reading, (2022).
Mingay, G.E., (ed.), Arthur Young and His Times, (1975).
Moncrieff, E., Joseph S., & Joseph, I., Farm Animal Portraits, (1996)
Monk, J., General View of the Agriculture of the County of Leicester, (1794).
Namier, L., & Brooke, J., (eds.) The History of Parliament: The House of Commons 1754-1790.
Nichols, J., History of Leicestershire, Sparkenhoe Hundred, Vol IV, Part 2, (1811).
Orde, A., Matthew and George Culley Travel Journals and Letters 1765-1798, (2002).
Orwin, C.S., A History of English Farming, (1949).

Orwin, C.S., Johnson's England, Agriculture and Rural Life, (1954).
Parkinson on the Breeding and Management of Live Stock, Vol II., p.59, quoted in Library of Useful Knowledge, The Horse, (1831).
Parkinson, R., Treatise on the Breeding and Management of Livestock, Vols I & II, (1810).
Parry, R., *Particulars of the breeding stock, late the property of Mr Robert Fowler, of Little Rollright, in the county of Oxford, deceased*, (1791).
Pawson, H.C., Robert Bakewell Pioneer Livestock Breeder, (1957).
Pitt, W., General View of the Agriculture of the County of Leicestershire, including a survey of the County of Rutland, (1809).
Pitt, W., The General View of the Agriculture of the County of Staffordshire, (1796 & 1813).
Postles, D.F., 'Trading beyond his Capital: The Business of Bankruptcy and Local Economies in Leicestershire, 1706-1842', in *LAHS Transactions*, Vol 96, (2022).
Potts, T., The British Farmers Cyclopaedia, (1809).
Pringle, R.O., 'The Roscommon Sheep', in Coleman, J., (ed.) The Sheep and Pigs of Great Britain, (1977).
Rowlin, J., The Complete Cow-Doctor; or Farmer's Companion, 2nd edn., (1799).
Royal Agricultural Society of England, Royal Show Catalogues, various dates.
Royal Agricultural Society of England, *Journal* (2nd series), Vol 12, Part II, No. XXIV, (1876).
Royal Agricultural Society of England, *Journal* (3rd series), Vol 5, Part 1, No. 17, (1894).
Rudge, T., General View of the Agriculture of the County of Gloucester, (1807).
Ruffhead, O., (compiler), The Statutes at Large.
Russell, N., Like Engend'ring Like – Heredity and animal breeding in early modern England, (Cambridge University Press, 1986).
Scarfe, N., Innocent Espionage the La Rochefoucauld Brothers' Tour of England in 1785, (1995).
Sheldon, J.P., Dairy Farming, (c1883).
Smith, B., 'The Leicester Sheep', *The Gardeners' Chronicle and Agricultural Gazette*, Vol XXXI, Part 2, (1871) [online facsimile https://books.google.co.uk/books?id=9I8SGsPl5JAC].
Society for the Encouragement of Arts, Manufacture, and Commerce, The Complete Farmer: or, a General Dictionary of Husbandry, (1789).
Spavold, J., (ed.) In the Name of God Amen. Everyday Life in South Derbyshire 1535-1700, (1992).
State Papers Domestic 1642-1660, Government Papers from the Civil Wars and Interregnum.

Stirling, A.M.W., Coke of Norfolk and his Friends, (1912).
Temple Patterson, A., Radical Leicester, (1954).
The Agricultural Gazette, (1876).
The Agricultural Magazine, Vol III & Vol V, (1809).
The Annual Necrology, (1797-8).
The Annual Register, (1773).
The British Farmer's Magazine, Vol 1, (1827).
The Farmer's Magazine, Vol VII, (1806).
The Farmer & Stockbreeder Yearbook, (1902).
The Gentleman's Magazine, Part II, (1795).
Throsby, J., Leicestershire Views Supplementary Volume, (1790).
Trow-Smith, R., A History of British Livestock to 1700, (1957).
Trow-Smith, R., A History of British Livestock 1700-1900, (1959).
Usher, J., Border Leicesters in Coleman, J., (ed.) The Sheep and Pigs of Great Britain, (1877).
Wallace, R., Farm Live Stock of Great Britain, 4th edn. (1907) & 5th edn., (1923).
Weaver, L.T., Painter of Pedigree: Thomas Weaver of Shrewsbury, (2017).
Wedge, J., General View of the Agriculture of the County of Warwick, (1794).
White's Directory of Leicestershire and Rutland, (1863).
Wilson, J., The Principles of Stock-Breeding, (1912).
Wrightson, J., Sheep, Breeds and Management, (1919), No. 1 of Sinclair, J., (ed.) Live Stock Handbooks.
Wykes, D.L., 'Robert Bakewell (1725-1795) of Dishley: farmer and livestock improver', in *Agricultural History Review* 52, I.
Youatt, W., Sheep, Their Breeds, Management and Diseases, (1837).
Youatt, W., Cattle, Their Breeds, Management and Diseases, (1838).
Youatt, W., The Pig, (1847).
Youatt, W., Fream, W., & Bear, W.E., The Complete Grazier and Farmers' and Cattle Breeders' Assistant, 15th edn., (1908).
Young, A., The Farmer's Tour Through the East of England, Vols I, II, III & IV, (1771).
Young, A., The Annals of Agriculture, Vols 6 & 7 (c1786-8), 10 (c1788), 27 (1796), 28 (1797), 29 (1797), 36 (1800), 41 (1805).
Young, A., General View of the Agriculture of the County of Sussex, (1793).
Young, A., On the Husbandry of Three Celebrated British Farmers, (1811).
Young, A., Tours in England and Wales, selected from the Annals of Agriculture, No. 14 in a series of reprints of scarce tracts in *Economic and Political Science*, published by the L.S.E., (1932).

Primary Sources

Record Office of Leicester, Leicestershire & Rutland (ROLLR):
Act Book1682/3
Archdeaconry Court, Box 4/XVI, 1, 1634/5, MF 609
Archdeaconry Court, Box XVII, 1635-6, 95/98, MF 610
Wills, Inventories and Administrations proved at Leicester.
DE 1061
DE 2178
DE 365/226
DE 5021
DE 7454/1, 1795-1817
DE 7454/3, 1810
DE 7454/4
DE 7474/22, 1732-53
DE 7454/23, 1739
DE 7454/24, 1748
DE 7454/25, 1751-2
DE 7454/26, 1752-3
DE 7454/35, 1775-6
DG 32/12
L 288
M 703
MF 154
RG 4/78

Nottinghamshire Archive Office (NAO):
Wills and Inventories held in Nottinghamshire Archive Office.
CA/TC/2/161
CA/TC/2/BOX/64/2
DD/BK/3/1-12
DD/588/11
M/7537
M/16889
M/16890
M/18835
M/18836
M/18837

The National Archives (TNA)
Pamphlet 994.02 TUR: Turner, H.N., 'Early Sheep in Australia, Results of Enquiries in Britain' (1978).

Prerogative Court of Canterbury Wills.
B 1/73, 1773
B 1/74, 1784
C 2/ChasI/H61/24
C 6/222/7
C 8/40/9
C 10/357/7
E 179/134/205
E 179/134/254
E 179/134/315
E 179/134/324
E 179/251/4 Part 8
FEC 1/1194
HO 107/191/16
HO 107/2011
HO 107/595/11
HO 107/593/27
HO 107/599/2
IR1 58, 1773 National Register of Apprentices.
PROB 11/657/190
PROB 22/816/98
RG 4/1586
RG 4/31/89
RG 4/3633
RG 4/4408
SP 23/161/407

Staffordshire Record Office (SRO)
Wills and Inventories proved at Lichfield.
D661/1/7/Parcel 5
D661/1/7/Parcel 6
D661/1/7/Parcel 7

The National Archives of Scotland (NAS)
GD224/1040/62
GD224/31/10

Derbyshire Record Office (DRO)
D2010/CT/10-11, 1775
D2534
D754/T/11
DE2375

Holkham Hall Archives
F/TWC2, 1803, 1805, 1811
E/X3, F/TWC2, 1804-7

London Metropolitan Archive
Freedom of the City Admission Papers 1712

Northamptonshire Archives
Administrative and biographical background on the Viscounts Cullen.

Nottingham University Manuscripts & Special Collections
Mi 4Da 2/8

India Office Library
Public Proceedings 1262, Original Consultations Range 4, Vol 11, No. 11, 16 March 1792

Digital Sources

www.academic-accelerator.com/encyclopedia/galway-sheep
www.ancestry.co.uk
www.bahs.org.uk/AGHR/ARTICLES/17n2a4.pdf
www.britishnewspaperarchive.co.uk
www.doi.org/10.3389/fgene.2019.00927
www.gale.com/intl/c/19th-century-us-newspapers
www.thegazette.co.uk
www.historyofparliamentonline.org
www.measuringworth.com/ppoweruk/
www.oldbaileyonline.org version 9, August 2023
www.openlibrary.org/books/Observations-on-the-manufactures-trade-and-present-state-of-Ireland
www.surman.english.qmul.ac.uk
www.genuki.org.uk/big/wal/Archives/NLWjournals/CattleDrovers1,
www.doi.org/10.48683/1926.00111432
www.merchantstaplers.co.uk & www.archiveshub.jisc.ac.uk/data/gb193-staple

General Index

Animal Index

[ ] = Illustration